Veronica Ambrogi, Pierfrancesco Cerruti, Valentina Marturano (Eds.)
Sustainability of Polymeric Materials

Also of interest

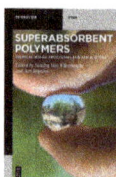

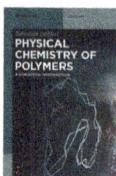

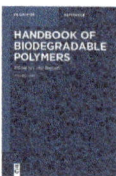

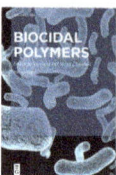

Sustainability of Polymeric Materials

Edited by
Veronica Ambrogi, Pierfrancesco Cerruti,
Valentina Marturano

DE GRUYTER

Editors

Prof. Veronica Ambrogi
University of Naples "Federico II"
Dep. of Chemical, Materials and
Production Engineering
Piazzale Tecchio 80
80126 Napoli
Italy
ambrogi@unina.it

Dr. Valentina Marturano
National Council of Research
Inst. for Polymers, Composites and
Biomaterials
Via Campi Flegrei 34
80078 Pozzuoli
Italy
valentina.marturano@ipcb.cnr.it

Dr. Pierfrancesco Cerruti
National Council of Research
Inst. for Polymers, Composites and
Biomaterials
Via Campi Flegrei 34
80078 Pozzuoli
Italy
cerruti@ipcb.cnr.it

ISBN 978-3-11-059093-7
e-ISBN (PDF) 978-3-11-059058-6
e-ISBN (EPUB) 978-3-11-059069-2

Library of Congress Control Number: 2020936251

Bibliographic information published by the Deutsche Nationalbibliothek
The Deutsche Nationalbibliothek lists this publication in the Deutsche Nationalbibliografie;
detailed bibliographic data are available on the Internet at http://dnb.dnb.de.

© 2020 Walter de Gruyter GmbH, Berlin/Boston
Cover image: Dr. Giuseppe Cesare Lama, Institute for Polymers,
Composites and Biomaterials, National Council of Research
Typesetting: Integra Software Services Pvt. Ltd.
Printing and binding: CPI books GmbH, Leck

www.degruyter.com

Preface

In 1987, Gro Harlem Brundtland, ex-Prime Minister of Norway, introduced the concept of sustainable development as "addressing the needs of the present without compromising the ability of the future generations to meet their needs," at the UN Commission on Sustainable Development. Nowadays, the technological development is globally improving the quality of human life, but the impact on the environment is increasingly destructive as the progress we are experiencing is not sustained by the limited natural resources of our limited planet. Therefore, an inversion on our daily approach is mandatory and the linear economy based on the take–make–dispose concept must be replaced by a circular economy model.

In the field of plastic materials, current growing concern on waste management and plastic pollution is pushing international institutions, as well as scientific and industrial community, toward a more sustainable economy for plastic products. In this frame, the design and development of a nature-mimicking circular system, where waste is transformed back to resources through the production of biobased and waste-derived precursors and polymers, represent a significant evolution toward a sustainable approach. In this respect, the optimization of innovative, low-cost, and low-energy-consuming processes for plastic production, and the definition of new strategies for the management of end-of-life oil-based plastics are becoming key importance for bioinnovation.

This book aims to offer a comprehensive overview on the green approaches to the research and technology of plastic materials. A critical perspective concerning both oil-based plastics and novel biobased and waste-derived polymer formulations is provided, with a special focus on the application fields in which biopolymers have the potential to compete with traditional plastics. This book is intended as a complimentary textbook for students as well as a reference for researchers and practitioners working on sustainable polymers and their applications.

Chapter 1 illustrates the key principles of the circular economy in comparison to linear approach and analyzes the new business, industry, and consumer models in the frame of a circular sustainable model. Chapter 2 discusses on the impact of microplastics on aquatic environments, providing remarkable data on the current scenarios in freshwater, sea, oceanic, and coastal environments, and the consequences that microplastics are having particularly on the fauna. Chapter 3 describes the current research trends on polyhydroxyalkanoates (PHAs), biopolyesters synthesized by a variety of microorganisms. This chapter also illustrates the potential applications PHAs can have due to their biodegradable and biocompatible nature, and their inherent properties. Chapter 4 provides an overview on chitosan and alginate polymers, and summarizes the most important novel concepts and inventions regarding marine biopolymer engineering. Chapter 5 emphasizes knowledge and recent advances in terrestrial biopolymers, such as cellulose, starch, and lignin, providing information on the current market and commercial distribution of this class

https://doi.org/10.1515/9783110590586-202

of bioplastics. Chapter 6 gives an insight on biobased thermosetting polymers, obtained by novel building blocks based on bioderived molecules designated to substitute petroleum-derived precursors. It also describes the properties of biobased thermosets in comparison with the traditional systems. Chapter 7 describes the most important synthetic methods and degradation mechanism of aliphatic biodegradable polyesters, such as polylactic acid, as well as current and promising applications in various fields. Chapters 8 and 9 provide a comprehensive description on natural and biobased functional and structural additives for polymers. Finally, chapter 10 provides an interesting insight on the use of additive manufacturing for the processing of biopolymers.

We would like to gratefully acknowledge all the authors for their eager and notable contribution in the preparation of this book, and to Dr. Vivien Schubert of DeGruyter Publisher, Germany, for her kind, continuous, and efficient support.

Veronica Ambrogi
Pierfrancesco Cerruti
Valentina Marturano

Contents

List of contributors

Francesco Abate
University of Naples Federico II
Department of Chemical, Materials and
Production Engineering (DICMAPI)
Piazzale Tecchio 80
80125 Naples
Italy
francesco.aba@libero.it

Sarai Agustin-Salazar
Institute for Polymers, Composites and
Biomaterials (IPCB-CNR)
Via Campi Flegrei 34
80078 Pozzuoli (Napoli)
Italy
iq.saraiasalazar@gmail.com

Veronica Ambrogi
University of Naples Federico II
Department of Chemical, Materials and
Production Engineering (DICMAPI)
Piazzale Tecchio 80
80125 Naples
Italy
ambrogi@unina.it

Paolo Aprea
University of Naples Federico II
Department of Chemical, Materials and
Production Engineering (DICMAPI)
Piazzale Tecchio 80
80125 Naples
Italy
paolo.aprea@unina.it

Maurizio Avella
Institute for Polymers, Composites and
Biomaterials (IPCB-CNR)
Via Campi Flegrei 34
80078 Pozzuoli (Napoli)
Italy
maurizio.avella@ipcb.cnr.it

Pierfrancesco Cerruti
Institute for Polymers, Composites and
Biomaterials (IPCB-CNR)
Via Campi Flegrei 34
80078 Pozzuoli (Napoli)
Italy
cerruti@ipcb.cnr.it

Amparo Chiralt
Polytechnic University of Valencia
Research Institute of Food Engineering for
Development
Camino de Vera s/n.
46022 Valencia
Spain
dchiralt@tal.upv.es

Mariacristina Cocca
Institute for Polymers, Composites and
Biomaterials (IPCB-CNR)
Via Campi Flegrei 34
80078 Pozzuoli (Napoli)
Italy
mariacristina.cocca@ipcb.cnr.it

Iuliana Cota
Université de Rennes I,
Institut des Sciences Chimiques de Rennes,
Centre of Catalysis and Gree Chemistry –
Team "Organometallics: Materials and
Catalysis"
Campus de Beaulieu, 243 Av. du Général
Leclerc
35042 Rennes Cedex
France
iuliana.cota2010@gmail.com

Cristina De Capitani
Institute of Polymers, Composites and
Biomaterials (IPCB-CNR-CNR)
via Previati n. 1/E
23900 Lecco
Italy
cristina.decapitani@polimi.it

https://doi.org/10.1515/9783110590586-204

Francesca De Falco
Institute for Polymers, Composites and
Biomaterials (IPCB-CNR)
Via Campi Flegrei 34
80078 Pozzuoli (Napoli)
Italy
francesca.defalco@ipcb.cnr.it

Emilia di Pace
Institute for Polymers, Composites and
Biomaterials (IPCB-CNR)
Via Campi Flegrei 34
80078 Pozzuoli (Napoli)
Italy
emilia.dipace@ipcb.cnr.it

Valentina Marturano
Institute for Polymers, Composites and
Biomaterials (IPCB-CNR)
Via Campi Flegrei 34
80078 Pozzuoli (Napoli)
Italy
valentina.marturano@ipcb.cnr.it

Alice Mija
Université Côte d'Azur, Université Nice-
Sophia Antipolis
Institut de Chimie de Nice, UMR CNRS 7272
06108 Nice Cedex 02
France
Alice.mija@unice.fr

Arash Moeini
University of Naples Federico II
Department of Chemical Sciences
Via Cintia 4, Complesso Universitario Monte
S. Angelo
80126 Naples
Italy
arash.moeini65@gmail.com

Isabel Moreno
Grupo de Química Macromolecular
(LABQUIMAC)
Universidad del País Vasco UPV/EHU
Departamento de Química Orgánica II,
Facultad de Ciencia y Tecnología
48940 Leioa
Spain

BC Materials – Basque Centre for Materials,
Applications and Nanostructures,
UPV/EHU Science Park
48940 Leioa
Spain
mariaisabel.moreno@ehu.eus

Aleksandra Nesic
University of Belgrade
Vinca Institute for Nuclear Sciences
12-14 Mike Petrovića Street
11158, Belgrade
Serbia
anesic@vin.bg.ac.rs

Federico Olivieri
Institute of Polymers, Composites and
Biomaterials (IPCB-CNR)
P.le E. Fermi 1,
80055 Portici,
Italy
federico.olivieri.89@gmail.com

Leyre Pérez-Álvarez
Grupo de Química Macromolecular
(LABQUIMAC)
Universidad del País Vasco UPV/EHU
Departamento de Química Física, Facultad de
Ciencia y Tecnología
48940 Leioa
Spain
BC Materials – Basque Centre for Materials,
Applications and Nanostructures,
UPV/EHU Science Park
48940 Leioa
Spain
leyre.perez@ehu.eus

Corinna Ponti
University of Naples Federico II
Department of Chemical, Materials and
Production Engineering (DICMAPI)
Piazzale Tecchio 80
80125 Naples
Italy
corinnaponti.na@gmail.com

Raquel Requena
Polytechnic University of Valencia
Research Institute of Food Engineering for
Development
Camino de Vera s/n.
46022 Valencia
Spain
rarepe@upv.es

Leire Ruiz-Rubio
Grupo de Química Macromolecular
(LABQUIMAC)
Universidad del País Vasco UPV/EHU
Departamento de Química Física, Facultad de
Ciencia y Tecnología
48940 Leioa
Spain
BC Materials – Basque Centre for Materials
Applications and Nanostructures
UPV/EHU Science Park
48940 Leioa
Spain
leire.ruiz@ehu.es

Julia Sánchez-Bodón
Grupo de Química Macromolecular
(LABQUIMAC)
Universidad del País Vasco UPV/EHU
Departamento de Química Física, Facultad de
Ciencia y Tecnología
48940 Leioa
Spain
julia.sanchez@ehu.eus

Gabriella Santagata
Institute for Polymers, Composites and
Biomaterials (IPCB-CNR)
Via Campi Flegrei 34
80078 Pozzuoli (Napoli)
Italy
gabriella.santagata@ipcb.cnr.it

Gennaro Scarinzi
Institute for Polymers, Composites and
Biomaterials (IPCB-CNR)
Via Campi Flegrei 34
80078 Pozzuoli (Napoli)
Italy
gennaro.scarinzi@ipcb.cnr.it

Andrea Sorrentino
Institute of Polymers, Composites and
Biomaterials (IPCB-CNR)
via Previati n. 1/E
23900 Lecco
Italy
andrea.sorrentino@cnr.it

Jean-Philippe Steyer
INRAe
Univ. Montpellier, LBE
102 avenue des Etangs
11100 Narbonne
France

María Vargas
Polytechnic University of Valencia
Research Institute of Food Engineering for
Development
Camino de Vera s/n.
46022 Valencia
Spain
mavarco@tal.upv.es

José Luis Vilas-Vilela
Grupo de Química Macromolecular
(LABQUIMAC)
Universidad del País Vasco UPV/EHU
Departamento de Química Física, Facultad de
Ciencia y Tecnología
48940 Leioa
Spain
BC Materials – Basque Centre for Materials
Applications and Nanostructures
UPV/EHU Science Park
48940 Leioa
Spain
joseluis.vilas@ehu.es

Gaetano Zuccaro
INRA – Laboratoire de Biotechnologie de
l'Environnement (LBE)
102 Avenue des Etangs
11100 Narbonne
France
gaetano.zuccaro@inra.fr

Francesco Abate, Corinna Ponti, Paolo Aprea

1 Feasibility of the circular economy and plastic pollution

Abstract: The last years have seen a general rise in individual awareness on the environmental issues that our society is about to face. The almost total scientific agreement over Global Warming and the day-by-day growing evidences of plastic pollution have fed requests worldwide to public institutions to take actively part in a radical change in today conception of the global economy.

This work retraces the steps that led to the recognition of the unsustainability of the current economic model and depicts the Circular Economy (CE) as a structured and promising approach to achieve a more sustainable development. Key principles of CE are extensively illustrated, along with examples of their practical implementation.

Plastics is the class of materials that, more than others, is undergoing to severe popular blame for its environmental impact. Bioplastics are commonly seen as a promising alternative to fossil based ones that perfectly fits within the CE approach. However if and how their use can lead to real environmental benefits has not been cleared yet.

Here we investigate this issue, providing a review of Life Cycle Assessment (LCA) studies relative to biobased and biodegradable polymers. What emerges is the need of a stronger standardization for LCAs to be truly comparable and significant. Apart of this, it appears clear that the mere substitution of a material with another, when not accompanied by a radical shift in the consumption model, can not be a final answer to the environmental question.

Finally, the current status of public policies regarding plastic pollution is discussed. If the European Union appears to be quite active on the topic, the same cannot be said for most of the extra-EU countries and a lot of work is still to be done in order to drew up a sustainable future.

Keywords: Circular Economy, Life Cycle Assessment, Biobased, Biodegradable, Biopolymer, Plastic Pollution, EU Policy

Francesco Abate, Corinna Ponti, Paolo Aprea, Department of Chemical, Materials and Production Engineering (DICMAPI), University of Naples Federico II, Naples, Italy

https://doi.org/10.1515/9783110590586-001

1.1 Transition from linear to circular economy

1.1.1 The need for a circular economy

The model of production and consumption of the industrial society has been based since its origins on the so-called *take–make–dispose* paradigm: raw materials are extracted and processed into consumer products, which are disposed once their end of life (EOL) comes. One assists the conversion of natural resources into waste through the use of large quantities of fossil energy. This approach depletes natural capital (through mining/unsustainable harvesting) and reduces its value due to pollution (waste dispersion and emissions) [1].

One must also consider that, although decreasing or stagnant in most OECD[1] countries [2], the world human population is continuing to rise. The current population of 7.6 billion people is expected to reach 8.6 billion by 2030, 9.18 billion by 2050, and 11.2 billion by 2100, and 97% of this growth will occur in developing countries such as China and India (Figure 1.1) [3].

The evaluation of the relationship between human population and environmental problems in terms of energy, resources demand, and waste production is not trivial. Ehrlich and Holdren in 1971 described the negative impact I of a society on the environment in terms of a simple equation: $I = P * F$, where P is the population and F is a function expressing the per capita impact [4].

In this perspective, the data previously reported seem even more alarming, also considering that the world gross domestic product (GDP) is also expected to increase by 57% in 2030, which would produce, especially in developing countries, a growing demand for health, food, education, and services. The subsequent growth of the middle class (particularly in Asia), that is estimated to reach 4.9 billion in 2030, would lead to an increase in terms of either consumptions or emissions [3].

The publication of the report *Limits to Growth* by the club of Rome[2] in 1972 significantly contributed to the spread of awareness of rethinking human activity in respect of the environment. It assumes that if human activity kept growing over the years as it did between 1900 and 1972 in terms of population growth, industrialization, pollution, food production, and waste of resources, the maximum carrying capacity of the planet would have been reached in the following 100 years [5].

Later in 1987, the report "Our common future," also known as Brundtland report, gave the first and much acknowledged definition of sustainable development as a development that "meets the needs of the present without compromising the ability of future generations to meet their own needs." This definition emphasizes

---

1 Organization for Economic Cooperation and Development.
2 Founded in 1968 at Accademia dei Lincei in Rome, Italy, the Club of Rome consists of current and former heads of state, UN bureaucrats, high-level politicians and government officials, diplomats, scientists, economists, and business leaders from around the globe.

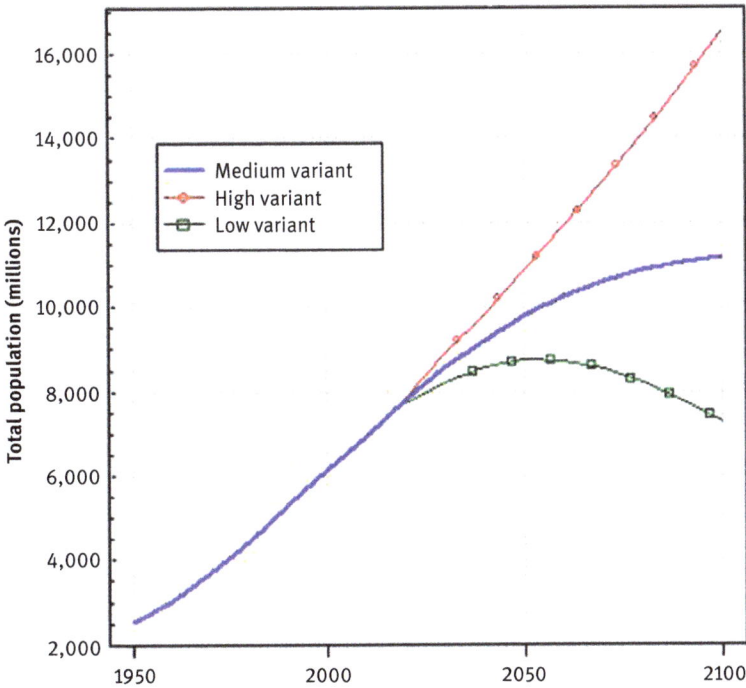

Figure 1.1: World's demographic profile for the period 1950–2100. Data from 1950 to 2015 are estimates, while from 2015 to 2100 are medium variant projections. Source: United Nations, Department of Economic and Social Affairs, Population Division (2017). World Population Prospects: The 2017 Revision, Online Demographic Profiles. Available from https://population.un. org/wpp/Graphs/DemographicProfiles/, Accessed on [23/04/2019].

the intergenerational aspect of sustainability, but the report also expresses clearly the need to find a harmony between development, environment, and social equity: "A world in which poverty is endemic will always be prone to ecological and other catastrophes" [6].

The world is still very far from achieving sustainable development in all its dimensions. Accounting the social dimension, despite the rise in the world GDP, chronic poverty will still affect more than 1.8 billion people in 2030 [3].

According to the report Emission World Gap, even if all the countries of the world respected their international agreements on carbon dioxide emissions, by the end of the century, there would be a global warming of around 3 °C compared to the preindustrial era, which would be catastrophic [7].

Further alarming data come from a study conducted by OECD about the projections on material resources world demand in 2060: without a substantial change in economic policies, global materials use could rise from 89 Gt in 2017 to 167 Gt in 2060. This positive consumption trend would affect all main categories of materials (the report refers to materials as the set of physical resources involved in production and consumption): metallic ores (9 to 20 Gt), nonmetallic minerals (44 to 86 Gt), biomass (22 to 37 Gt), and fossil fuels (15 to 24 Gt) [8].

During the last 10 years, the concept of circular economy (CE) is raising attention as a new, more structured approach to achieve and implement sustainable development. Several definitions have been formulated for the CE. Among them, the most diffused and quoted, provided by the Ellen MacArthur foundation[3] in 2012, asserts that "A circular economy is an industrial system that is restorative or regenerative by intention and design." Such an economy is based on facilitating the transition to renewable resources and renewable sources of energy, on reducing the use of toxic substances and, above all, on designing materials, products, and systems that avoid waste production and enhance materials reuse, recovery, and recycling [9]. The CE ultimate goal is decoupling economic growth from resource depletion: this requires a completely new function for resources, aiming to maintain their value as long as possible along the economy [10].

1.1.2 Origins of circular economy

According to different reviews on the subject [1, 11] the concept of CE is not new and can be traced back to different schools of thought. In 1966, Kenneth Boulding first recognized the need for a cyclical system for human activity, given the intrinsic finitude of the earth's resources. He suggested the transition from a linear system to a "spaceship economy," a closed system "which is capable of continuous reproduction of material form even though it cannot escape having inputs of energy" [12]. Under Boulding's influence, Pearce and Turner in their essay *Economics of Natural Resources and the Environment*, first introduced the term circular economy [11, 13]. They identify three environmental economic functions:
1. to provide multifunctional resources, either renewable or nonrenewable;
2. to assimilate waste and emissions;
3. to ensure direct utility through amenity values.

The previous functions can be summarized in a fourth function defining the environment as a life-supporting system.

These environmental economic functions set specific limits that the linear economic system does not consider. The scheme shown in Figure 1.2 illustrates the connections between economic activities and the environmental functions described by Pearce and Turner in a circular system: the economic activity creates utility (U)

---

3 A charity founded in 2010. It collaborates with business, government, and academia to accelerate the transition to a circular economy.

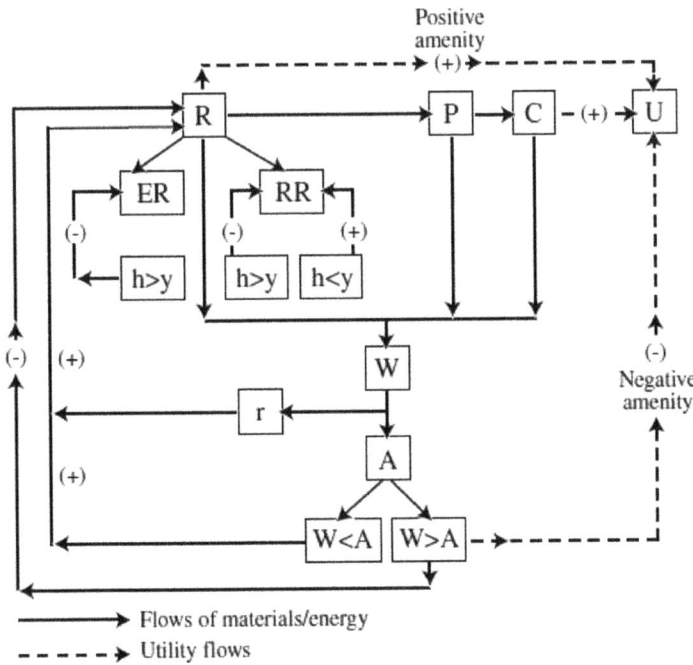

Figure 1.2: The circular economic system. Source: [14].

converting natural resources (R) in consumer goods (C) through production (P). Waste and emissions (W) come from both extraction of resources, production, and consumption. Once produced, waste can be, in principle, assimilated by the environment (A). In the diagram, the authors represent the conservation/improvement of a function and its depletion with the plus (+) and minus (−) signs, respectively.

It can be noted that nonrenewable resources (ER) always involve a depletion, as the production rate (y) is zero and the consumption rate (h) is positive. On the other hand, to not alter the amount of renewable resources (RR), it is important that the consumption rate stays always lower than the renewal rate (h < y), otherwise renewable stocks will be transformed in nonrenewable resources. Moreover, to preserve the assimilating function of the environment, it is of primary importance to not exceed its capacity, otherwise, if W > A, the landscape value (U) and the ability to ensure the availability of natural resources (R) will be compromised. Through recycling (r), a process output becomes the input of another process: this can help to preserve the environmental functions, but there is no possibility to 100% recycle resources due to both entropic reasons and technological limitations [14].

1.1.3 Key principles of the circular economy

CE has a twofold meaning: literally, it has a descriptive meaning that emphasizes the circular character of materials and resources flows, but also expresses a clear contrast to the traditional Linear Economy, which relies on the aforementioned *take–make–dispose* model. In the CE, there are two types of flows pertaining biological nutrients and technical nutrients, respectively [9].

The idea is directly derived from Braungart and McDonough, which define the *biological nutrients* as a set of natural, plant-based, or synthetic materials that are not dangerous for the environment and can be safely returned to the biosphere to renew the natural capital. On the contrary, the *technological nutrients* are minerals or synthetic materials that, after their first use as a consumer good, must maintain their maximum value in a closed-loop system consisting in remanufacture, reuse, or recovery (Figure 1.3). The substantial difference between the two cycles is that consumption should only be allowed in the first one, because it does not develop waste but further nutrients for biological processes and no value is lost. This is a completely new way to conceive resources management that goes beyond the classical approach to sustainability based on the maximization of efficiency, which aims to reduce the energy and resources input in a product or system process, but it is still grounded on a linear, *cradle-to-grave* point of view. The system must instead switch to a *cradle-to-cradle* design, which is a means of achieving effectiveness, and which implies, but is not limited to, the maximization of efficiency [15].

Design of products and systems plays a fundamental role in the CE with the purpose to slow, close, and narrow resource loops: slowing resource loops consists in designing long-lasting goods preferentially made of durable and reliable materials, and in enhancing the reuse and recovery of such materials, thus ensuring a high standardization of the components and degree of disassembly; closing resource loops means to guarantee a high amount of high-quality recycled materials, and narrowing resource loops aims at using the lowest quantity of materials and energy for a certain purpose (involving the efficiency maximization *stricto sensu*). Disassembly not only helps to extend the life of a product but also helps to keep the biological and technological cycles separated, avoiding irreversible mixes, which is against the CE principles [16].

The previous concepts define a sort of hierarchy of CE actions: *reduce, reuse, recover, and recycle.*

To understand this hierarchy, it is important to specify that a product has both a physical- and an economic value.

The physical value relays on the raw materials from which the product is made of, and on the energy that has been necessary to the production. A useful parameter to quantify the energy required to produce a material is the *embodied energy:* it expresses (in MJ/kg or MJ/item) the amount of energy required to extract the raw materials, to transport, and to process them. This definition, typical of a *cradle-to-gate* approach, may appear limited if applied to a final product: in this case, best practises

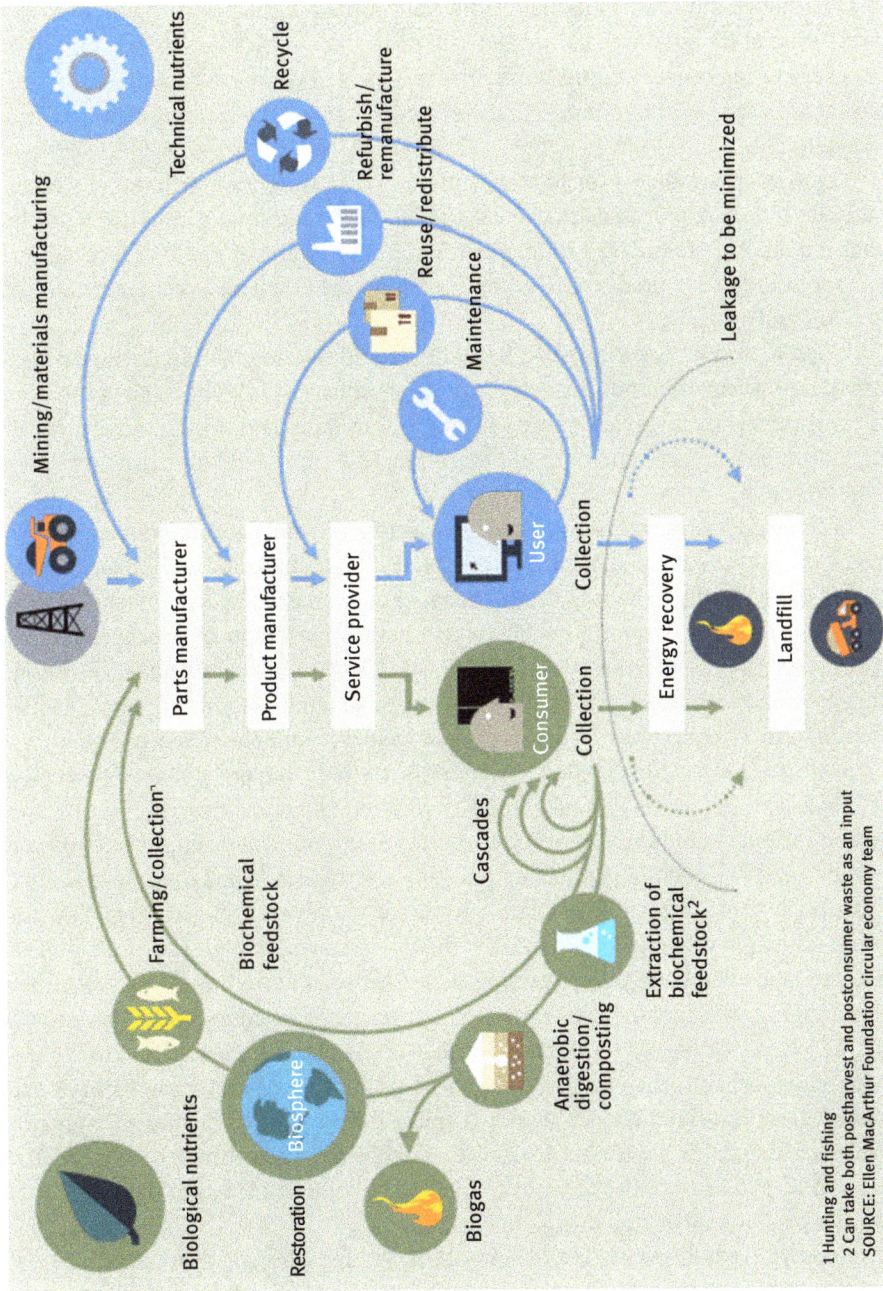

Figure 1.3: Circular economy flow diagram. Source: [9].

also include energy contributions for the transport from the firm (necessary to the distribution of the product) and for the EOL phase (i.e., recycling energy) [17, 18]. In any case, to make meaningful comparisons between different products, it is important to know the boundaries that have been set for the quantification of this parameter.

The economic value takes into account the costs associated with the raw materials, the different processes that have transformed them into a product, and the labor required to produce the item. The CE objective is, therefore, to *reduce* resources inputs and waste production, while *maximizing* physical and economic value of products and systems.

When a product is disposed in landfills, in addition to creating environmental damage, its economic and physical value is completely lost. The same applies to disposal processes involving energy recovery (like incineration), although heat or energy can be produced there: in CE, these processes are therefore considered only a last option [9].

Recycling activities consist in recovering waste materials and reprocessing them in new products, materials, or substances [19]. Different materials and products have a different possibility to be recycled, depending on both economic and technological reasons. A first and essential clarification concerns the origin of waste: industrial scraps are very easy to be recycled (often in situ) because they are only slightly contaminated and dispersed, while municipal waste generally requires a much deeper effort in terms of collection and separation and a lower purity level is achieved.

A further distinction regards materials nature: paper, polymers, metals, ceramics, and glasses have substantially different technological properties and require different amount of energy to be reprocessed. For example, metals are characterized by peculiar values of density, electrical and magnetic properties, which allow them to be easily separated from other metals and materials. For this reason, although the energy required for the process is quite high (i.e., for stainless steel the embodied energy for recycling varies between 11 and 13 MJ/kg [20]), they have a high recycling rate [20]. On the other side, polymers share similar properties, and their color features are determined by marketing choices. This makes their separation more difficult, therefore new design strategies are of primary importance to improve their recyclability. Furthermore, unlike thermoplastic polymers, thermosetting polymers cannot be recycled using traditional mechanical methods (currently, they can be reduced to smaller parts and used as fillers) and chemical recycling procedures are still only explored experimentally.

Moreover, while metals could potentially be recycled an infinite number of times [21], thermoplastic polymers are affected by a degradation in mechanical properties at each cycle, which compromises their processability too [22]. The latter is, therefore, ascribable not as a real recycling process but as downcycling: a process in which there is a degradation of material properties limiting its use only to lower value applications. It does not concern only thermoplastic polymers but also other materials such as paper and even metals when they are reprocessed in multicomponent metal alloys. Often, secondary generation products are not recyclable anymore, and such a paradigm does not alter the linear character of waste production, but only delays the product transition from resource to waste [23]. These facts emphasize the urgency to achieve a higher quality recycling through conscious design and technological progress. However, even in the best scenario, recycling can only preserve the economic and physical value of the materials a product is made of, while the value associated with the processing is lost. Besides, recycling requires energy and involves CO_2 emissions, although generally such energy and emissions are lower than those associated with virgin materials [20]. For these reasons, in the CE hierarchy mentioned previously, recycling should be only a last step to close materials cycle after that a product or its component have already been used many times, through reuse and additional forms of recovery such as refurbishment and remanufacturing.

Reuse means to use a product again for the same or a different purpose without or with just minimum modifications. Reuse can be practiced directly by consumers (i.e., reuse of bottles, plastic bags, second hand sales of books and clothes) or by companies that design their products according to a *cradle-to-cradle* vision (i.e., refillable toner cartridges) [10].

Refurbishing a product consists in replacing or repairing faulty components and updating the product in terms of aesthetics, while remanufacturing lies in disassembly and assembly process by which components that are still good are recovered and placed in another new product. These three alternatives allow to maintain product and materials at a higher level and with lower environmental impacts compared to recycling [9].

1.1.4 New business, industry, and consumer models

To put into practice CE principles, new business, industry, and consumer models must be spread. It is a complex process of which only a few fundamental aspects are briefly analyzed here.

To improve reuse practices, it is essential creating a demand for reused, remanufactured, and refurbished objects among consumers which is also economically attractive for companies. Regarding the last point, an important tool is the *extended producer responsibility* (EPR) (European Commission – DG environment, 2014 [16]). EPR is an environmental policy stating that the producer is responsible for the environmental impact of a product, from the design to the disposal phase. Its implementation involves, for example, taxation on waste, emissions, use of virgin materials or nonrecyclable materials, and direct incentives for a correct design of products [24].

Regarding consuming reduction, an effective way to shift to CE is the diffusion of collaborative consumption models as sharing, renting, and lending. This concept is based on Stahel's conception of the economy as a system that should sell services instead of products [25].

Another crucial concept, regarding industrial development, is the *industrial symbiosis* (IS), defined by Chertow as a system "engaging traditionally separate industries in a collective approach to *competitive advantage involving physical exchange of materials, energy, water, and by-products. The keys to IS are collaboration and the synergistic possibilities offered by geographic proximity*" [26]. IS has been already seen, starting from 1990s, as a means to reduce the production of waste, emissions, and the use of raw materials and energy in industrial process developing so-called ecoindustrial parks [10], where the output of one firm becomes the input for a different one. IS can be considered as a subfield of industrial ecology [11, 27], a discipline that studies the interactions between the industrial activities and the biosphere in order to find a way to reshape the former to make it compatible with the latter [28].

1.2 The role of polymers in the context of the CE

1.2.1 Traditional polymers: main environmental concerns

Polymers show peculiar properties that have motivated their rapid development over the last 70 years, making them a fundamental part of our everyday life. Among their many features, they can be easily processed in different shapes and dimensions, have generally low density (and, consequently, reasonably high specific mechanical properties), and can be chemically and physically engineered to perform a certain function (e.g., in terms of mechanical, optical, wettability, and permeability properties). These properties are coupled with low cost, thanks to their easy processability.

According to a study conducted by Geyer, Jambeck & Law, the world annual production of polymers (resins and fibers) has grown from 2 Mt in 1950 to 380 Mt in 2015, and simultaneously their contribution to solid urban waste, in middle- and high-income countries, has increased from 1% to 10% by mass from 1960 to 2005. During the same time period, 8.3 Gt of plastics from virgin raw materials have been produced. The most widely ever produced nonfiber polymers are polyethylene (PE), polypropylene (PP), and polyvinylchloride (PVC), accounting for 36%, 21%, and 12% respectively, followed by polyethylene terephthalate (PET), polyurethanes (PUR), and polystyrene (PS), each contributing less than 10%. Among the synthetic fibers, polyesters (especially PET) cover 70% of the total production ever made. Packaging sector demands the highest amount of plastic: in 2015, it brought to the market 42% of primary nonfiber polymers produced (146 Mt), and the majority left the market, becoming waste, in the same year [29].

Almost the total of plastics (99%) are produced from fossil raw materials and, if the trends in oil and plastic consumption remained the same as today, in 2050 the production of plastic would require one-fifth of the cumulative oil consumption [30]. Such a strong dependence on nonrenewable raw materials, for products whose majority, as mentioned previously, has a very short life, is probably the most evident ensign of a wicked model of linear economy.

Regarding material efficiency, plastics show good performances compared to other materials. According to a study dating back to 2010, if plastics were substituted, whenever possible, by other conventional materials, there will be a negative impact in terms of greenhouse gas (GHG) emissions and energy demand either for their production, usage, or recovery [31]. This suggests that plastics are not materials to be banned at all costs, a widely held opinion today, but that there is an urgent need to completely rethink their use and disposal paradigm.

The most serious and perceptible environmental problem associated with plastics occurs at their EOL, due to the long time in which they remain unchanged in the environment if they are dispersed in it [32]. Polymers' high-molecular weight, hydrophobicity, and cross-linked chemical structure are factors that guarantee stability and durability, but also prolong degradation time [33]. For instance, PE biodegradation can take decades, or more, depending on the type and on the environment [32]. Plastics in the environment can be classified according to the size as nanoplastics (<1 μm), microplastics (<5 mm), mesoplastics (<2.5 cm), macroplastics (<1 m), and megaplastics (>1 m) [34].

Plastic litter has been clearly noticed in the most different environments as terrestrial, freshwater, estuarine, coastal, marine, and also in remote zones of the globe as deep-sea sediments, submarine canyons, and in Arctic sea ice [35]. The most abundant data currently available are relative to marine littering: every year, eight million tons of plastic end up in the oceans [36].

The first source of plastic for the oceans is the transport from the mainland through the rivers. A large part of the dispersion occurs due to incorrect waste management but can also occur near the production plants, during the transportation phase and usage phase (for accident or intension). Microplastics are produced by erosion and degradation of polymers during their transportation in water but are also released during the washing of synthetic fabrics, the rinsing of cosmetics (at the first place), and the abrasion of tires and fishing gears. The latter are also responsible for trapping animals, a phenomenon known as "ghost fishing" [37].

Entanglements, caused by macroplastics, are the most visible negative impact of the marine litter on marine fauna [37] but evidences about microplastics are also emerging: they enter the food chain and their assimilation may be causing an increase in complications in diseases such as inflammation, cancer in animals, and toxic effects (especially humans) due to bioaccumulation [38].

1.2.2 CE-inspired strategies for plastics: the role of biopolymers

Previous section showed how complex it is to deal with plastics: they are useful materials, which may be preferable to others as they have less impact on the environment if properly used, but they constitute a serious threat to the ecosystem when dispersed in it. CE strategies regarding polymers management must therefore include different levels of action.

Biopolymers (also commonly referred as nonrenewable resources) can play a really important role in this scenario and are regarded as a possibility to solve or limit the main problems associated with the consumption of conventional fossil-based and with the production of nonbiodegradable plastic mentioned previously. According to *European bioplastics,* a plastic material is referred to as a bioplastic if it is either biobased, biodegradable, or both [European bioplastics, 2018].

Biobased plastics are polymers wholly or partially made from renewable biomass sources (EN 16575). The international energy agency defines the biomass as "any organic matter, i.e. biological material, available on a renewable basis. It includes feedstock derived from animals or plants, such as wood and agricultural crops, and organic waste from municipal and industrial sources" [39]. Biobased plastics can be directly derived from natural polymers (for example produced from plants, microorganisms, and algae), or manufactured from natural monomers [40].

Biodegradability is a property very important for the disposal phase of a material: biodegradation is a biochemical process in which, thanks to microbial activity, materials break down into natural compounds (carbon dioxide, water, and methane) [38, 41]. Such mechanism is strictly dependent on polymeric chain structure and environmental biotic- and abiotic conditions: for example, if the process occurs in an oxygen rich environment (as for composting process), the source material partially degrades into carbon dioxide, while in presence of anaerobic conditions methane is produced [European bioplastics, 2018].

Main bioplastics are starch-based polymers, polyhydroxyalkanoates (PHAs), polylactic acid (PLA), cellulose-based polymers, biobased polyethylene (PE), biobased PVC, and protein-based polymers [38].

One of the environmental benefits associated with biobased polymers, compared to conventional plastics, is connected with the renewable nature of the raw materials they are made from [42], but there is still plenty of room for progress, since today their most efficient source of raw materials are carbohydrate-rich plants (i.e., corn, sugar cane) [European bioplastics, 2017], that are disputed by the food industry [38].

Regarding the GHG emissions, the switch from the use of traditional to biobased polymers could also be highly beneficial: for instance, polymers that are both biobased and biodegradable can be used to produce CO_2 neutral products: vegetables provide raw materials while sequestering CO_2 through photosynthesis and, when biodegradation occurs, the product is partially converted back to it, while CO_2 can be then again fixed by plants [43]. This can be a desirable option for short-to-medium shelf life products. Biobased but nonbiodegradable products can also play a role as long term carbon sink [38, 44].

Biodegradable polymers are at the center of CE biological nutrients flow, as they can safely return to the environment, rebuilding natural capital: they fit with the cradle-to-cradle design approach and are also claimed to play an important role in facing up plastic littering, even if a further technological improvement is needed, because, as previously mentioned, biodegradability strongly depends on the environmental conditions: industrial composting plants, soils, freshwaters and marine water evidently are different kind of environments, so biodegradation can occur at different rates and with different results. While standards concerning industrial composting and anaerobic digestion are available, biodegradation at home compost and marine environment are still not fully covered by the international standards and legislation [European bioplastics, 2017].

1.3 Evaluation of sustainability – the life cycle assessment approach

Potential benefits of a large-scale adoption of biopolymers in place of their conventional counterparts should be now conceptually clear, but a quantitative methodical analysis is required in order to verify this prospect.

Sustainability of a product is commonly valued by means of its environmental impact, measured through a life cycle assessment (LCA). An LCA is an articulated procedure that aims to evaluate the quantitative impact of a product on the environment analyzing the resources consumed and the emissions produced during its life cycle. The computation can be made over the entire life cycle (cradle-to-grave) or only on a part of it (gate-to-gate). The life cycle of a product consists of at least five steps (Figure 1.4): raw material extraction and material production, product manufacturing, transport and logistics, product use and, finally, product disposal. If along with disposal options recycling routes are also available, the correspondent LCA type is called cradle-to-cradle.

Two standards by the International Organization for Standardization [45, 46] define the minimal scheme of an LCA as follows:

– *Definition of goal and scope*: specification of the product and the functional unit (i.e., the reference quantity in mass, volume or, for example, number of items) chosen for the study; also here are specified system boundaries, simplification assumed, and limits of the study; finally, motivation and goals of the work are explained.
– *LCI*: declaration of data sources and flows inside the system. LCI includes writing off the bill of materials required for assembling the final product but also throughout the entire manufacturing process, determining how they combine into one or more new materials until the final product is obtained (system flows) and then

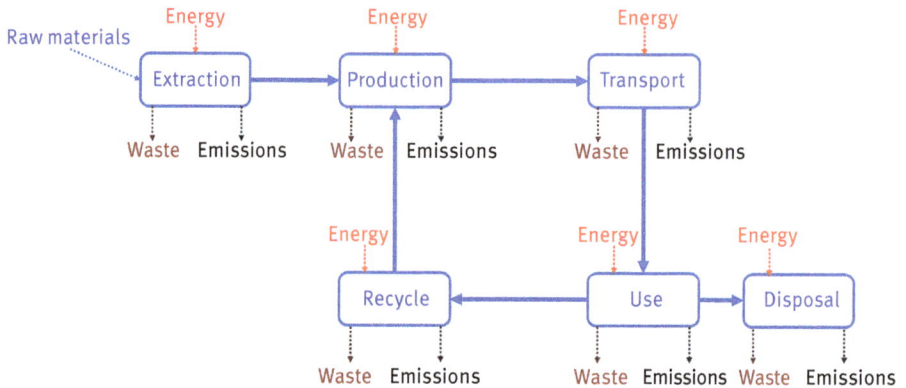

Figure 1.4: A basic example of an LCA scheme.

selecting the sources for the environmental data of each material. Usually, data are taken from dedicated databases and/or directly from manufacturers.

– *Life cycle impact assessment*: methods and algorithms used to elaborate data from LCI into quantitative results, expressed through specific impact categories such as global warming potential (GWP), nonrenewable resource depletion, eutrophication, acidification, and so on.

– *Interpretation*: evaluation and discussion of results obtained throughout the analysis, herein verification of the fulfillment of initial goals and identification of significant outcome about single contributions, single stages or the whole cycle, all aimed to drawing conclusions and recommendations.

LCA approach still presents some critical issues [47]. For example, the standards mentioned previously still leave too high a level of arbitrariness to the operator, thus LCA results objectivity can often undergo to critiques, while lack of uniformity causes problems on comparing two or more works.

LCA is nevertheless recognized as a unique and powerful tool and nowadays is widely adopted by companies to assess environmental performance of their products and by institutions to build up their policies.

1.3.1 LCA of biopolymers – an overview

The driving force to biopolymers development and diffusion clearly relies on their role as an environmental-friendly alternative to conventional polymers, so one can ask how strong is that force nowadays. In the following, a brief overview of significant results from LCA analysis and reviews provided by scientific literature to date is offered. In order to examine the question properly, biobased polymers and biodegradable polymers are discussed separately.

1.3.2 Biobased polymers

For the former category stands a tradeoff situation. Data suggest that switching from fossil-based to fully biobased polymers would decrease the depletion of nonrenewable resources and GWP in most of the cases [48–50], but unfortunately this is only part of the full picture. First of all, this outcome is shown to be greatly dependent on not only the material or product investigated but also from the specific production system and technology adopted, implying that general conclusions cannot be drawn. Furthermore, the analysis of other impact factors related to land use (like eutrophication, acidification, photochemical ozone formation) show that biopolymers impact more than their conventional counterparts [48, 50–56], mostly due to feedstock *production/cultivation*. To switch from fossil to biobased plastics without incurring in these drawbacks would thus be necessary to work on farming practices, avoiding tillage farming [57], improving fertilizers management, and exploiting of degraded soils. Reaching a higher production efficiency would also help to obtain the high amount of biomass required without increasing the environmental impact of the process. A significant contribution in achieving this goal can be given by the use of biomass derived from agricultural residues (also defined as second-generation feedstock, as opposed to first-generation feedstock, which is produced directly from agriculture) or waste (third-generation feedstock) can significantly contribute to lowering the original impact by simply avoiding new crops and instead profiting on the ongoing productions, reducing waste and at the same time increasing productivity, thus improving the overall efficiency of the system [48, 50, 56].

This is a clear example of how the CE principles, specially integrating the management of different product systems, can effectively shape the production patterns into more sustainable forms, extracting as much value as possible from raw materials by making use of every by-product of each phase and thus integrating all the system flows. Furthermore, investing on second- and third-generation feedstock can ensure less competition with the food production chain. Finally, the synthesis of chemicals and materials from biomass are up to biorefineries, so the continuous progress on synthesis methods and plants efficiency will positively affect the biobased scenery.

A significant source of variability in LCAs' results arises from the energy mix parameter. Every single process the material goes through requires some energy in one of its forms (electrical, mechanical, etc.). Manufacturers usually obtain this energy from the electric grid and then convert it in the desired form, but any way in which electricity is generated upstream, in fact, has a different environmental footprint that must be taken into consideration considering the so-called energy mix. The production process can also involve other forms of energy source: manufacturers can, for example, produce the energy by themselves with a gasoline or carbon powered generator instead of fastening to the local electric grid. Thus, the energy mix more generally describes how energy consumed in each step of the life cycle is generated. Typically, the energy mix is strongly dependent from the geographical localization of the LCA study [53, 56, 58]. The relevance of the energy mix over the whole life cycle stems up from the fact that the biobased production is highly energy-intensive, even more than fossil-based one [48, 52, 56]. Thus, even though biobased polymers do not require petrochemical raw material to be incorporated in the final product, the energy consumed by the production process is usually higher, which implied a still significant depletion of nonrenewable resources, because most of the energy production is still fossil-based. Switching to the biobased philosophy needs a contextual transition to renewable energy sources to get a real improvement in reducing both the use of natural resources and greenhouse gas emissions [56]. This point appears evident when considering that production of synthetic materials represents just about the 6% of global fossil fuel depletion, while the remaining part is related to energy conversion somewhere else in the economy [50]: transition to renewable energy sources will be unavoidable in future, even if 100% of synthetic materials are made from renewable resources. On the bright side, the energy surplus can eventually be supplied by other agricultural residues, not otherwise exploitable, or by biomass from biodegradable waste disposal (gerngross).

1.3.3 Biodegradable polymers

Making net statements about the influence of biodegradability on the environmental footprint of polymers is hard, even more than what has been observed for the biobased ones. The main reason is the lack of literature on the subject [55], with the few existing studies seeming to be widely heterogeneous and dispersive. In fact, when EOL options are taken into account, system complexity increases extremely, introducing inventory problems concerning the availability, sensitivity, and reliability of data [50, 52, 54, 59]. Also, modelling problems arise: for example, not every EOL option is available for each product because of technological limits, or lack of sufficient market volumes and subsequent economical unsustainability [60]; this aspect is also geographical sensitive since the local availability of specific disposal facilities can vary greatly. The high complexity of such systems and the poor reliability of inventory data, together with the lack of severe LCA standards and the consequent large methodological variety, result in poor significance of single studies and in a high-case specificity. Moreover, drawing general conclusions from comparisons among different analysis becomes extremely troublesome [55, 56, 60]. The only way left is to define indicative trends, but when more specific answers are required it is not possible to prescind from a case-by-case analysis [50, 52].

Impact assessment analysis on biodegradable polymers has been focusing on the three major products currently available on the market: PLA, PHA, and starch blends. Regarding the EOL options, as a global trend, mechanical and chemical recycling options show the best performance when they are considered, because on one hand they minimize greenhouse gases emissions and then GWP, and on the other hand maximize second raw material recovery thus abating the depletion of additional resources [54, 56, 60]. Apparently, even incineration with energy recovery performs better than composting or anaerobic digestion [56]. This paradox arises from the inability of current models to properly estimate a number of factors, including the facilitation of food waste collection, which avoids landfilling [54], the environmental credits acquired from the use of compost and digestate as soil conditioners [59], and the positive impact on littering phenomenon [60] and downcycling of polymers (i.e., loss of material quality) through the recycling cascade [54, 60]. Speaking of EOL, there is a point that everyone agrees on: the need to reach the zero-waste-to-landfill target. Landfilling of biodegradable products shows a high impact on GWP due to emission of harmful greenhouse gases (CO_2 and, even worse, CH_4) by the decomposition process. For this reason, landfill is the least desirable EOL option and should always be avoided [53–56, 60]. Finally, biopolymer technology is still immature compared to conventional plastics, whose productive processes have been studied and optimized for decades. Research is running fast on this field and impact assessment analysis may already be obsolete when they get published. A general overestimation of impact assessment results for biodegradable polymers is thus expected [55, 56].

In summary, biopolymers have gained high visibility as a potential answer to plastics sustainability question but, although a very interesting prospect for the immediate future, as of today they cannot be regarded as a universal solution to this problem, and any kind of general conclusion cannot be drawn in the absence of a prioritization of goals. Surely, their potential as a major weapon to struggle with depletion of natural resources gets confirmed, but some hidden costs [already reported by 58] cannot be neglected: in the first place, agricultural production is not as low impact as it might seem; second, biodegradation processes provide some impactful by-products as well. As a consequence, in order to wield the full potential of biopolymers from an environmental point of view, the development of efficient agricultural strategies and adequate disposal practices is strictly required [49, 53, 56, 60]. Furthermore, more effort should be put on outlining more rigorous standards for environmental LCA of biopolymers, soon allowing the availability of a base of more solid and comparable (hence, meaningful) data [49, 53, 56]. In conclusion, despite to date, those materials do not represent the final answer to plastics environmental problem, they still show a lot of room for improvement, so further research on this field appears promising and thus recommendable [51, 52, 55, 56].

1.4 Dealing with plastic pollution: the policy-making process

Since public attention on plastic pollution is increasing day by day, the interest of governments and small-to-big companies in the development of new strategies toward a less impacting plastics economy is increasing accordingly, so that several initiatives have been launched in recent years.

One of the most widespread is the new plastics economy global commitment, launched in 2018 by Ellen MacArthur foundation in collaboration with UN environment [61] following the two reports "The new plastics economy: rethinking the future of plastics" [41] and "The new plastics economy: catalyzing action" [36]. It involves over 150 companies worldwide, representing about 20% of the global plastic packaging production, along with numerous organizations like sector associations, universities, and governments. The aims of the initiative range from stimulating industry into investing in ecodesign, to making a strong network among companies that allows them to share their information and data in order to encourage transparency and simplify communication, collaboration, and the creation of a true CE framework.

European Union is devoting much effort to building strong policies, as proved by the plastics strategy launched by European Commission in 2018 [62]. This document stems up from the CE action plan [63] and represents a comprehensive scheme to face the challenge of leading the worldwide transition to a more sustainable plastics economy from an environmental, economic and social point of view. In the plastics strategy, the key questions, the vision proposed by the European Union and the actions necessary to implement this vision are outlined, including driving investments and innovation toward circular solutions through direct funding of research and infrastructure and promoting impact assessment of biodegradable and biobased plastics during the entire life cycle.

In the document is also discussed how to improve the plastics recycling chain, from an industrial and economic perspective. Part of the answer to that question is left to the circular plastics alliance [64], a voluntary platform that aims to connect the main stakeholders of the entire plastics value chain (from plastics manufacturers to the recycling industry, the waste collection infrastructures, etc.), in order to coordinate investments and actions, report the main obstacles faced by each stakeholder, and monitor progresses.

Finally, the plastics strategy underlines also the need of curbing plastic waste and littering. The core of this section has been developed to become the single-use plastics (SUP) directive, voted by European Parliament in March 2019 [65]. Attempts to curb littering problem taking aim at SUP are the focus of most of the policy-making activities around the world and this topic is so hot that *Collins Dictionary* named "single-use" word of the year in 2018.

The EU SUP directive obliges each member state to take actions about key issues regarding SUP, oxodegradable plastics, and fishing gears, like:

- obtain a sensible reduction in consumption of SUP;
- ban the oxodegradable and several SUP items (cotton bud sticks, cutlery, plates, straws, beverage stirrers, balloon sticks, polystyrene food containers);
- impose the production of caps and lids that stay attached to bottles;
- adopt a mandatory marking on some SUP items (sanitary towels, tampons, wipes, filters for tobacco and tobacco products with filters, cups for beverage) that inform of appropriate waste management and negative impact of inappropriate disposal;
- establish the extended producer responsibility (EPR) on all previously mentioned items (including fishing gears), intended to cover educational and informational, waste collection and cleaning up litter costs;
- set severe targets for SUP bottle collection and recycling (77% by 2025, 90% by 2029);
- promote educational campaigns aimed to push responsible consumer behaviors;
- encourage the use of sustainable alternatives for SUP items for food use;
- report annually on data and actions taken.

Each point will have to be discussed and received on a national legislative level within 2 years from formal approval by the European Council.

Others initiative regarding the plastic waste issues are under development: a microplastics ban is currently under study of European Chemicals Agency and it will be discussed by the European Commission in the next future. Some member states already planned a partial ban: for example, cosmetics containing microplastics will be banned from the Italian market by 2020 [66].

First attempts on making structured policies are coming to life, but a lot of work is still to be done. As pointed out by Stahel [67], a sustainable future is achievable only if a serious change in perspective is made: growth measured in stock quality instead of flow of resources (i.e., GDP), internalization of external costs, stewardship over ownership are just some actions the policies should foster through promotions, punishments and bans, guiding the society toward a more balanced equilibrium with the environment and the resources it donates to us.

References

[1]   Murray, A., Skene, K., and Haynes, K. (2017) The circular economy: An interdisciplinary exploration of the concept and application in a global context. J. Bus. Ethics, 140(3), 369–380.
[2]   OECD (2018), Meeting Policy Challenges for a Sustainable Bioeconomy, OECD Publishing, Paris, p. 18.
[3]   UN (2017), World population projected to reach 9.8 billion in 2050 and 11.2 billion in 2010, New York.
[4]   Ehrlich, P., and Holdren, J. (1971). Impact of population growth. Science, 171(3977), 1212–1217.

[5] Meadows, D. H., Meadows, D. L., Randers, J., and Behrens, W. (1972), The limit to growth, Universe Books, New York.

[6] WCED (UN) (1987) Our Common Future, the World Commission on Environment and Development of the United Nations.

[7] UN (2018), Emissions World Gap, New York.

[8] OECD (2019), Global Materials Resources Outlook to 2060, OECD Publishing, Paris, p. 5.

[9] Ellen MacArthur Foundation (2012) Towards the Circular Economy: Economic and Business Rationale for an Accelerated Transition.

[10] Preston, F. (2012). A global redesign?: Shaping the circular economy. London: Chatham House.

[11] Ghisellini, P., Cialani, C., and Ulgiati, S. (2016). A review on circular economy: the expected transition to a balanced interplay of environmental and economic systems. J. Clean. Prod., 114, 11–32.

[12] Boulding, K. (1966). E., 1966, The Economics of the Coming Spaceship Earth. New York.

[13] Andersen, M. S. (2007). An introductory note on the environmental economics of the circular economy. Sust. Sci., 2(1), 133–140.

[14] Pearce, D. W., and Turner, R. K. (1990). Economics of Natural Resources and the Environment. JHU Press, chap. II.

[15] McDonough, W., and Braungart, M. (2010). Cradle to cradle: Remaking the way we make things. North point press.

[16] Bocken, N. M., de Pauw, I., Bakker, C., and van der Grinten, B. (2016). Product design and business model strategies for a circular economy. J. Ind. Prod. Eng., 33(5), 308–320.

[16] European Commission – DG Environment (2014). Development of Guidance on Extended Producer Responsibility (EPR) FINAL REPORT.

[17] Hammond, G.P., and Jones, C.I, (2006) Inventory of (Embodied) Carbon & Energy (ICE), Department of Mechanical Engineering, University of Bath, United Kingdom

[18] Haynes, R. (2010). Embodied energy calculations within life cycle analysis of residential buildings. Etet1812. Staging-Cloud. Netregistry, 1–16.

[19] EU. (2008). Official Journal of EU, L 312, 19.11.2008. Directive 2008/98/EC of the European Parliament and of the Council of 19 November 2008 on waste and repealing certain directives. Available: http://eur-lex.europa.eu/LexUriServ/LexUriServ.do?uri¼OJ: L:2008:312:0003:0030:en:PDF.

[20] Ashby, M. F. (2012). Materials and the Environment: Eco-Informed Material Choice. Elsevier.

[21] Birat, J. P. (2015). Life-cycle assessment, resource efficiency and recycling. Metal. Res. Technol., 112(2), 206.

[22] Ignatyev, I. A., Thielemans, W., and Vander Beke, B. (2014). Recycling of polymers: a review. ChemSusChem, 7(6), 1579–1593.

[23] Braungart, Michael, McDonough, William, and Bollinger, Andrew. (2007). "Cradle-to-cradle design: creating healthy emissions – a strategy for eco-effective product and system design." J. Clean. Prod. 15 (13–14): 1337–1348.

[25] Stahel, W. R. (1998). From products to services: selling performance instead of goods. IPTS Rep., 27(1998), 35–42.

[26] Chertow, M. R. (2007). "Uncovering" industrial symbiosis. J. Ind. Ecol., 11(1), 11–30.

[27] Yazan, D. M., Romano, V. A., and Albino, V. (2016). The design of industrial symbiosis: an input–output approach. J. Clean. Prod., 129, 537–547.

[28] Erkman, S. (1997). Industrial ecology: an historical view. J. Clean. Prod., 5(1–2), 1–10.

[29] Geyer, R., Jambeck, J. R., and Law, K. L. (2017). Production, use, and fate of all plastics ever made. Sci. Adv., 3(7), e1700782.

[30] CIEL (2017) Fossils, plastics, & petrochemical feedstocks. In Fueling Plastics, pp. 1–5, Center for International Environment Law.

[31] Pilz, H., Brandt, B., and Fehringer, R. (2010). The impact of plastics on life cycle energy consumption and greenhouse gas emissions in Europe. Summary report June.

[32] Gewert, B., Plassmann, M. M., and MacLeod, M. (2015). Pathways for degradation of plastic polymers floating in the marine environment. Environ. Sci. Process Impacts, 17(9), 1513–1521.

[33] Gautam, R., Bassi, A. S., and Yanful, E. K. (2007). A review of biodegradation of synthetic plastic and foams. Appl. Biochem. Biotechnol., 141(1), 85–108.

[34] GESAMP. (2016). Sources, fate and effects of microplastics in the marine environment: part two of a global assessment. IMO/FAO/UNESCO-IOC/UNIDO/WMO/IAEA/UN/ UNEP/UNDP Joint Group of Experts on the Scientific Aspects of Marine Environmental Protection.

[35] Horton, A. A., Walton, A., Spurgeon, D. J., Lahive, E., and Svendsen, C. (2017). Microplastics in freshwater and terrestrial environments: evaluating the current understanding to identify the knowledge gaps and future research priorities. Sci. Total Environ., 586, 127–141.

[36] Ellen MacArthur Foundation (2017). The New Plastics Economy: Rethinking the future of Plastics and Catalysing Action.

[37] Crippa, M., De Wilde, B., Koopmans, R., Leyssens, J., Muncke, J., Ritschkoff, A-C., Van Doorsselaer, K., Velis, C., and Wagner, M. (2019). A Circular Economy for Plastics – Insights from Research and Innovation to Inform Policy and Funding Decisions, (M. De Smet & M. Linder, Eds.). European Commission, Brussels, Belgium.

[38] Karan, H., Funk, C., Grabert, M., Oey, M., and Hankamer, B. (2019). Green bioplastics as part of a circular bioeconomy. Trends Plant Sci.

[39] IEA, Bioenergy and Biofuels, 2017. https://www.iea.org/topics/renewables/bioenergy/.

[40] Meraldo, A. (2016). Introduction to bio-based polymers. In Multilayer Flexible Packaging (pp. 47–52). William Andrew Publishing.

[41] Ellen MacArthur Foundation, World Economic Forum and McKinsey & Company, 2016. The New Plastics Economy: rethinking the future of plastics. Ellen MacArthur Foundation: https://www.ellenmacarthurfoundation.org/assets/downloads/The-New-Plastics-Economy-Rethinking-the-Future-of-Plastics.pdf.

[42] Yu, J., and Chen, L. X. L. 2008. "The greenhouse gas emissions and fossil energy requirement of bioplastics from cradle to gate of a biomass refinery." Environ. Sci. Technol. 42: 6961–6.

[43] Sudesh, K., and Iwata, T. (2008). Sustainability of biobased and biodegradable plastics. CLEAN–Soil, Air, Water, 36(5–6), 433–442.

[44] Chen, Y. J. 2014. "Bioplastics and their role in achieving global sustainability." J. Chem. Pharm. Res. 6 (1): 226–31.

[45] ISO Environmental management, 2006. ISO Environmental management - Life cycle assessment - Principles and framework; ISO 14044: 2006. International Organization for Standardization, Switzerland.

[46] ISO Environmental management, 2006. ISO Environmental management - Life cycle assessment - Principles and framework; ISO 14040: 2006. International Organization for Standardization, Switzerland.

[47] Reap, J., Roman, F., Duncan, S., and Bras, B., 2008. A survey of unresolved problems in life cycle assessment Part 1: goal and scope and inventory analysis. Int. J. Life Cycle Assess 13: 290–300.

[48] Chen, L., Pelton, R.E.O., and Smith, T.M., 2016. Comparative life cycle assessment of fossil and bio-based polyethyleneterephthalate (PET) bottles. J. Clean. Prod. 137: 667–676.

[49] Spierling, S., Knüpffer, E., Behnsen, H., Mudersbach, M., Krieg, H., Springer, S., Albrecht, S., Herrmann, C., and Endres, H.-J., 2018a. Bio-based plastics - A review of environmental, social and economic impact assessments. J. Clean. Prod. 185: 476–491.

[50] Weiss, M., Haufe, J., Carus, M., Brandão, M., Bringezu, S., Hermann, B., and Patel, M.K., 2012. A review of the environmental impacts of bio-based materials. J. Ind. Ecol. 16 (s1): S169–S181.

[51] Guo, M., and Murphy, R.J., 2012. Is there a generic environmental advantage for starch–PVOH biopolymers over petrochemical polymers? J. Polym. Environ. 20: 976–990.

[52] Hermann, B.G., Blok, K., and Patel, M.K., 2010. Twisting biomaterials around your little finger: environmental impacts of bio-based wrappings. Int. J. Life Cycle Assess. 15: 346–358.

[53] Hottle, T.A., Bilec, M.M., and Landis, A.E., 2013. Sustainability assessments of bio-based polymers. Polym. Degrad. Stab. 98: 1898–1907.

[54] Hottle, T.A., Bilec, M.M., and Landis, A.E., 2017. Biopolymer production and end of life comparisons using life cycle assessment. Resour. Conserv. Recycl. 122: 295–306.

[55] Patel, M., 2003. Do Biopolymers Fulfill Our Expectations Concerning Environmental Benefits? Biodegradable Polymers and Plastics: 83–102. Springer, Boston.

[56] Yates, M.R., and Barlow, C.Y., 2013. Life cycle assessments of biodegradable, commercial biopolymers — A critical review. Resour. Conserv. Recycl. 78: 54–66.

[57] Kim, S., and Dale, B. E., 2008. Energy and greenhouse gas profiles of polyhydroxybutyrates derived from corn grain: A life cycle perspective. Environ. Sci. Technol. 42(20): 7690–7695.

[58] Gerngross, T.U., and Slater, S.C., 2000. How green are green plastics? Sci. Am. 283(2): 37–41.

[59] Hermann, B.G., Debeer, L., De Wilde, B., Blok, K., and Patel, M.K., 2011. To compost or not to compost: Carbon and energy footprints of biodegradable materials' waste treatment. Polym. Degrad. Stab. 96: 1159–1171.

[60] Spierling, S., Röttger, C., Venkatachalam, V., Mudersbach, M., Herrmann, C., and Endres, H.-J., 2018b. Bio-based plastics – a building block for the circular economy? Proc. CIRP 69: 573–578.

[61] Ellen MacArthur Foundation and UN Environment, 2019. New Plastics Economy Global Commitment: Spring 2019 Report. Ellen MacArthur Foundation: https://newplasticseconomy. org/assets/doc/GC-Spring-Report.pdf.

[62] European Commission, 2018. A European Strategy for Plastics in a Circular Economy. COM (2018) 28 final, 16 January 2018.

[63] European Commission, 2015. Closing the loop - An EU action plan for the Circular Economy. COM(2015) 614 final, 2 December 2015.

[64] European Commission, 2018. Commission launches Circular Plastics Alliance to foster the market of recycled plastics in Europe. IP/18/6728 (Press Release), 11 December 2018.

[65] European Parliament, 2019. Reduction of the impact of certain plastic products on the environment. P8_TA-PROV(2019)0305 (Provisional Edition), 27 March 2019.

[66] Italian Budget Law 2018. GU Serie Generale n.302 - Suppl. Ordinario n. 62, legge n.205 art. 1 comma 546.

[67] Stahel, W. R., 2016. The circular economy. Nature 531: 435–438.

[68] OECD (2016), Extended Producer Responsibility: Updated Guidance for Efficient Waste Management, OECD Publishing, Paris, https://doi.org/10.1787/9789264256385-en.

[69] Ellen MacArthur Foundation, World Economic Forum and SYSTEMIQ, 2017. The New Plastics Economy: Catalysing action. Ellen MacArthur Foundation: https://www.ellenmacarthurfounda tion.org/assets/downloads/New-Plastics-Economy_Catalysing-Action_13-1-17.pdf.

[70] European Chemical Agency (ECHA), 2019. Annex XV restriction report, proposal for a restriction on intentionally added microplastics: https://echa.europa.eu/documents/10162/ 0724031f-e356-ed1d-2c7c-346ab7adb59b.

[71] Jameel, F., Daystar, J., and Venditti, R.A., n.d. Environmental Life Cycle Assessment. North Carolina State University.

[72] World Economic Forum, Ellen MacArthur Foundation and McKinsey & Company. (2016). The New Plastics Economy — Rethinking the future of plastics. http://www.ellenmacarthur foundation.org/publications.

[73] European Bioplastics (2018) Fact Sheets What are Bioplastics? Available at: https://docs.euro peanbioplastics.org/publications/fs/EuBP_FS_What_are_bioplastics Acessed on 15/05/20.

Francesca De Falco, Emilia Di Pace, Maurizio Avella,
Mariacristina Cocca

2 Impact of plastics on marine environments: from macro- to microplastic pollution

Abstract: Since 1950s, the introduction of plastic materials in the global markets has led to important technological advances and brought many advantages in our life. Synthetic polymers are durable, lightweight, moldable and for these properties widely used in different sectors. However, the main disadvantage of the worldwide diffusion and the increased demand of plastics is represented by their disposal, which occurs too often in an uncontrolled way and ending up in the natural environment. Combining this aspect with the durability of plastics, the consequence is the problem of plastic pollution. In recent years, great attention and concern has been arising on the presence and consequent impact of plastics in marine environments. Plastic litter has been reported to interact by entanglement or ingestion with several species, from seabirds to crustaceans, fish, turtles and even big marine mammals. The problem of plastic pollution have become even more worrisome and challenging, when it was discovered that it involves not only the macro scale, but also the micro dimensions. Microplastics, plastic particles smaller than 5 mm, have been found in seas and oceans, rivers, lakes and remote places like the artic and the deep sea. These particles can be ingested and interact with marine fauna and potentially can leach toxic compounds or accumulate pollutants already present in the water. Currently, it is important to provide more researches and data on the impact of microplastics on the environment, particularly to better understand the possible consequences on humans.

Keywords: plastics, plastic pollution, microplastics, marine environment, marine litter

2.1 Diffusion of plastics

Strong, lightweight, moldable plastics are used in several products for different application sectors such as food and nonfood packaging, automotive, thermal and acoustic insulation, energy efficiency, sports, medicine, and electronics.

The diffusion of plastics started with the use of natural polymers, such as shellac and natural latex. The chemical modification of these substances represents the

Francesca De Falco, Emilia Di Pace, Maurizio Avella, Mariacristina Cocca, Institute for Polymers, Composites and Biomaterials, National Research Council of Italy, Pozzuoli, NA, Italy

https://doi.org/10.1515/9783110590586-002

starting point of the evolution of plastic materials, which have led to rubber, nitrocellulose, collagen, and galalith. Finally, the wide range of completely synthetic materials, recognized as modern plastics, has been developed during the last century [1].

In 1862, Alexander Parkes presented at the Great International Exhibition in London, a new material called "Parkesine," an organic material obtained by cellulose nitrate and suitable for molding. Treating cellulose nitrate with camphor, John Wesley Hyatt discovered, in 1870, a plastic that could be molded into various shapes and can substitute natural substances like tortoiseshell, horn, and ivory. The new horn-like material known as "celluloid" was a commercial success.

The first fully synthetic resin was invented in 1907 by the Belgian chemist Leo Hendrik Baekeland, named "Bakelite." Bakelite is a thermosetting resin that was synthetized by heating a mixture of phenol and formaldehyde together with a catalyst. The resin was a good insulator, durable, heat resistant, and suitable for mechanical mass production [2].

In 1945, at the beginning of the "plastic age," Yarsley and Couzens, in their book *Plastics*, described with optimism a world where man could make things and satisfy needs using synthetic materials from universally distributed substances [3].

With the introduction of new classes of synthetic polymers after the Second World War and later in the 1950s and 1960s, as a consequence of economic development, polymeric materials have become objects of everyday life [4]. During this period, the plastic industry starts its growth, new plastics were synthetized and produced such as polystyrene in 1929, polyester in 1930, polyvinylchloride and polythene in 1933, and nylon in 1935. The Ziegler–Natta catalysts allowed the massive industrial productions of polyolefins [5]. From 1950 to 2012, plastic grew at a rate of 8.7% per year, replacing materials like glass, metal, and paper [6]. It has been estimated that the amount of plastics (resins, fibers, and additives) produced from 1950 to 2015 was 8,300 million tons (Mt), around half of that was produced in the last 13 years [7].

The success of plastics is mainly due to the properties of such materials, suitable for the most disparate applications. Their properties and their low-cost production are responsible for the annual worldwide demand for plastics [8]. The benefits deriving from the use of plastics are globally accepted. For example, the introduction of plastics, in the sector of food and water packaging, have given the possibility to have a safe storage with the ability to control the inside atmosphere and temperature. In addition, reductions of material and energy consumption were obtained after the application of plastics in construction and transportation sectors [9]. The presence of about 50% of composite materials in the new Boeing 787 has generated a reduction of 20% in fuel costs. The improvements in materials science has allowed technological advances, the development of nanocomposites with nanofillers like graphene or carbon nanotubes have improved the application range of composites to different fields ranging from electronics (i.e., sensors), to aerospace, actuators, sporting goods, and fuel cells. Finally, the compatibility of polymers with biological tissues, have allowed biomedical applications such as drug delivery, orthodontic therapy, vascular stents, orthopedic implants, and so on [10, 11].

The increasing plastic demand is clearly visible from analyzing the amount of the global production of plastics that in 2016 was esteemed to be 335 Mt with 60 Mt produced in Europe. Moreover, in 2016, the total converter demand of plastics was 49.9 Mt, of which 39.9% was the demand for packaging sector, 19.7% for building and construction, 16.7% for other applications (i.e., furniture, mechanical engineering, appliances, medical, etc.), 10% for automotive and the rest for electrical and electronics, household, leisure, sports, and agriculture. The European converter demand by polymer types in 2016 is represented in Figure 2.1 [12]. This data does not consider the world consumption of synthetic fibers.

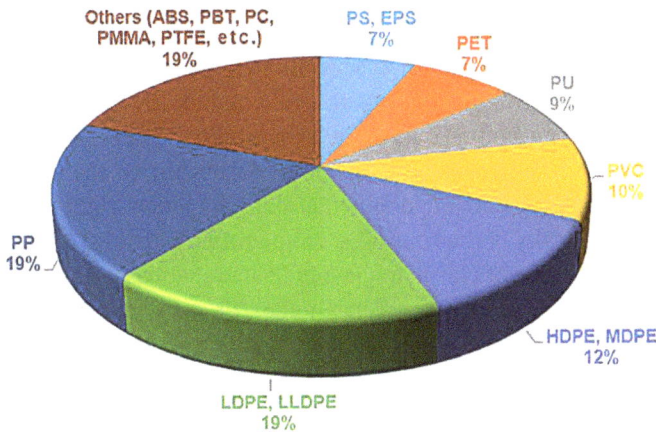

Figure 2.1: European plastics converter demand by polymer types in 2016. Source: Plastics Europe 2017 [12].

The disadvantage of this wide plastic diffusion, large demand, and their intrinsic high durability is their disposal. Actually, plastics after their usage and at the end of their life can be recycled, thermally destructed, or disposed in landfills. The landfill disposal represents an environmental problem due to uncontrolled discarding in dumps or in the environment. According to the European Plastic Industry Association, in 2016, 27.1 Mt of plastic wastes were collected through official schemes: 41.6% was destined to energy recovery, 27.3% was disposed in landfills, and 31.1% was recycled both inside and outside Europe [12].

It was esteemed that from 1950 to 2015, 6,300 Mt of plastics have been produced and that an amount of 12% (800 Mt) was recycled, 9% (600 Mt) was incinerated, and the other amounts were disposed in landfills or in the natural environment.

Clearly, taking into account the plastic disposal in landfill or in the environment and considering the plastic durability, the main drawback is plastic pollution.

Nowadays, plastic pollution represents a serious threat for the marine environment and it is still growing. In fact, plastic waste can be adopted as an indicator of Anthropocene era. Moreover, the current trends of plastic production and waste management could lead to the disposal of about 12,000 Mt of plastic waste in landfills or in the natural environment by 2050 [7].

2.2 Plastic as pollutant of marine environment

The absence of a correct waste management strategy in many countries and the massive plastic production has allowed the diffusion of plastic debris over the world, also in remote places. The accumulation of plastic debris is particularly visible, becoming a relevant problem, in the marine environment. The presence of plastic debris in marine ecosystems is documented in several scientific papers [13].

The ingestion of plastic fragments in albatross was reported for the first time in 1960 [14]. From this date, a lot of scientific data have been acquired on the accumulation and impact these plastic fragments have on the environment, on the living organisms and on human health [15]. Using debris distribution models and taking into account the population density of the countries, the maritime activities, and so on, it was esteemed that 5.25 trillion of plastic fragments floated the sea in 2014 [16]. Due to polymer density, polyethylene, polypropylene, and expanded polystyrene are the main floating debris while PVC, polyamides, and PET tend to sink. Moreover, it is to be highlighted that the fouling, sedimentation, and entanglement phenomena can favor the sinking of the plastic debris to deep sea and to the sediment. Once sunk, the degradation rates of the polymers slow down due to the low temperatures and the absence of light [15]. The fouling of plastic fishing gear induces their going down below the photic zone where the fouling colony is unable to survive, allowing the plastic to float again inducing the phenomenon defined as marine "snow" with a consequent alteration of the ecosystem [17].

Plastic debris recovered in marine environment are produced on the land or directly in the sea. These pollutants can be intentionally discarded or abandoned, and can reach the oceans through wastewater treatment plants, can be lost by ships, and so on [18]. The 18% of the marine plastic debris found in the ocean environment is attributed to the fishing industry that use polyolefin and nylons for fishing gears. Marine plastic debris produced in land is of about 80% [19]. The annual input of plastic to the ocean from waste generated by coastal populations worldwide was recently esteemed taking into account the mass of waste annually produced per person, the plastic waste, and the mismanaged plastic waste analyzing 192 coastal countries with at least 100 residents. Results showed that 275 Mt of plastic waste was generated in 2010 with 4.8 to 12.7 Mt entering the ocean. Without changing the waste management strategies, it has been predicted that by 2025 the amount of plastic waste reaching the ocean will increase by an order of magnitude. The waste produced directly in the sea are mainly due to the waste loss from ships notwithstanding that the dumping was forbidden by the international shipping regulation MARPOL Annex V [20].

Plastic debris remains unaltered in seas and oceans for many years due to their durability [21]. Degradation of plastics can be slowed by fouling effects and by low exposure to UV light. As a matter of fact, plastic fragments ingested by an albatross was related to an airplane shot down 9,600 km away and 60 years ago [22].

Two hundred and fifty species have been listed as species that have met plastics via entanglement or via ingestion, among that turtles; penguins; albatrosses, petrels, and shearwaters; shorebirds, skuas, gulls, and auks; coastal birds other than seabirds; baleen whales, toothed whales, and dolphins; earless or true seals, sea lions, and fur seals; manatees and dugong; sea otters; fish and crustaceans [23]. A lot of scientific data have been published on this field [13, 24, 25]. Recently, the impact of marine plastic pollution on the loggerhead turtle (*Caretta caretta*), is reported to lead to serious injuries and even the death of the animal. Three specimens of this turtle were collected stranded in Terceira Island, Azores (NE Atlantic): the first had to be amputated due to a piece of a nylon longline that was strangling its right forelimb causing a necrotic process; the second had swallowed a piece of swordfish longline which caused its death; and the third had its left forelimb entangled in a bowl of floating debris and was already amputated when found (Figure 2.2) [26].

Even though it might seem unexpected, plastic debris may also affect bigger marine animals, like sharks or whales. For instance, 25.16% of blue sharks caught in the Northwestern Mediterranean Sea, resulted positive to the ingestion of plastic litter, whose majority was composed of polypropylene (PP) sheet-like user plastic, the one commonly used as packaging material [27]. Moreover, the analysis of the gastrointestinal tracts of sperm whales found stranded along the North Sea coast in 2016, revealed that some of them ingested a wide range of different plastic debris, from netting to a part of a car [28]. In this last case, it is to be noted that the ingestion of plastic was not the cause of death of the mammals, but certainly indicate the wide presence and potential risk for the fauna of these pollutants.

Unfortunately, the risks related to the presence of plastic litter in marine environments are related to possible entanglement or ingestion, as reported above. The manufacturing process of plastic materials involves the use of additives to modify the bulk properties of a polymer (i.e., plasticizers, antiaging, flame retardants, UV stabilizers, colorants, fillers, etc.) [29]. Investigations have been carried out to assess if these toxic chemicals could be leached from plastic to marine ecosystems and organisms [30, 31]. Besides this possibility, there is also evidence that plastic can adsorb and accumulate other types of pollutants already present in the surrounding marine environment.

Figure 2.2: A specimen of *Caretta caretta* found in 2 April, 2008 in Terceira Island, Azores, entangled in a bowl of plastic lines and with its left forelimb already amputated and cicatrized [26].

The impact that plastic litter has on the environment is, of course, strongly correlated to the dimensions of these pollutants. When plastic waste is introduced in the environment, it does not degrade but is subjected to the action of phenomena like weathering and abrasion, which causes the fragmentation of the plastic debris [32], which could be easily spread by wind and wave action [33, 34].

In this respect, nowadays, more concern is arising in the scientific community regarding a new dimensional level of plastic pollution, represented by plastic particles called "microplastics."

2.3 Microplastics

The first time the term "microplastic" appeared was in a study in 1990 about the presence of plastic litter on beaches in South Africa. The term was used to indicate debris with a diameter less than 20 mm [35].

Actually, the first encounters with plastic fragments with submillimetric dimensions, occurred in the 1970s. During a sampling campaign in 1971 to assess possible environmental impacts of a nuclear power station in the Niantic Bay, New England, USA, marine scientists found in their plankton tows spherical polystyrene particles with a mean diameter of 0.5 mm. Further investigations revealed that some species of fish were ingesting them and hypotheses arose that such particles could be "suspension beads," normally molded into a pellet shape before being sold to plastic producers [36]. Then, in 1973, another plankton survey in North Atlantic also found particles of different polymer type and shape but all characterized by dimensions smaller than 5 mm. Most of the items recovered were polystyrene spherules and polyethylene (PE) cylinders or disks, both employed in plastic manufacturing [37].

Nevertheless, the first study that turned the spotlight on the problem of microplastic pollution was carried out in 2004 by researchers of the University of Plymouth, UK. They reused the term microplastic to describe the presence of microscopic plastic debris in sediments collected from beaches and from estuarine and subtidal sediments around Plymouth, UK [33]. After these findings, the scientific community has focused its attention on the investigation of the presence and environmental impact of these pollutants. In fact, at the International Research Workshop on the Occurrence, Effects, and Fate of Microplastic Marine Debris in 2009, a common definition of the term "microplastic" was decided: plastic particles smaller than 5 mm. Though the prefix "micro" suggests the use of microscopy to view these plastic pieces, due to the early state of research, it was chosen not to exclude visible components of the small plastic spectrum and thus set the upper limit at 5 mm. The workshop also introduced the differentiation between two main types of microplastics, on the basis of their sources: "primary" and "secondary" microplastics. The first are intentionally produced in submillimetric dimensions, either for direct use or as precursors to other products; the latter are formed in the environment from breakdown of larger plastic materials [23]. This terminology was also adopted by the Joint Group of Experts on the Scientific Aspects of Marine Environmental Protection (GESAMP) and by the United Nations Environment Program [38, 39]. While the term "secondary" generically comprises any microplastics deriving from the fragmentation or weathering of larger plastic debris (i.e., bags, packaging, etc.), the term "primary" includes more specific types of microplastics.

A recent report written by the International Union for Conservation of Nature (IUCN) expanded a bit the definition of primary microplastics, indicating in general plastics that are released directly into the environment in the form of submillimetric particles. Such definition includes not only plastics that are produced on purpose in these dimensions, but also particles that reached such dimensional scale due to the abrasion of other plastic items before they enter the environment. For this purpose, the IUCN identified the main sources of primary microplastics, estimating their impact on marine environments (Figure 2.3). According to their results, synthetic textiles are responsible for 35% of the global release of primary microplastics to the world oceans. Synthetic fibers like polyester and polyamide represent almost 60% of the annual global consumption of fibers, that is 69.7 Mt, used in the apparel

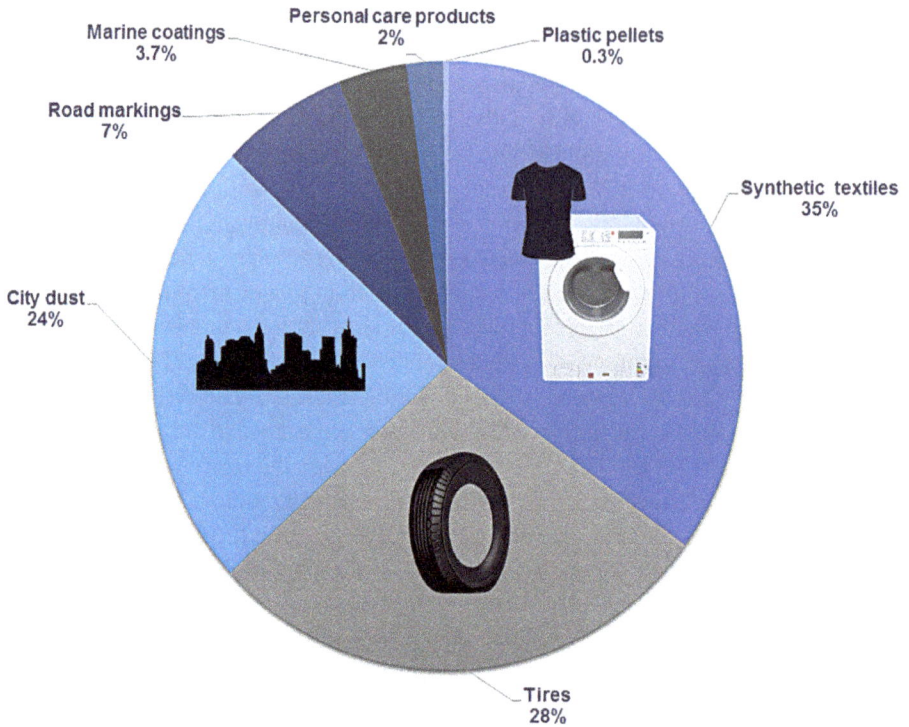

Figure 2.3: Global releases of primary microplastics to the world oceans by source (in %) [40].

industry, and their usage is constantly increasing [40]. During the washing, fabrics are subject to chemical and mechanical stresses that cause the detachment of microfibers from the yarns. Due to their dimensions, such microfibers cannot be completely blocked by wastewater treatments plants, and eventually end up in marine ecosystems [41]. Until now, this source of microplastics is one of the most investigated, with several works that have tried to evaluate the parameters that influence the release [42–45] in order to identify mitigation strategies [46, 47].

The second and third main sources of primary microplastics the IUCN identified are respectively tires and city dust. In fact, the outer parts of tires are made of a matrix of styrene butadiene rubber mixed with additives; the wear and tear of these parts cause the release of particles that are spread by the wind or washed off the road by rain [48]. Such source still needs more studies on the occurrence of tire particles and their possible impact on the environment. The expression "city dust" includes nine sources that are grouped together since their contribution is small, but they all have been found in urban environments. They are the abrasion of infrastructure (household dust, city dust, artificial turf, harbors and marine, building coating), the abrasion of objects (synthetic soles of footwear, synthetic cooking utensils), the blasting of abrasives, and intentional pouring (detergents) [40]. Also, in this case, more research is needed since this assessment is mainly based on the extrapolation of data from Nordic countries' studies [48–50]. Finally, other sources that contribute by less than 10% to the release of primary microplastics are: road marking, which are mainly made of paint and thermoplastics and can release microplastics from weathering or abrasion by vehicles; marine coatings, whose particles (i.e., polyurethane, epoxy, vinyl, and lacquers) may be released during building, maintenance or use of boats [40]; plastic pellets that can be spilled during manufacturing, processing transport or recycling [36]; and personal care products which has synthetic "microbeads" (mainly made of polyethylene) [51]. This last source has also been the first type of microplastics that has led to legislation actions by several governments worldwide [52].

2.4 Environmental impact of microplastics

Marine environments have been the first ecosystems investigated for the presence of microplastic pollution. First accounts of the presence of microplastics come from experimental activities performed in marine, oceanic, and coastal environments [33, 53]. Microplastics have been widely detected in subtidal sediments and on beaches worldwide [41, 54, 55], with great concentrations found in the subtropical gyres, already known as accumulation points of real "plastic islands" [56, 57], and in seas like the Mediterranean. In fact, recent works have shown that an average of 26,898 particles/km^2 are present in the South Pacific gyre [57], whereas an average density of plastic of 1 item per 4 m^2 was detected in the Mediterranean Sea, whose 83% of the total number of items were microplastics [58]. Regarding the type of polymers mostly found, for instance in the Mediterranean Sea, microplastics collected were composed by 52% of polyethylene, 16% of polypropylene, followed by minor percentages of synthetic paints (7.7%), polyamides (6.6%), and epoxy resins (5%) [59]. Recently, several researches have also highlighted how microplastics are present not only in marine environments, but also in freshwater systems. Their presence was reported in rivers like the Rhine and the Meuse in the Netherlands and Germany [60], in the Danube river [61], in estuaries in the UK and in China [34, 62], in French rivers [63]; in several lakes in Canada [64], in northern and central Italy [65, 66], in Mongolia [67], and in the African Great Lakes [68]. Nevertheless, an even more concerning situation arises from studies that have detected microplastics in remote locations, like the Arctic and the deep sea [69–71]. According to the modelling of van Sebille et al., from 15 to 51 trillion microplastic particles accumulated in 2014 globally, with the greatest quantities estimated in the North Pacific and in the Mediterranean Sea [72].

The inevitable consequence of such ubiquitous presence of microplastics in the most different ecosystems, is the impact they may have on the fauna, since they may be ingested by marine organisms like plankton, fish and so on, with the possibility to end up in the human food web [73]. In fact, several experiments have demonstrated that different species of zooplankton can ingest microplastics, since the dimensions of these particles are extremely similar to those of their natural food [74–76]. Passing to greater organisms, Lusher et al. analyzed 504 fish of 10 different species front the English Channel and found microplastics in 36.5% of them [77].

Fourteen species of fish from the Amazon River Estuary were also examined by Pegado et al., detecting 228 microplastics in the gastrointestinal tracts of 26 fish, with sizes ranging from 0.38 to 4.16 mm. Of these, 97.4% of the microplastics found were pellets, while much smaller percentages were recorded for sheets (1.3%), threads (0.9%), and fragments (0.4%), and the main polymers recognized were polyamide, rayon, and PE [78]. Several works has pointed out how the ingestion of these pollutants may have a negative impact on filtration rates, reproductive capacity, immunology, and respiration rates of filter-feeders such as mussels and oysters [79–81].

Another aspect to take into account in the investigation of the effects of microplastics on marine organisms is that they can transfer toxic substances to the fauna. First of all, plastics may leach compounds already present in their chemical structure, like nonreacted monomers or oligomers and additives. Moreover, due to their intrinsic hydrophobicity, microplastics may concentrate persistent organic pollutants (POPs), already present in water [82]. The first study that reported the presence of microplastics, in 1972, already detected the presence of polychlorinated biphenyls (PCBs) on the surface of the polystyrene pellets recovered and, since PCBs are not involved in the production of polystyrene, it was hypothesized that they came from the surrounding seawater [36]. Later, another research in 2001, pointed out the presence of PCBs, dichlorodiphenyldichloroethylene and nonylphenols on polypropylene resin pellets collected on four Japanese coasts [83]. After these first accounts, more studies have investigated the mechanisms that lead to the sorption/desorption of hydrophobic organic pollutants by microplastics in marine environments.

Teuten et al. compared the capacity of polyvinyl chloride (PVC), PP, PE, and of natural sediments of sorption and desorption of phenanthrene (PHNH). Results showed that the sorption capacity of PHNH for the polymers was much higher than for sediments, while desorption rates of PHNH in seawater were significantly lower than for natural sediments [84]. Regarding the types of plastic more worrisome, several works have shown that PE seems to accumulate more organic pollutants than other plastics like PVC and PP [85–87].

In 2012, Bakir et al. investigated the simultaneous sorption of two model POPs, PHNH and dichlorodiphenyltrichloroethane (DDT) onto PVC and ultrahigh molecular weight polyethylene, finding an antagonistic effect of DDT which interfere with the sorption of PHNH onto plastic [88]. Then, they carried out a broader analysis considering the potential for PVC and PE to sorb and desorb DDT, PHNH, ^{14}C-perfluorooctanoic acid and ^{14}C-di-2-ethylhexyl phthalate (DEHP). They evaluated the desorption rates of POPs both in seawater and under simulated gut conditions, studying the influence of pH and temperature to represent cold- and warm-blooded organisms. Their outcomes showed that desorption rates were faster with gut surfactant and simulating warm-blooded organisms, with desorption under gut conditions up to 30 times greater than in seawater alone [89].

Such findings highlight another key question about the ability of plastics to concentrate POPs in marine environment, that is, if these synthetic polymers can actually transfer such contaminants to marine organisms. At the moment, there is no clear answer. Rochman et al. analyzed fish exposed to a mixture of PE that absorbed chemical pollutants from the marine environment, finding that the fish was able to bioaccumulate the contaminants and then suffered liver toxicity and pathology [90]. In other experiments, lugworms exposed to contaminated microparticles of PVC, a transfer of pollutants from plastics to the organisms was also recorded, with consequent negative biological effects [91, 92]. In contrast with these works, other studies pointed out that the concentration of microplastics generally used in most experiments are much higher than those commonly reports in the natural environment. As a consequence, doubts arise if the transfer of contaminants from microplastics is a quantitatively important route in comparison with other pathways like respiratory or uptake from food, as also suggested by calculations made with mathematical models [93–95].

In conclusion, the impact of microplastics on marine fauna is still an open matter that strongly needs more scientific data, particularly in light of possible consequences on humans, since microplastics have been found in bivalves cultured and sold for human consumption [96], in fishes [97] and even in sea salt [98].

References

[1]   Richardson, T. L., and Lokensgard, E. (2004). Industrial Plastics: Theory and Applications. Thomson Delmar Learning, Albany.
[2]   Mossman, S. T. I., and Morris, P. J. T. (1994). The Development of Plastics, Special Publication No. 141. Royal Society of Chemistry, Cambridge.
[3]   Yarsley, V. E., and Couzens, E. G. (1945). Plastics. Middlesex: Penguin Books Limited.
[4]   Albus, S., Bonten, C., Keßler, K., Rossi, G., and Wessel, T. (2007). Plastic Art: A Precarious Success Story. AXA Art, Cologne.
[5]   Cecchin, G., Morini, G., and Piemontesi, F. (2003). Ziegler–Natta catalysts. In: KirkOthmer Encyclopedia of Chem Technology. Wiley, New York.
[6]   Gourmelon, G. Global Plastic Production Rises, Recycling Lags – New Worldwatch Institute Analysis Explores Trends in Global Plastic Consumption and Recycling. http://www.world watch.org/global-plastic-production-rises-recycling-lags-0.

[7] Geyer, R., Jambeck, J., and Law, K. L. (2017). Production, use, and fate of all plastics ever made. Sci. Adv. 3, e1700782.

[8] Thompson, R. C., Swan, S. H., Moore, C. J., and vom Saal, F. S. (2009). Our plastic age. Phil. Trans. R. Soc. B 364, 1973–1976.

[9] Andrady, A. L., and Neal, M. A. (2009). Applications and societal benefits of plastics. Phil. Trans. R. Soc. B 364, 1977–1984.

[10] Mittal, G., Dhand, V., Rhee, K. J., Park, S.-J., and Lee, W. L. (2015). A review on carbon nanotubes and graphene as fillers in reinforced polymer nanocomposites. J. Ind. Eng. Chem. 21, 11–25.

[11] Tian, H., Tang, Z., Zhuang, X., Chen, X., and Jing, X. (2012). Biodegradable synthetic polymers: preparation, functionalization and biomedical application. Prog. Polym. Sci. 37, 237–280.

[12] Europe, Plastics. (2017). An analysis of European plastics production, demand and waste data. https://www.plasticseurope.org/en/resources/publications/274-plastics-facts-2017.

[13] Moore, C. J. (2008). Synthetic polymers in the marine environment: a rapidly increasing, long-term threat. Environ. Res. 108, 131–139.

[14] Kenyon, K. W., and Kridler, E. (1969). Laysan Albatross swallow indigestible matter. The Auk 86, 339–343.

[15] Thompson, R. C., Moore, C. J., vom Saal, F. S., and Swan, S. H. (2009). Plastics, the environment and human health: current consensus and future trends. Phil. Trans. R. Soc. B 364, 2153–2166.

[16] Eriksen, M., Lebreton, L. C. M., Carson, H. S., Thiel, M., Moore, C. J., Borerro, J. C., Galgani, F., Ryan, P. G., and Reisser, J. (2014). Plastic pollution in the World's oceans: more than 5 trillion plastic pieces weighing over 250,000 tons afloat at sea. PLoS One 9, 1e15.

[17] Andrady, A. L. (2000). Plastics and their impacts in the marine environment. In: Proceedings of the International Marine Debris Conference on Derelict Fishing Gear and the Ocean Environment, 6–11 August 2000, Honolulu, Hawaii.

[18] Cesa, F. S., Turra, A., and Baruque-Ramos, J. (2017). Synthetic fibers as microplastics in the marine environment: a review from textile perspective with a focus on domestic washings. Sci. Total Environ. 598, 1116–1129.

[19] Andrady, A. L. (2011). Microplastics in the marine environment. Mar. Pollut. Bull. 62, 1596–1605.

[20] Jambeck, J. R., Geyer, R., Wilcox, C., Siegler, T. R., Oerryman, M., Andrady, A., Narayan, R., and Law, K. L. (2015). Plastic waste inputs from land into the ocean. Science 347, 768–771.

[21] Sudhakar, M., Trishul, A., Doble, M., Kumar, K. S., Jahan, S. S., Inbakandan, D., Viduthalai, R. R., Umadevi, V. R., Murthy, P. S., and Venkatesan, R. (2007). Biofouling and biodegradation of polyolefins in ocean waters. Polym. Degrad. Stab. 92, 1743–1752.

[22] Barnes, D. K. A., Galgani, F., Thompson, R. C., and Barlaz, M., (2009). Accumulation and fragmentation of plastic debris in global environments. Phil. Trans. R. Soc. B 364, 1985–1998.

[23] Arthur, C., Baker, J., and Bamford, H. (2009). Proceedings of the International Research Workshop on the Occurrence, Effects and Fate of Microplastic Marine Debris, 49. NOAA Technical Memorandum NOS-OR&R-30.

[24] Gregory, M. R. (2009). Environmental implications of plastic debris in marine settings – entanglement, ingestion, smothering, hangers-on, hitch-hiking and alien invasions. Phil. Trans. R. Soc. B 364, 2013–2025.

[25] Gall, S. C., and Thompson, R. C. (2015). The impact of debris on marine life. Mar. Poll. Bull. 92, 170–179.

[26] Barreiros, J. P., and Raykov, V. S. (2014). Lethal lesions and amputation caused by plastic debris and fishing gear on the loggerhead turtle Caretta caretta (Linnaeus, 1758). Three case reports from Terceira Island, Azores (NE Atlantic). Mar. Pollut. Bull. 86, 518–522.

[27] Berdardini, I., Garibaldi, F., Canesi, L., Fossi, M. C., and Baini, M. (2018). First data on plastic ingestion by blue sharks (Prionace glauca) from the Ligurian Sea (North-Western Mediterranean Sea). Mar. Pollut. Bull. 135, 303–310.

[28] Unger, B., Bravo Rebolledo, E. L., Deaville, R., Gröne, A., Ijsseldijk, L. L., Leopold, M. F., Siebert, U., Spitz, J., Wohlsein, P., and Herr, H. (2016). Large amounts of marine debris found in sperm whales stranded along the North Sea coast in early 2016. Mar. Pollut. Bull. 112, 134–141.

[29] Brydson, J. A. (1999). Plastics materials. Butterworth-Heinemann, Oxford.

[30] Teuten,, et al. (2009). Transport and release of chemicals from plastics to the environment and to wildlife. Phil. Trans. R. Soc. 364, 2027–2045.

[31] Koelmans, A. A., Besseling, E., and Foekema, E. M. (2014). Leaching of plastic additives to marine organisms. Environ. Pollut. 187, 49–54.

[32] Thompson, R., Moore, C., Andrady, A., Gregory, M., Takada, H., and Weisberg, S. (2005). New directions in plastic debris. Science 310, 1117–1119.

[33] Thompson, R. C., Olsen, Y., Mitchell, R. P., Davis, A., Rowland, S. J., John, A. W. G., McGonigle, D., and Russell, A. E. (2004). Lost at sea: where is all the plastic? Science 304, 838.

[34] Browne, M. A., Galloway, T. S., and Thompson, R. C. (2010). Spatial patterns of plastic debris along estuarine shorelines. Environ. Sci. Technol. 44, 3404–3409.

[35] Ryan, P., and Moloney, C. (1990). Plastic and other artefacts on South African beaches: temporal trends in abundance and composition. S. Afr. J. Sci. 86, 450–452.

[36] Carpenter, E. J., Anderson, S. J., Harvey, G. R., Miklas, H. P., and Peck, B. B. (1972). Polystyrene spherules in coastal waters. Science 178, 749–750.

[37] Colton, J. B., Jr., Knapp, F. D., and Burns, B. R. (1974). Plastic particles in surface waters of the Northwestern Atlantic. Science 185, 491–497.

[38] GESAMP. (2015). Sources, fate and effects of microplastics in the marine environment: a global assessment. In: IMO/FAO/UNESCO-IOC/ UNIDO/WMO/IAEA/UN/UNEP/UNDP Joint Group of Experts on the Scientific Aspects of Marine Environmental Protection (P. J. Kershaw, Ed., 96p.). Rep Stud GESAMP No. 90.

[39] UNEP. (2016). Marine Plastic Debris and Microplastics: Global Lessons and Research to Inspire Action and Guide Policy Change. United Nations Environment Programme (UNEP), Nairobi. http://ec.europa.eu/environment/marine/good-environmental-status/descriptor-10/pdf/Marine_plastic_debris_and_microplastic_technical_report_advance_copy.pdf.

[40] Boucher, J., and Friot, D. (2017). Primary Microplastics in the Oceans: A Global Evaluation of Sources (43 pp.). IUCN, Gland, Switzerland. https://doi.org/10.2305/IUCN.CH.2017.01.en.

[41] Browne, M. A., Crump, P., Niven, S. J., Teuten, E., Tonkin, A., Galloway, T., and Thompson, R. C. (2011). Accumulations of microplastic on shorelines worldwide: sources and sinks. Environ. Sci. Technol. 45, 9175–9179.

[42] Napper, I. E., and Thompson, R. C. (2016). Release of synthetic microplastic plastic fibres from domestic washing machines: effects of fabric type and washing conditions. Mar. Pollut. Bull. 112, 39–45.

[43] Carney Almroth, B., Aström, L., Roslund, S., Petersson, H., Johansson, M., and Persson, N., 2018. Quantifying shedding of synthetic fibers from textiles; a source of microplastics released into the environment. Environ. Sci. Pollut. Res. 25, 1191.

[44] De Falco, F., Gullo, M. P., Gentile, G., Di Pace, E., Cocca, M., Gelabert, L., Brouta-Agnésa, M., Rovira, A., Escudero, R., Villalba, R., Mossotti, R., Montarsolo, A., Gavignano, S., Tonin, C., and Avella, M. (2018). Evaluation of microplastic release caused by textile washing processes of synthetic fabrics. Environ. Pollut. 236, 916–925.

[45] De Falco, F., Di Pace, E., Cocca, M., and Avella, M. (2019). The contribution of washing processes of synthetic clothes to microplastic pollution. Sci. Rep. 9, 1–11.

[46] De Falco, F., Gentile, G., Avolio, R., Errico, M. E., Di Pace, E., Ambrogi, V., Avella, M., and Cocca, M. (2018). Pectin based finishing to mitigate the impact of microplastics released by polyamide fabrics. Carbohydr. Polym. 198, 175–180.

[47] De Falco, F., Cocca, M., Guarino, V., Gentile, G., Ambrogi, V., Ambrosio, L., and Avella, M. (2019). Novel finishing treatments of polyamide fabrics by electrofluidodynamic process to reduce microplastic release during washings. Polym. Degrad. Stab. 165, 110–116.

[48] Sundt, P., Schultze, P., and Syversen, F. (2014). Sources of microplastic pollution to the marine environment. Norw. Environ. Agency 108, 1. http://www.miljodirektoratet.no/Documents/publikasjoner/M321/M321.pdf.

[49] Lassen, C., Foss Hansen, S., Magnusson, K., Noren, F., Bloch Hartmann, N. I., Rehne Jensen, P., Gisel Nielsen, T., and Brinch, A. (2015). Microplastics: Occurrence, Effects and Sources of Releases to the Environment in Denmark. The Danish Environmental Protection Agency.

[50] Magnuson, K., Eliason, K., Frane, A., Hakonen, K., Hulten, J., Olshammar, M., Stadmark, J., and Voisin, A. (2016). Swedish Sources and Pathways for Microplastics to the Marine Environment. www.ivl.se/webdav/files/Rapporter/C183 pdf.

[51] Fendall, L. S., and Sewell, M. A. 2009. Contributing to marine pollution by washing your face: microplastics in facial cleansers. Mar. Pollut. Bull. 58, 1225–1228.

[52] Campaign "Beat the microbead". http://www.beatthemicrobead.org/results-so-far/.

[53] Barboza, Luís Gabriel Antão, and Gimenez, B.C.G. (2015). Microplastics in the marine environment: current trends and future perspectives. Mar. Pollut. Bull. 97, 5–12.

[54] de Lucia, G. A., Caliani, I., Marra, S., Camedda, A., Coppa, S., Alcaro, L., Campani, T., Giannetti, M., Coppola, D., Cicero, A. M., Panti, C., Baini, M., Guerranti, C., Marsili, L., Massaro, G., Fossi, M. C., and Matiddi, M. (2014). Amount and distribution of neustonic micro-plastic off the western Sardinian coast (Central-Western Mediterranean Sea). Mar. Environ. Res. 100, 10–16.

[55] Song, Y. Y., Sang, H. H., Jang, M., Kang, J. H., Kwon, O. Y., Han, G. M., and Shim, W. J., (2014). Large accumulation of micro-sized synthetic polymer particles in the sea surface microlayer. Environ. Sci. Technol. 48, 9014–9021.

[56] Moore, C. J., Moore, S. L., Leecaster, M. K., and Weisberg, S. B. (2001). A comparison of plastic and plankton in the North Pacific Central Gyre. Mar. Pollut. Bull. 42, 1297–1300.

[57] Eriksen, M., Maximenko, N., Thiel, M., Cummins, A., Lattin, G., Wilson, S., Hafner, J., Zellers, A., and Rifman, S. (2013). Plastic pollution in the South Pacific subtropical gyre. Mar. Pollut. Bull. 68, 71–76.

[58] Cózar, A., Sanz-Martín, M., Martí, E., González-Gordillo, J. I., Ubeda, B., Gálvez, J. A., Irigoien, X., and Duarte, C. M. (2015). Plastic accumulation in the Mediterranean Sea. PLoS One 10(4), e0121762.

[59] Suaria, G., Avio, C. G., Mineo, A., Lattin, G. L., Magaldi, M. G., Belmonte, G., Moore, C. J., Regoli, F., and Aliani, S. (2016). The Mediterranean Plastic Soup: synthetic polymers in Mediterranean surface waters. Sci. Rep. 6, 37551.

[60] Leslie, H. A., Brandsma, S. H., van Velzen, M. J. M., and Vethaak, A. D. (2017). Microplastics en route: field measurements in the Dutch river delta and Amsterdam canals, wastewater treatment plants, North Sea sediments and biota. Environ. Int. 101, 133–142.

[61] Lechner, A., Keckeis, H., Lumesberger-Loisl, F., Zens, B., Krusch, R., Tritthart, M., Glas, M., and Schludermann, E. (2014). The Danube so colourful: a potpourri of plastic litter outnumbers fish larvae in Europe's second largest river. Environ. Pollut. 188, 177–181.

[62] Zhao, S., Zhu, L., Wang, T., and Daoji, L. (2014). Suspended microplastics in the surface water of the Yangtze Estuary system, China: first observations on occurrence, distribution. Mar. Pollut. Bull. 86, 562–568.

[63] Sanchez, W., Bender, C., and Porcher, J. (2014). Wild gudgeons (Gobio gobio) from French rivers are contaminated by microplastics: preliminary study and first evidence. Environ. Res. 128, 98–100.

[64] Driedger, A. G. J., Dürr, H. H., Mitchell, K., and Van Cappellen, P. (2015). Plastic debris in the Laurentian Great Lakes: a review. J. Great Lakes Res. 41, 9–19.

[65] Sighicelli, M., Pietrelli, L., Lecce, F., Iannilli, V., Falconieri, M., Coscia, L., Di Vito, S., Nuglio, S., and Zampetti, G. (2018). Microplastic pollution in the surface waters of Italian Subalpine Lakes. Environ. Pollut. 236, 645–651.

[66] Fisher, E. K., Paglialonga, L., Czech, E., and Tamminga, M. (2016). Microplastic pollution in lakes and lake shoreline sediments – a case study on Lake Bolsena and Lake Chiusi (central Italy). Environ. Pollut. 213, 648–657.

[67] Free, C. M., Jensen, O. P., Mason, S. A., Eriksen, M., Williamson, N. J., and Boldgiv, B. (2014). High-levels of microplastic pollution in a large, remote, mountain lake. Mar. Pollut. Bull. 85, 156–163.

[68] Biginagwa, F. J., Mayoma, B. S., Shashoua, Y., Syberg, K., and Khan, F. R. (2016). First evidence of microplastics in the African Great Lakes: Recovery from Lake Victoria Nile perch and Nile tilapia. J. Great Lakes Res. 42, 146–149.

[69] Obbard, R. W., Sadri, S., Wong, Y. Q., Khitun, A. A., Baker, I., and Thompson, R. C. (2014). Global warming releases microplastic legacy frozen in Arctic Sea ice. Earth's Future 2, 315–320.

[70] Lusher, A. L., Tirelli, V., O'Connor, I., and Officer, R. (2015). Microplastics in Arctic polar waters: the first reported values of particles in surface and sub-surface samples. Sci. Rep. 5, 14947.

[71] Van Cauwenberghe, L., Vanreusel, A., Mees, J., and Janssen, C. J. (2013). Microplastic pollution in deep-sea sediments. Environ. Pollut. 182, 495–499.

[72] van Sebille, E., Wilcox, C., Lebreton, L., Maximenko, N., Hardesty, B. D., van Franeker, J. A., Eriksen, M., Siegel, D., Galgani, F., and Law, K. L. (2015). A global inventory of small floating plastic debris. Environ. Res. Lett. 10, 124006.

[73] Wright, S. L., Thompson, R. C., and Galloway, T. S. (2013). The physical impacts of microplastics on marine organisms: a review. Environ. Pollut. 178, 483–492.

[74] Cole, M., Lindeque, P., Fileman, E., Halsband, C., Goodhead, R., Moger, J., and Galloway, T. S. (2013). Microplastic ingestion by zooplankton. Environ. Sci. Technol. 47, 6646–6655.

[75] Setälä, O., Fleming-Lehtinen, V., and Lehtiniemi, M. (2014). Ingestion and transfer of microplastics in the planktonic food web. Environ. Pollut. 185, 77–83.

[76] Krogh Frydkjær, C., Iversen, N., and Roslev, P. (2017). Ingestion and egestion of microplastics by the Cladoceran Daphnia magna: effects of regular and irregular shaped plastic and sorbed phenanthrene. Bull. Environ. Contam. Toxicol. 99, 655–661.

[77] Lusher, A. L., McHugh, M., and Thompson, R. C. (2013). Occurrence of microplastics in the gastrointestinal tract of pelagic and demersal fish from the English Channel. Mar. Pollut. Bull. 67, 94–99.

[78] Pegado, S., de Souza, T., Schmid, K., Winemiller, K. O., Chelazzi, D., Cincinelli, A., Dei, L., and Giarrizzo, T. (2018). First evidence of microplastic ingestion by fishes from the Amazon River estuary. Mar. Pollut. Bull. 133, 814–821.

[79] von Moos, N., Burkhardt-Holm, P., and Koehler, A. (2012). Uptake and effects of microplastics on cells and tissue of the blue mussel Mytilus edulis L. after an experimental exposure. Environ. Sci. Technol. 46, 11327–11335.

[80] Van Cauwenberghe, L., Claessens, M., Vandegehuchte, M. B., and Janssen, C. R. (2015). Microplastics are taken up by mussels (Mytilus edulis) and lugworms (Arenicola marina) living in natural habitats. Environ. Pollut. 199, 10–17.

[81] Green, D. S. (2016). Effects of microplastics on European flat oysters, Ostrea edulis and their associated benthic communities. Environ. Pollut. 216, 95–103.

[82] Andrady, A. L. (2017). The plastic in microplastics: a review. Mar. Pollut. Bull. 119, 12–22.

[83] Mato, Y., Isobe, T., Takada, H., Kanehiro, H., Ohtake, C., and Kaminuma, T. (2001). Plastic resin pellets as a transport medium for toxic chemicals in the marine environment. Environ. Sci. Technol. 35, 308–324.

[84] Teuten, E. L., Rowland, S. J., Galloway, T. S., and Thompson, R. C. (2007). Potential for plastics to transport hydrophobic contaminants. Environ. Sci. Technol. 41, 7759–7764.

[85] Teuten, E. L., Saquing, J. M., Knappe, D. R. U., Barlaz, M. A., Jonsson, S., Björn, A., Rowland, S. J., Thompson, R. C., Galloway, T. S., Yamashita, R., Ochi, D., Watanuki, Y., Moore, C., Viet, P. H., Tana, T. S., Prudente, M., Boonyatumanond, R., Zakaria, M. P., Akkhavong, K., Ogata, Y., Hirai, H., Iwasa, S., Mizukawa, K., Hagino, Y., Imamura, A., Saha, M., and Takada, H. (2009). Transport and release of chemicals from plastics to the environment and to wildlife. Philos. Trans. R. Soc. B Biol. Sci. 364, 2027–2045.

[86] Fries, E., and Zarfl, C. (2012). Sorption of polycyclic aromatic hydrocarbons (PAHs) to low and high density polyethylene (PE). Environ. Sci. Pollut. Res. 19, 1296–1304.

[87] Rochman, C. M., Hoh, E., Hentschel, B. T., and Kaye, S. (2013). Long-term field measurement of sorption of organic contaminants to five types of plastic pellets: implications for plastic marine debris. Environ. Sci. Technol. 47, 1646–1654.

[88] Bakir, A., Rowland, S. J., and Thompson, R. C. (2012). Competitive sorption of persistent organic pollutants onto microplastics in the marine environment. Mar. Pollut. Bull. 64, 2782–2789.

[89] Bakir, A., Rowland, S. J., and Thompson, R. C. (2014). Enhanced desorption of persistent organic pollutants from microplastics under simulated physiological conditions. Environ. Pollut. 185, 16–23.

[90] Rochman, C. M., Hoh, E., Kurobe, T., and Teh, S. J. (2013). Ingested plastic transfers hazardous chemicals to fish and induces hepatic stress. Sci. Rep. 3, 1–7.

[91] Browne, M. A., Niven, S. J., Galloway, T. S., Rowland, S. J., and Thompson, R. C. (2013). Microplastic moves pollutants and additives to worms, reducing functions linked to health and biodiversity. Curr. Biol. 23, 2388–2392.

[92] Wright, S. L., Rowe, D., Thompson, R. C., and Galloway, T. S. (2013). Microplastic ingestion decreases energy reserves in marine worms. Curr. Biol. 23, R1031–R1033.

[93] Koelmans, A. A., Bakir, A., Allen Burton, G., and Janssen, C. R. (2016). Microplastic as a vector for chemicals in the aquatic environment: critical review and model-supported reinterpretation of empirical studies. Environ. Sci. Technol. 50, 3315–3326.

[94] Bakir, A., O'Connor, I., Rowland, S. J., Jan Hendriks, A., and Thompson, R. C. (2016). Relative importance of microplastics XE "microplastics" as a pathway for the transfer of hydrophobic organic chemicals to marine life. Environ. Pollut. 219, 56–65.

[95] Lohmann, R. (2017). Microplastics are not important for the cycling and bioaccumulation of organic pollutants in the oceans – but should microplastics be considered POPs themselves? Integr. Environ. Assess. Manage. 13, 460–465.

[96] Van Cauwenberghe, L., and Janssen, C. R. (2014). Microplastics in bivalves cultured for human consumption. Environ. Pollut. 193, 65–70.

[97] Rochman, C. M., Tahir, A., Williams, S. L., Baxa, D. V., Lam, R., Miller, J. T., Teh, F. C., Werorilangi, S., and Teh, S. J. (2015). Anthropogenic debris in seafood: plastic debris and fibres from textiles in fish and bivalves sold for human consumption. Sci. Rep. 5, 1–10.

[98] Yang, D. Q., Shi, H. H., Li, L., Li, J. N., Jabeen, K., and Kolandhasamy, P. (2015). Microplastic pollution in table salts from China. Environ. Sci. Technol. 49, 13622–13627.

Raquel Requena, María Vargas, Amparo Chiralt

3 Plastics from bacteria (polyhydroxyalkanoates)

Abstract: The development of environmentally friendly alternatives to limit the pollution associated with the massive use of synthetic plastics, such as bio-based and/or biodegradable polymer materials, is one of the current research trends. Polyhydroxyalkanoates (PHAs) are biopolyesters synthesized by a variety of microorganisms as energy reserve materials. Given their biodegradable and biocompatible nature and their inherent properties, PHAs have potential applications in different areas, such as the packaging and biomedical sectors. However, adapting the properties of these polymers to different use requirements is necessary. Different approaches aimed at improving the PHA functional properties as packaging materials have been reviewed. Thus, the use of additives, such as plasticizers or compatibilizers, the combination with other biodegradable polymers in blending or multilayer assemblies, as well as the development of biocomposites have rendered greatly improved functional properties. However, the high production cost compared to the synthetic plastics is still an obstacle to the growth of the PHA applications and their expansion in the plastics market. In this sense, different strategies to produce these polyesters at competitive costs are being developed. These focus both on the development of superior-bacterial strains and transgenic plants by recombinant DNA technology which permits the use of waste as growth feedstock, as well as on the optimization of the PHA extraction process.

Keywords: biodegradable polymers, biopolymer, bioplastics, polyhydroxyalkanoates (PHAs), poly(3-hydroxybutyrate), poly(3-hydroxybutyrate-co-3-hydroxyvalerate)

3.1 Introduction

Polyhydroxyalkanoates (PHAs) are a family of linear polyesters of (R)-3-hydroxy fatty acid monomers with the general structural formula shown in Figure 3.1. They are the only family of polymers that act as energy/carbon reservoirs for more than 300 species of gram-negative and gram-positive bacteria, as well as a broad variety of archaea [1]. PHAs are synthetized intracellularly as discrete inclusions when an

Raquel Requena, María Vargas, Amparo Chiralt, Instituto de Ingeniería de Alimentos para el Desarrollo, Universitat Politècnica de València, Valencia, Spain

https://doi.org/10.1515/9783110590586-003

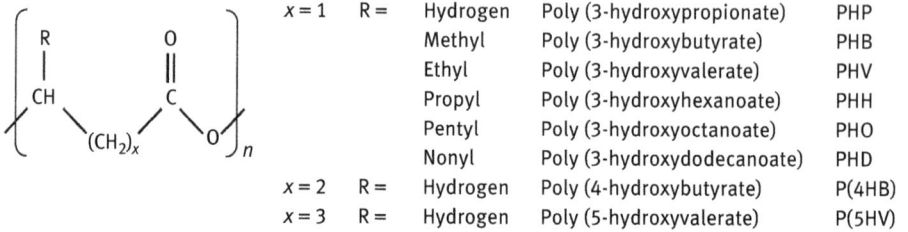

$x = 1$	$R =$	Hydrogen	Poly (3-hydroxypropionate)	PHP
		Methyl	Poly (3-hydroxybutyrate)	PHB
		Ethyl	Poly (3-hydroxyvalerate)	PHV
		Propyl	Poly (3-hydroxyhexanoate)	PHH
		Pentyl	Poly (3-hydroxyoctanoate)	PHO
		Nonyl	Poly (3-hydroxydodecanoate)	PHD
$x = 2$	$R =$	Hydrogen	Poly (4-hydroxybutyrate)	P(4HB)
$x = 3$	$R =$	Hydrogen	Poly (5-hydroxyvalerate)	P(5HV)

Figure 3.1: The general molecular structure of PHAs. In the figure, R-side chains consist of hydrogen or alkyl groups up to 13 carbon atoms in length, $x = 1$–4, $n = 600$–$35,000$. Adapted from [4].

essential nutrient, such as nitrogen, phosphorous, magnesium, sulfate, or phosphate, is available in limiting concentrations in the presence of an excess of carbon source. These inclusions of 0.2–0.5 µm in diameter are localized in the cell cytoplasm and can comprise up to 80% of the dry cell weight. It is known that PHA in vivo is not a crystalline solid, indeed, it is a mobile amorphous polymer. However, what hinders PHA crystallization still remains doubtful. Some authors support the hypothesis of the plasticizing effect of the water present in the PHA inclusions, which could form hydrogen bonds with the carbonyl groups of the polyester backbone between adjacent polymer chains [2]. Owing to their high refractivity, native PHAs can be visualized with a phase contrast light microscope and stained with Sudan Black B, which indicates that they are lipid in nature [2, 3].

The molecular weight of these polymers ranges from 200,000 to 3,000,000 Da, depending on the microorganism and growth conditions. Thus, PHAs can be divided into two main groups depending on their chain length. Short-chain-length PHAs (scl-PHAs), such as poly(3-hydroxybutyrate) (P(3HB) or PHB), contain monomer units with less than five carbon atoms, whereas the medium-chain-length PHAs (mcl-PHAs) contain monomer units of 6–14 carbon atoms. There is also a third group with monomer units containing more than 14 carbon atoms, which is uncommon and little studied, called long-chain-length PHAs (lcl-PHAs) [5]. This difference with regard to the chain length is caused by the substrate specificity of the PHA synthases [1–3].

PHAs are also classified depending on the types of monomers: homopolymers (only one type of monomer) or copolymers or heterocopolymers (different monomers in the chain). More than 150 different (R)-3-hydroxy fatty acid monomers have been identified in this heterogeneous family of homo- or copolyesters, thus resulting in polymer materials with a wide spectrum of properties. While scl-PHAs lead to thermoplastic materials that are highly crystalline and, therefore, brittle and barely stretchable, mcl-PHA are more flexible and less crystalline materials with lower melting temperatures [6, 7]. The most common PHA are PHB and poly(3-hydroxybutyrate-co-3-hydroxyvalerate) (P(3HB-3HV) or PHBV), since they possess properties that are similar to conventional polymers, such as polypropylene and polyethylene [1].

3.2 PHA synthesis

PHAs are naturally occurring polymers, which are synthetized by bacteria utilizing common carbon sources (i.e., glucose). Based on the requirements for PHA production, bacteria are divided into two different groups. The first group consists of bacteria such as *Alcaligenes eutrophus*, *Protomonas extorquens*, and *Protomonas oleovorans* requires the presence of excess carbon, while an essential nutrient is available only in limiting concentrations. This group can only accumulate PHAs during starvation periods, since once the bacteria are provided with the limiting nutrient, the reservoirs are depolymerized to recover the stored carbon. In contrast, the second group of bacteria, which includes *Alcaligenes latus* (a mutant strain of *Azotobacter*) and recombinant *Escherichia coli*, does not need nutrient limitation for PHA synthesis and can accumulate PHAs during growth [3]. In its simplest approach, PHA production by pure cultures, with bacteria belonging to the first group, consists of a two-stage batch cultivation process, in which the bacterial inoculum is introduced into a sterile solution containing an available carbon source (first stage). After a desired biomass concentration is obtained without nutrient limitation, the carbon source and an essential nutrient in limiting concentrations should be fed at an optimal ratio to allow an efficient PHA synthesis (second stage) [1].

In order to reduce the PHA production costs, several alternative approaches have been proposed, such as the use of mixed cultures adapted to complex waste feedstock, which does not require reactor sterilization. In this way, real fermented agricultural or industrial waste can be used as low-cost feedstock, thus reducing the high cost of the bacterial fermentation associated with the substrate cost. In this way, many different fermented wastes, such as effluents from fruit and tomato cannery, sugar cane molasses, effluents from olive oil, saponified sunflower oil, bi-ooil (a pyrolysis by-product), and waste waters from the food industry and households, have been used as substrate for the PHA production [1]. Alcohols, such as methanol, are also cheap available carbon sources, which can be used to synthetize PHAs by using methylotrophs, although their productivity still needs to be enhanced [1, 3]. Bacteria can use different carbon sources to synthetize PHAs, since the PHA synthase enzyme shows a broad substrate spectrum, polymerizing a wide variety of monomers. The main factor that determines PHA composition is the type of carbon source used [2, 5].

The use of recombinant strains that utilize cheap carbon sources is another alternative means of lowering the cost of the PHA production. PHA biosynthesis in *E. coli* by introducing PHA synthase genes allows for a cost-effective production, since many different cheap carbon sources can be used as substrate and the extraction process is easier and less costly. Moreover, because *E. coli* is not a natural PHA producer, it does not require the limitation of essential nutrients. In addition, since *E. coli* does not have PHA depolymerase, PHA accumulation could reach 80–90% of the dry cell weight [2, 3].

Transgenic plants expressing bacterial genes have also been considered to synthetize PHAs in a more cost-effective way. Novel metabolic pathways either in the cytoplasm, plastid or peroxisome have been engineered in the model plant *Arabidopsis thaliana* to obtain different PHAs. Moreover, PHA production has been successfully performed in agriculturally relevant crops, such as tobacco, *Brassica napus*, cotton, and corn [8]. However, adverse effects, like stunted growth and infertility, have been observed because of high expression levels of the acetoacetyl-CoA reductase, which hinders their large-scale cultivation. Nonetheless, this problem can be minimized by regulating the tissue specificity, the timing of expression and the cellular location of the PHA-producing enzymes [2, 3, 9]. High levels of PHB (up to 18% PHB of the cellular dry weight) in plants with fertile offspring are obtained in the plastid of *Nicotiana tabacum* [10]. For large-scale production, oilseed crops, such as rapeseed, have been proposed as ideal PHA producers, since the results obtained in the closely related species *Arabidopsis thaliana* are directly applicable [2, 3].

Despite the high interest of the plant-based expression systems that do not need external organic sources, the use of these systems is limited because of competition with subsistence crops in cultivation areas. Moreover, the application of transgenic plants is restricted in many countries due to ethical concerns and strict legislation regarding their dissemination. In this sense, algae or microalgae, which have emerged as a by-product of the biorefinery processes, show the same advantages as transgenic plants: high growth rates, ease of handling, and ability to grow in a wide range of environments [9]. Furthermore, this approach enables the use of carbon, thus neutralizing greenhouse gas emissions. According to [11], two stages are distinguished for the algae polymer production: a first stage, in which the algae growth is initiated, and a second stage, in which PHA accumulation is promoted [9] and demonstrated for the first time that the polyhydroxybutyrate (PHB) production is feasible in a microalgal system by introducing the bacterial PHB pathway of *Ralstonia eutropha* H16 into the diatom *Phaeodactylum tricornutum* (PHB levels up to 10.6% of algal dry weight).

PHAs can also be produced by chemical synthesis or by chemical modifications of natural PHAs, as described by Abe et al. (1995) for the chemical synthesis of PHB by the ring-opening polymerization of a mixture of (R)- and (S)-β-butyrolactone in the presence of aluminum or zinc alkyl catalysts. This group of synthetic PHAs cannot be produced by microorganisms, since the monomer units are toxic or cannot be taken from the culture broth by bacteria, due to the PHA synthase specificity. Nevertheless, this approach is not economically profitable for large-scale production and requires the use of toxic reagents [1, 7].

The PHA recovery process should also be taken into account, since it is a determining factor in PHA production costs. Owing to its simplicity and ease of operation, solvent extraction has been the most commonly used PHA recovery method. Solvent extraction consists of a previous treatment based on cell disruption to liberate the PHA granules contained within bacteria, followed by the solubilization of PHA granules with organic solvents, and a final precipitation with nonsolvents [5, 12]. This is the preferred extraction method when high purity is required, since it does not degrade the polymer and removes cell endotoxins [12, 13]. The flotation method, a modification of the solvent method that avoids the last step, is also used to extract PHAs by the shelf-flotation of the cell debris with high purity and recovery efficiency [5]. Enzymatic or chemical (hypochlorite or surfactants) digestion methods have emerged as alternatives to solvent methods with good PHA recovery. Two-phase aqueous extraction and supercritical fluid extraction are the latest and most innovative strategies for PHA extraction, due to their sustainability and low cost. Aqueous two-phase system is comprised of two polymers at low concentrations (or of one polymer and an inorganic salt) that display incompatibility and coexist as two immiscible phases when mixed [14]. Meanwhile, the supercritical fluid extraction uses supercritical fluids for the isolation, purification and recovery of PHAs, such as carbon dioxide ammonia and methanol [5, 12]. For in vivo PHA applications, some impurities, like endotoxins, can cause inflammation even when PHAs are obtained by solvent extraction [15, 16]. Supercritical solvent extraction is recommended in these cases since it allows impurity-free PHAs to be obtained for medical applications [5].

3.3 PHA applications

Of the different biodegradable polymers, PHAs are those gaining attention due to the broad chemical variety of their radicals, which provides physical properties as good as the ones exhibited by conventional polymers [17]. In this sense, tailor-made PHAs can be produced by selecting the proper bacteria, as well as the appropriate culture conditions and carbon sources. Moreover, PHA variability can be increased by the chemical modification of the natural PHAs [18]. Thus, depending on the PHA monomer composition, these biopolymers could vary from rigid and brittle plastics to soft elastomers, rubbers, and adhesives [17].

PHA applications in the material industry have been related with the development of cosmetics containers, shampoo bottles, shopping bags, cups, pens, combs, bullets, and disposable items such as razors, feminine hygiene products, moisture barriers for nappies, and sanitary towels [7, 19, 20]. Moreover, in the last three decades, numerous patented applications for the PHAs have been developed such as flavor delivery agents in foods [21], dairy cream substitutes [22], fibers for nonwoven fabrics [23], latex for paper coating [24], and hot-melt adhesives [25].

3.3.1 Food packaging materials

Although some mcl-PHAs can be elastomeric materials in a narrow temperature window, scl-PHAs, such as PHB, PHBV, or poly(3-hydroxyvalerate) (P(3HV) or PHV), are thermoplastic polyesters and are used as bioplastics, which can replace petroleum-based polymers in packaging and coating applications, with the food sector being the principal target [17, 20, 26]. Table 3.1 shows several studies that analyzed the impact of packaging materials based on PHB, the most common PHA, on the quality of the packaged foodstuffs.

Table 3.1: Impact of polyhydroxybutyrate (PHB)based packaging on food quality.

Food	Impact on food quality	Reference
Sour cream	PHB-based films were suitable for maintaining the quality of the dairy product in terms of color, pH, and secondary lipid oxidation products.	[27]
Mayonnaise, margarine and cream cheese	PHB films were found suitable for storage of fat-rich products, according to physical, mechanical, sensorial, and dimensional analyzes.	[28]
Meat salad in mayonnaise	PHB films could be successfully used for sous vide thermal food treatment at a temperature no higher than 63 °C.	[29]
Orange juice simulant and a dressing	PHB-based cups were equivalent to those of high-density polyethylene in terms of color changes, primary and secondary lipid oxidation products and reduction of α–tocopherols, during storage of samples with light exposure.	[30]
Tomatoes	PHB-coated paperboard trays maintained tomato quality as good as that of those stored in low-density polyethylene bags, in terms of weight loss, moisture content, color, firmness, and flavor.	[31]

Despite the acceptable results regarding the impact on the food quality of PHB-based materials, its application as food packaging has been hindered because of its brittleness, stiffness, thermal instability, limited gas barrier properties as well as high cost and limited availability [17, 32]. In this sense, different strategies have been investigated to improve PHA functional properties as packaging material in order to extend their applicability. Copolymerizing approaches have been effective alternative means of improving thermal stability, mechanical performance, gas and water vapor barrier capacity of these biopolymers, through decreasing both their brittleness and melting temperatures [12, 17, 26, 32–36]. In this sense, several copolymers have been developed using different monomer units, such as 3-hydroxyvalerate, 4-hydroxybutyrate, or 3-hydrohexanoate, thus improving the physical properties of the PHB homopolymer [37]. PHBV obtained by different levels of copolymerization of 3-hydroxyvalerate leads to less stiff, tougher materials with improved thermal stability and barrier properties, with a broadened application range in the food packaging sector [12, 26, 32, 34, 35, 38].

In the field of food applications, PHA-based packaging materials have also been used as carriers of antimicrobial or antioxidant compounds, which can extend the shelf life of the packaged foodstuffs, maintaining their quality. To impart antimicrobial action to PHA materials, antimicrobial agents can be incorporated into the packaging structure during manufacturing. Thus, the antimicrobial activity can be achieved by direct contact with the foodstuff (the antimicrobial compound can be absorbed or immobilized into the polymeric matrix), or by modifying the headspace of the package by the controlled release of the active compound from the packaging structure [39]. In this sense, many studies have been performed aiming to provide this added functionality to PHA-based materials (Table 3.2).

3.3.2 Medical applications

The biocompatibility and the biodegradable nature of PHAs have attracted much interest in the medical sector [5, 7, 12]. Thus, PHB and PHBV have been widely used in tissue engineering as matrices (scaffolds) for the proliferation of different human cells, such as endothelium cells, isolated hepatocytes, fibroblasts [51] and neurons [52]. PHAs have also been used for the tissue regeneration of bones [53] and eyelids [54]. Based on the results obtained for orthopedic implants and bone replacement in animals, PHAs could be used as biomedical implant materials in the human body [5]. PHAs have also been used in repairing damaged nerves in rates as an alternative to the conventional nerve graft [11] and for wound dressing [55] and scaffolds [56].

The use of PHAs as drug delivery carriers is also receiving more and more attention, since the micro- or nanospheres of PHAs containing drugs allow for a prolonged release of the encapsulated compounds, as the polymer degrades [5, 7, 12]. Thus, PHAs have been tested as drug or vaccine carriers on several animals, such as mice, rabbits, sheep, or dogs, and even in the gingivitis treatment in humans [57, 58]. Specific drug targeting systems based on PHA nanoparticles for cancer therapy have been successfully developed by the linkage of specific ligands, whose effectiveness has been proven in both in vitro and in vivo studies [59–61].

Both D-3-hydroxybutyrate (D-3HB), which is the most common PHA degradation product, and D-3HB derivatives showed inhibitory effects on cell apoptosis, which explains the PHA biocompatibility and suggests their application as neural protective agents [7, 62]. These products improved neural capacities, such as learning and memory, since they enhanced neural communication [63].

Table 3.2: Antimicrobial packaging materials based on polyhydroxyalkanoate (PHAs).

Active agent	Film processing	Main results	Reference
Silver nanoparticles (AgNPs) obtained in an aqueous solution of PHBV	Melt blending of the PHBV/AgNPs nanocomposites (8%)	PHBV-AgNPs films showed a strong and sustained (even after 7 months) antibacterial activity against *Salmonella enterica* and *Listeria monocytogenes*.	[40]
Hexagonal-pyramid zinc oxide nanoparticles (ZnONP)	Nanocomposites based on PHBV and ZnONP are obtained by three different methods: (i) direct-melt blending, (ii) melt blending of preincorporated ZnO into PHBV electrospun fiber mats and, (iii) as a coating of the annealed electrospun PHBV/ZnO fiber mats on the compression molded PHBV.	ZnO is contained in electrospun fiber mats, as coatings of the film outer layer increased the Zn availability and, thus, its antimicrobial performance (reduction of 3 CFU log units).	[41]
ZnONP	Nanocomposites are based on ZnONP (1, 2, 5, and 10 wt %) and PHB, obtained by casting method from chloroform solutions.	PHB-ZnONP films showed antibacterial activity against both *Staphylococcus aureus* and *Escherichia coli*, which was progressively improved upon increasing ZnO concentration.	[42]
Eugenol	PHB active films were obtained by casting from the chloroform solutions, with eugenol at different concentrations (10–200 µg/g of PHB)	PHB films containing eugenol (80 µg/g) inhibited the growth and the sporulation of molds, i.e., *Aspergillus niger*, *Aspergillus flavus*, *Penicillium* sp., and *Rhizopus* sp., until 7 days. Significant antibacterial activity was observed against *Salmonella typhimurium*, *Staphylococcus aureus*, *Escherichia coli*, and *Bacillus cereus*	[43]

Vanillin	PHB active films were obtained by casting from chloroform solutions with vanillin at different concentrations (10–200 µg/g of PHB)	The minimum concentration of vanillin required to exhibit antimicrobial activity was ≥80 µg/g PHB for bacteria and ≥50 µg/g PHB for fungi.	[44]
Cinnamaldehyde	Electrospun zein- cinnamaldehyde layers were applied to obtain PHBV multilayer films by compression molding: (i) PHBV + zein-cinnamaldehyde; (ii) PHBV + zein-cinnamaldehyde + PHBV; (iii) PHBV + zein-cinnamaldehyde + alginate.	The active multilayer structures showed antibacterial activity against *L. monocytogenes*, the PHBV/zein-cinnamaldehyde/PHBV system being the one that showed the greatest antibacterial activity.	[45]
Cinnamaldehyde	Trilayer structures with PHB-outer layers were obtained by compression molding including an electrospun inner layer of zein-cinnamaldehyde.	Multilayer films containing cinnamaldehyde (2.60 mg/cm^2) completely inactivated feline calicivirus, while murine norovirus titers were reduced by 3 log. Both are enteric viruses, causing foodborne illnesses.	[46]
Catequin (antioxidant)	PLA-PHB (75:25) films containing 5 wt% catequin were prepared by melt blending.	Polyester films containing catequin showed significant antioxidant activity coherent with the catechin release.	[47]
N-Halamine precursor	Electrospun fiber membranes based on PHB and the *N*-halamine precursor were chlorinated to obtain antimicrobial materials.	The chlorinated membranes had excellent antimicrobial functions, which could inactivate 92% *S. aureus* and 85% *E. coli* within 30 min of contact time.	[48]
Nisin	PHB-PCL (50:50) films were obtained by melt blending and compression molding. Antimicrobial films were obtained by nisin adsorption.	PHB-PCL nisin activated films were effective against *Lactobacillus plantarum* inoculated on sliced ham, thus extending its shelf-life	[49]
Montmorillonite organoclays (Cloisite® 30B and Claytone APA)	P(3HB-4HB) nanocomposites with 5 and 10 wt% of each type of organoclays were obtained by casting.	P(3HB4HB) exhibited antimicrobial activity against *S. aureus* but not against *E. coli*. Films containing Claytone APA showed higher antimicrobial properties compared to those containing Cloisite®30B. Stronger antimicrobial effects were observed as the clay concentration in the film increased.	[50]

3.3.3 Biofuel

The use of PHAs as precursors of green biofuels has recently been suggested, since the hydrolysis of PHAs followed by methyl esterification gives 3-hydroxyalkanoates methyl esters, whose combustion energy is comparable to that of bioethanol [64, 65]. However, PHA conversion into biofuel is more expensive than the direct fermentation of sugars into bioethanol [12]. Nevertheless, the interest in PHA-based biofuels is enhanced by the fact that they are biodegradable and produce low greenhouse gas emissions, in contrast with fossil fuels. Moreover, other biofuels are obtained from biomass, whereas PHAs can be produced from waste water or activated sludge, dealing with the controversial use of agricultural crops as fuels [7, 66].

3.3.4 Other industrial applications

PHB and PHBV can be used as carriers of herbicides, such as ametryn, for agricultural applications, due to their capacity to deliver the agents contained inside micro or nanoparticles progressively, thus reducing the requirement of successive applications as well as the negative effects on nontarget species [12, 67]. Due to their hydrophobic nature, PHA composites can be used as biomimetic adsorbents for the removal of lipid-soluble organic pollutants from water [68]. PHB granules in feed have also been applied as pathogenic bacterial growth inhibitors in some aquaculture applications [69, 70].

3.4 Properties of PHA-based packaging materials

The physical properties of PHA can be as good as those of synthetic plastics, due to the wide chemical variety of their radicals. Thus, depending on the monomer composition, PHAs can range from brittle, stiff plastics to soft elastomers, rubbers and adhesives. Moreover, some PHA packaging materials offer good resistance to moisture, because of their hydrophobic nature [11], and gas barrier properties comparable to polyvinyl chloride and polyethylene terephthalate [17], with the advantage of being biodegradable at worst in 6 weeks in a microbiologically active environment [26]. Compared to polysaccharide-based materials, PHAs films exhibit much better moisture barrier properties, whereas the gas barrier capacity is inferior [35].

3.4.1 Physical properties

PHB-homopolymer-based films are brittle and stiff, with poor impact resistance, due to their high degree of crystallinity and the secondary crystallization that occurs after film processing with aging [1, 11]. PHA film properties are known to be dependent on the polymer composition and film processing, particularly so when they are obtained by melt blending or solvent casting [1, 71]. Table 3.3 reviews the range of reported values for the properties of PHB-based films obtained by both processing methods. As shown in Table 3.3, PHB films prepared by solvent casting show higher elongation at break and greater impact resistance than those obtained by melt blending and compression molding, probably due to the finer spherulitic morphology associated with the low crystallization temperature, which results in high nucleation density [72]. Similar results have been observed for copolymers, such as poly(3-hydroxybutyrate-co-3-hydroxyhexanoate) (P(3HB-3HH) or PHBH), where the solvent cast films were more stretchable than those obtained by compression molding [1]. As regards thermal properties, PHB solvent cast films exhibit lower glass transition temperature (T_g) and melting temperature (T_m) values than those obtained by melt blending and thermocompression. However, the film barrier capacities are in the same order of magnitude for PHB films obtained either by solvent casting or thermoprocessing.

As previously mentioned, the brittleness and the thermal instability of PHB-based materials are what mainly limit practical applications. Therefore, different approaches, such as the use of copolymeric materials, have been investigated as a means of improving PHA applicability. As shown Table 3.3, compared to PHB, PHBV-based packaging materials exhibit less stiffness and brittleness, and improved stretchability and tensile strength as the content in the copolymer increases [1, 38], since 3-hydroxyvalerate units hinder the formation of hydroxybutyrate crystals [81]. Likewise, as the second monomer fraction increased in the copolymeric materials, such as PHBV, poly(3-hydroxybutyrate-co-3-hydroxypropionate) (P(3HB-3HP) or PHBP), poly(3-hydroxybutyrate-co-4-hydroxybutyrate) (P(3HB-4HB)) and poly(3-hydroxybutyrate-co-6-hydroxyhexanoate) (P(3HB-6HH)), their glass transition and melting temperatures dropped, thus broadening the processing window, since there was improved melt stability at lower processing temperatures, [2, 32, 38] reported that T_m values of PHBV films decreased from 180 °C to 123 °C and T_g from 3.1 °C to 6.3 °C as the 3-hydroxyvalerate content in the copolymers increased from 0 to 20 mol%, with elongation at break values of 180% for the PHBV film with 20 mol% HV.

As concerns the effect of the method used for the PHBV film processing, Fabra et al. [71] reported that the PHBV/PHBV bilayer films obtained by thermoprocessing exhibited greater mechanical resistance and lower water vapor permeability (WVP) and transparency, as compared to the corresponding multilayer obtained by casting.

Table 3.3: Range of typical properties of polyhydroxyalkanoate 9PHA) films obtained by solvent casting and compression molding.

Polymer	Method	T_g (°C)	T_m (°C)	EM (GPa)	TS (MPa)	E (%)	X_c (%)	WVP·10^{15}	OP·10^{19}	Reference
PHB	Solvent casting	-2–9	160–180	1.2–3.6	19–44	2–5	57–63	2–10	8	[42, 73–77]
	Compression molding	10–18	170–180	1.5–3.5	8–40	0.8–2.1	45–50	8	5–8	[1, 47, 78, 79]
P(3HB-3HV$_{12*}$)	Solvent casting	–	140–155	–	6.4	1.4	33	50	15	[71]
	Compression molding	-3–2	140–155	1.7	14.6	1.2	37	4	15–18	[1, 32, 71, 80]
P(3HB-3HV$_{50-55*}$)	Solvent casting	-10	77	–	16	1,200	–	–	–	Laycock et al., 2014
	Compression molding	0	162	0.4	13.4	230	–	–	–	Laycock et al., 2014

*Subscript indicate the mol% comonomer unit.

T_g: glass transition temperature; T_m: melting temperature; EM: elastic modulus; TS: tensile strength; E: elongation at break; X_c: degree of crystallinity; WVP: water vapor permeability (kg m/m^2 Pa s); OP: oxygen permeability (m^3 m/m^2 s Pa).

3.4.2 Biodegradability

PHA-based materials are degraded in biological media, giving rise to harmless products: carbon dioxide and water in aerobic conditions or methane and water in anaerobic environments [26, 82, 83]. Bacteria, actinomycetes, and micromycetes have been described as PHA degraders in soil, compost, activated sludge, and river and sea water, since they produce extracellular PHA depolymerases, which hydrolyze the polymer by surface erosion to water-soluble monomers and/or oligomers used as substrates by other microorganisms [82]. The main advantage of PHAs as compared to other biodegradable polymers is that they do not need specific environmental conditions to be degraded [4]. Moreover, PHAs can be recycled like petrochemical thermoplastics [20].

The degradation rate and the mechanism of PHA biodegradation are influenced by several factors, including PHA chemical composition and properties (crystallinity, molecular mass, and polydispersity), type of polymer and processing method, physicochemical conditions of the environment (oxygen availability, temperature, pH, moisture, acidity, and salinity), weather and climate and, to a larger extent, by the composition and metabolic activity of the microbial community [82, 83]. In this sense, Lee [84] reported that PHBV was completely degraded after 6 to 75 weeks in anaerobic sewage and soil, respectively, which reflects the great influence of the type of media where the biodegradation takes place. Regarding the molecular weight of the polymer, Thellen et al. [85] reported the fastest degradation rate for the PHB films with the lowest polymer molecular weight, whereas the film with the highest PHB molecular weight had the slowest degradation rate [83]. found a negative correlation between the degree of crystallinity of PHAs and their biodegradation rate: P(3HB-4HB), which had the lowest degree of crystallinity (50%), showed the highest degradation rate (1.63 mg/day) and the most crystalline (78%) PHB was degraded at the slowest rate (0.61 mg/day), whereas the polymer with intermediate crystallinity, PHBV, exhibited an intermediate degradation rate.

Compared to other biodegradable polyesters, such as PLA, PHBV-based packaging materials have been reported to degrade faster; and an improved biodegradability of PLA-PHBV blend films has been achieved by adding hydrophilic plasticizers, such as polyethyleneglycol (PEG). These kinds of hydrophilic substances promote the matrix swelling required for the hydrolysis reactions [86]. However, the addition of other plasticizers in PHA-based matrices, such as laprol, was reported to halve the biodegradation rate of PHB films compared to pure films, probably due to the fact that laprol partially encapsulated the PHB component, thus retarding the biodegradation of the blend [38].

3.5 PHA-based multiphase materials

A great deal of effort has gone into improving the brittleness and stiffness of PHB-based packaging materials that result from their high degree of crystallinity and ageing recrystallization [1, 11]; this effort has focused on different strategies aimed at reducing the crystallinity and improving the thermal stability and stretchability of the final packaging materials. The use of plasticizers has been investigated, as well as the blending with other polymers.

3.5.1 Plasticizers

In biopolymer film production, plasticizers are essential additives since they reduce the intermolecular forces along polymer chains, which improves the flexibility and chain mobility, at the same time as they provoke a decrease in the T_g and changes in the crystallization behavior, thus enhancing the functional properties of the packaging materials [87]. The effect of different kinds of plasticizers on the mechanical and thermal properties of PHA-based matrices has been studied in depth. Table 3.4 summarizes some of the different studies [88]. reported that as the

Table 3.4: Plasticizer effects on the functional properties of polyhydroxyalkanoate (PHA)based films.

Plasticizer	Polymer (process)	Main results	Reference
Propylene glycol (PG) Glycerol (G) Poly(ethyleneglycol) with M_w 1,000 Da (PEG1000) Castor oil (CO) Epoxidized soybean oil (ESO)	PHBV (extrusion)	In general, as the plasticizer concentration increased, T_m, EM, and TS decreased, whereas E and X_c of the plasticized film increased. PEG1000 and TEC effectively improved the films' stretchability. Conversely, glycerides with long fatty acids (PG and G) and low M_w substances with hydroxyl groups (CO and ESO) did not lead to a plasticizing effect in PHBV films. All the plasticizers significantly increased the WVP and OP of the films, but to a different extent.	[87]
ESO Soybean oil (SO) Triethyl citrate (TEC) Dibutylphthalate (DBP)	PHBV (casting)	TEC and DBP were more effective at plasticizing PHBV matrices, since they led to significantly lower T_m and T_g Likewise, stretchability and impact resistance of PHBV-TEC and PHBV-DBP blends were also greater than PHBV-SO and PHBV-ESO blends, since the solubility parameter, and the polar and hydrogen-bonding components of TEC and DBP were closer to the corresponding values of PHBV.	[37]

Table 3.4 (continued)

Plasticizer	Polymer (process)	Main results	Reference
Acetyl butyl citrate (ATBC)	PHBV (extrusion)	The addition of ATBC significantly shifted both T_g and T_m as the plasticizer concentration increased. Moreover, ATBC led to less crystalline films and less stable crystallites.	[89]
PEG300 Laprol Dioctylsebacate (DOS) Polyisobutylene (PIB) Dibutylsebacate (DBS)	PHB (casting)	All the additives progressively enhanced the E of the PHB film from concentrations that ranged from 15% to 33%. Laprol, DOS and DBS led to better mechanical behavior, whereas PIB showed the mildest plasticizing effect. Laprol was selected as plasticizer based on the greatest improvement of the PHB film stretchability.	[38]
Sophorolipid (SL)	PHB (casting)	SL, a kind of glycolipid produced by the fermentation broth of *Candida bombicola*, significantly hindered the PHB crystallization (decrease in ΔH_m and T_m) by the interruption of intermolecular interactions.	[90]
Dodecanol Lauric acid Tributyrin Trilaurin	PHB (casting)	All the compounds acted as plasticizers, since T_g and T_c of the mixtures were lower than those of the neat PHB films, but to a different extent. The miscibility between PHB and the different plasticizers was greater in the order of tributyrin > dodecanol~ lauric acid > trilaurin.	[91]
PEG200	PHB (casting)	The incorporation of PEG200 (15%) led to a reduction in both TS (30 to 22 MPa) and EM (1,180 to 537 MPa), and enhanced the film extensibility (from 6% to 12%). Moreover, T_m decreased from 175 °C to 156 °C.	[92]

T_g: glass transition temperature; T_m: melting temperature; EM: elastic modulus; TS: tensile strength; E: elongation at break; X_c: degree of crystallinity; WVP: water vapor permeability; OP: oxygen permeability.

plasticizer concentration increased, the T_m, the elastic modulus (EM) and the tensile strength (TS) decreased, whereas the elongation at break (E) and the degree of crystallinity (X_c) of the plasticized film increased. Moreover, the plasticizers significantly increased the water vapor and oxygen permeability, but to a different extent.

In general, medium molecular weight substances with oxygen atoms (e.g., ethers or ketones), which are accessible for interactions with the polymer matrix, such as PEG with M_w of 1,000 Da (PEG1000), triethyl citrate (TEC) or dibutylphthalate (DBP), have effectively improved the film stretchability [37, 88]. Conversely, glycerides with long fatty acids and low-M_w substances with hydroxyl groups do not lead to a plasticizing effect in PHBV films [88]. The addition of some plasticizers, such as acetyl butyl citrate (ATBC) and sophorolipid, hindered the PHBV and PHB crystallization, respectively, in line with the interruption of interactions among the polymer chains [89, 90]. In contrast, some other compounds, such as propylene glycol (PG), PEG1000 or epoxidized soybean oil (ESO), promoted molecular mobility, thus favoring the crystallization process [88].

There is increasing interest in the application of natural plasticizers, such as castor oil (CO), ESO, soybean oil (SO), DBP, or TEC. Natural plasticizers show low toxicity and migration and improve both the thermal and mechanical properties of PHBV films [37, 87, 88]. Some other biodegradable additives, such as dodecanol, lauric acid, tributyrin, and trilaurin, were miscible with PHB and enhanced its molecular mobility, thus decreasing both the glass transition and crystallization temperatures (T_c) of PHB films [91].

3.5.2 Polymer blends

Unlike the development of new polymer materials, the blending strategy is a cheap and rapid method to improve plastic functional properties. Therefore, this approach has been applied to obtain PHA-based blends with improved film properties as compared to that of the neat polymer films. Table 3.5 summarizes different studies dealing with the blending of PHAs with other polymers. PHAs have been widely used to improve both the mechanical properties and barrier capacity of PLA films [47, 93–95]. Nonetheless, Ferreira et al. [96] reported immiscibility between both polyesters, since PLA-PHBV blends showed two different T_g and T_m values, which did not vary in relation to the blend composition. Therefore, additives, such as plasticizers and/or compatibilizers, are necessary to improve the ductile properties of the polymer blends and to achieve the stretchability required for their application, as well as the compatibility of the polymer blends, which renders better functional film properties. As mentioned in Table 3.5, lactic acid oligomer, limonene, PEG300, acetyl tributyl citrate (ATBC) and poly(butylene adipate-co-terephthalate) (PBAT) have enhanced the functional properties of PLA-PHB blend films [47, 93–95]. As reported by Jost and Kopitzky [97] 20–35% of PHBV is the suitable concentration range for PLA-PHBV blend films in terms of polymer interactions, phase separation and barrier properties. WVP and OP decreased by 46% and 40%, respectively, in PLA-PHBV (75:25) films with respect to the neat PLA films.

PHA-starch blend films have been proposed as a means of lowering the costs associated with the use of neat PHA and accelerating their biodegradation, because starch is cheaper and faster to degrade than PHA. However, in order to use starch as packaging material, it should be gelatinized and plasticized, usually by using water and/or nonvolatile plasticizers, such as glycerol or other polyols, obtaining

Table 3.5: Polymer blends based on polyhydroxyalkanoates (PHAs) and other biodegradable polymers.

Polymers	Process (compatibilizing or plasticizing agent)	Main results	Reference
PLA-PHB (75:25)	Melt blending and compression molding (ATBC or PEG300 at 15%)	PLA-PHB blend exhibited improved barrier capacities compared to the neat PLA films, since PHB enhanced the blend crystallinity. ATBC showed a greater plasticizer effect than PEG300, because of the similar solubility parameters between ATBC and both polyesters, as well as the higher retention in the polymer matrix.	[94]
PLA-PHB (75:25)	Melt blending and compression molding (Limonene 15%)	The polymer blend was miscible, showing only one glass transition, whose T_g decreased after limonene addition. PHB reinforced the PLA matrix, thus improving its oxygen barrier properties and the surface water resistance, which was enhanced by the limonene addition. More limonene was retained in the polymer blend than in PLA films and it increased the film extensibility.	[78]
PLA-PHB (85:15)	Extrusion (OLA at different levels: 15%, 20%, and 30%)	Single T_g value was reported in all the formulations, proving the high degree of compatibility between the three components. Thus, improved barrier properties were achieved by adding both PHB and 30% OLA, due to the increase in the blend crystallinity. Moreover, OLA significantly plasticized the PLA-PHBV blend, since it lowered the T_g and improved its ductile properties.	[93]
PLA-PHB (70:30)	Extrusion (PBAT at different levels: 5%, 10%, 15%, and 20%)	PBAT improved the compatibility between PLA and PHB. Moreover, the compatibilizer acted as plasticizer, since it enhanced the E and decreased both TS and EM, 10% of PBAT being the more convenient concentration. Biodegradation rates improved as the PBAT concentration increased.	[95]
PLA-PHBV	Solvent casting (PEG-PE1400, PEGDO, and PMMA at 2%)	20–35% of PHBV is the suitable concentration range for PLA-PHBV films in terms of polymer interactions, phase separation and barrier properties, since the WVP and OP decreased 46% and 40%, respectively in PLA:PHBV (75:25) films with respect to the neat PLA films. The strongest influence was observed with PMMA used as surface-active compatibilizer.	[87]

Table 3.5 (continued)

Polymers	Process (compatibilizing or plasticizing agent)	Main results	Reference
PHB-TPS PHB-native starch	Solvent casting	Blend films had a single T_g for all the PHB: starch ratios tested. However, no interactions between either polymer were observed, since the T_m values of the blends did not change. PHB:TPS blends were more thermostable than those blends with starch powder. The TS was optimum for the 70PHB:30TS formulation.	[77]
PHB-TPS PHBV-TPS	Extrusion (PCL or PBAT)	TPS acted as compatibilizer between the PHA and the PBAT phases in the blend, thus minimizing the aging related with PHA- and TPS-based materials. PHBV and PBAT were able to hinder the moisture uptake of the starch and the glycerol leak. PHBV-PBAT-TPS (50:30:12) films with 70% biobased content showed outstanding mechanical features suitable for flexible packaging.	[98]
HPDSP-P (3HB-4HB)	Extrusion blowing (Glycerol, PEG200 and OMMT at 30%, 10%, and 10% of total starch/PHA mass, respectively)	Crystallinity of the films gradually increased as the polyester content increased, thus improving the WVP of the blends as compared to the HPDSP-based films. The addition of PHA to starch films enhanced their TS, but decreased both E, and thermal stability. The films containing 12% PHA exhibited both the highest light transmittance and TS.	[99]
HPDSP-PHA (80:20)	Extrusion blowing (Cross-linking agents: citric acid, adipic acid, borax, or boric acid at 1% of the total polymer mass). (Glycerol, PEG200 and OMMT at 24%, 4%, and 10% of total polymer mass, respectively)	The cross-linking agents significantly improved light transmittance, TS, E, thermal stability, and barrier capacities as well as compatibility between the starch and PHA phases of 80HPDSP:20PHA blend films. Citric acid and adipic acid were more suitable as cross-linking agents than boric acid and borax.	[100]

T_g: glass transition temperature; T_m: melting temperature; EM: elastic modulus; TS: tensile strength; E: elongation at break; X_c: degree of crystallinity; WVP: water vapor permeability; OP: oxygen permeability; TPS: thermoplastic starch; OLA: lactic acid oligomer; PBAT: poly(butylene adipate-co-terephthalate); PCL: poly(e-caprolactone) ; HPDSP: hydroxypropyl distarch phosphate; OMMT: organically modified montmorillonite; PEG: poly(ethylene glycol); ATBC: acetyl tributyl citrate; PEG-PE1400: poly-(ethylene-block-polyethylene glycol); PEGDO: polyethylene glycol 400 dioleat; PMMA: polymethyl methacrylate.

thermoplastic starch (TPS) [98]. Thus, PHB-TPS blends were more thermostable and exhibited better mechanical performance than those blends prepared with native starch granules [77]. Likewise, Parulekar and Mohanty [98] successfully developed PHBV-PBAT-TPS films by extrusion with 70% biobased content and outstanding mechanical features, which were suitable for flexible packaging, since both hydrophobic materials, PHBV and PBAT, were able to delay both the high moisture uptake of starch and the glycerol leach. Moreover, the aging process related to PHA- and TPS-based materials was significantly reduced due to the compatibilizing effect of the TPS in the PHBV-PBAT-TPS blend. Likewise, Sun et al. [99] recommended starch-PHA blend films based on hydroxypropyl distarch phosphate (HPDSP) and P(3HB-4HB) in a polymer ratio 88:12 by using extrusion blowing and glycerol, PEG200 and organically modified montmorillonite at 30%, 10%, and 10%, respectively, since this blend formulation showed improved TS and water vapor resistance. HPDSP-PHA (80:20) blend films were obtained by using different cross-linking agents, with improved light transmittance, TS, E, thermal stability and barrier capacity as well as compatibility between the starch and the PHA phases. The different cross-linking agents were citric acid, adipic acid, boric acid, and borax; the first two being more suitable than boric acid and borax [100].

3.5.3 Multilayer systems

Although improved film properties can be obtained by blending PHA with other biodegradable polymers, this strategy involves reactive processes, which, in turn, could lead to reactant residues in the films, whose food safety needs to be proven. Conversely, the bilayer or multilayer systems provide an easier approach to combine biopolymers with complementary properties and no great effort to enhance the interfacial compatibility between the polymers is needed, since the polymer interactions only take place at the layer interface. In addition, as regards the barrier capacity, multilayer assembly is more advantageous, since polymers with complementary barrier properties can be combined to obtain packaging materials that are highly water resistant (nonpolar sheet) as well as impermeable to gas (polar sheet), thus being better adapted to different food packaging applications [101]. In this sense, PHAs have been combined with biodegradable polymers with complementary properties by using different strategies, obtaining multilayer systems with improved capacities as compared to the neat polymer monolayers (Table 3.6). PHB and PHBV have been used as outer layers in three-layer systems to protect the inner water-soluble sheet by electrospinning the polyester on a preformed layer and the subsequent compression molding of the assembly [102–104] or by the coextrusion and compression molding of the different sheets [105]. PHAs have also been used to obtain moisture-resistant protein-based films by dip coating of the PHA monolayer with a filmogenic solution based on whey protein and gelatin [106].

Table 3.6: Polyhydroxyalkanoates (PHA)-based multilayer materials.

Polymer system	Process	Main results	Reference
PHBV[e]/WG/ PHBV[e] PHB[e]/WG/PHB[e]	WG monolayer plasticized with glycerol by compression molding Electrospinning of PHB or PHBV on both sides of the WG layer Compression molding of the monolayers	WVP of the multilayers depended on the deposited amount of the electrospun PHB and the temperature used for the assembly. OP was greatly influenced by the type and amount of the PHA. The lowest WVP and OP values were obtained for three-layer films prepared with 1 mg PHBV/cm^2 and processed at 160 °C.	[103]
PHBV/zein[e]/ PHBV PHB/zein[e]/PHB	PHBV and PHB monolayer by compression molding Electrospinning of zein nanofibers onto the PHA films Compression molding of the electrospun PHA monolayer with an identical PHA sheet	WVP and OP were improved in PHA multilayer systems by the addition of electrospun zein nanofibers. This strategy reduced the film mechanical resistance and improved their stretchability.	[79]
PHB[e]/TPS/ PHB[e]	TPS monolayer by compression molding Electrospinning of PHB fibers onto both sides of the TPS monolayer Compression molding of the monolayers	Improved water barrier capacity compared to the neat TPS monolayer by obtaining the multilayer system.	[104]
W-G/PHA/W-G	PHA monolayer by solvent casting Dip the PHA monolayer in a filmogenic solution plasticized with glycerol	The hydrocolloids decreased both the WVP and the opacity of the films compared to the neat PHA structure. Moreover, improved mechanical properties were achieved due to the hydrocolloids. Thermal analysis proved the polymeric miscibility.	[106]
PHB/PVOH/ PHB	Three-layer coextrusion Compression molding of the preformed layers	The grafted PHB improved the adhesion of the PVOH layer to the PHB layer. PHB protected the water-soluble PVOH core and greater film barrier capacities were achieved by using less-plasticized PVOH grade and more crystalline PHB.	[105]

Table 3.6 (continued)

Polymer system	Process	Main results	Reference
PHBV^e/CNF/ PHBV^e PHB^e/CNF/ PHB^e PHBV^e/LCNF/ PHBV^e PHB^e/LCNF/ PHB^e	Casting of the properly obtained CNF and LCNF solutions Electrospinning of PHB or PHBV fibers onto both sides of the CNF and LCNF films Compression molding of the monolayers	Electrospinning coating technique significantly decreased the nanopapers hydrophilicity and provided a more balanced mechanical performance. In addition, the PHBV^e coating enhanced the oxygen barrier capacity of the LCNF films.	[102]
PLA/PHBV multilayer films (3 to 2049 theoretical layers) (90/10 wt%)	Custom multilayer coextrusion	PLA/PHBV multilayer films exhibited improved mechanical and barrier properties compared to classical methods, such as melt blending or the three-layer coextrusion process, since the structure presented many long, thin, crystallized lamellas of PHBV dispersed within the PLA matrix, with increased compatibility.	[107]

WVP: water vapor permeability; OP: oxygen permeability; PVOH: polyvinyl alcohol; WG: wheat gluten; W-G: whey-gelatin; TPS: thermoplastic starch; CNFs: cellulose nanofibrils LCNFs: lignocellulose nanofibrils; superscript e represents electrospun layer.

Nonetheless, when incompatible polymers are used for the different layers, additional strategies could be required for adhesion purposes. For example, Thellen et al. [105] developed PHA tie layers, by grafting it with maleic anhydride (MA) using a dicumyl peroxide initiator to act as adhesive, thus more than doubling the peel strength with the core PVOH layer in three-layer systems. In this way, PHAs protected the water-soluble PVOH core, taking advantage of the high-oxygen barrier capacity of PVOH as well as the biodegradable and water barrier properties of PHA materials. Likewise, a custom multilayer coextrusion process was designed to develop PLA/PHBV multilayer films (90/10 wt%) containing from 3 to 2,049 theoretical layers, which exhibited improved ductility and barrier properties as compared to neat PLA films or melt blended films and three-layer coextruded films. These multilayer structures presented many long, thin, crystallized lamellas of PHBV dispersed within the PLA matrix, with increased compatibility [107].

3.5.4 Biocomposites

The incorporation of micro- and nanoreinforcing agents into the PHA polymer matrices to obtain biocomposites have also been considered to improve their functional properties and their competitiveness in the food packaging sector. Biocomposites are made up of a continuous biopolymer matrix in which organic or inorganic fillers of differing geometry (fibers, flakes, spheres, and particulates) are dispersed [108]. Natural fibers obtained from agro- and food waste have been studied as organic microfillers to develop PHA biocomposite films by using melt blending and compression molding (Table 3.7). Lignocellulosic fibers from different food by-products, such as wheat straw [109, 110], olive mills [109], almond shell, rice husk, and seagrass [111], have exhibited reinforcing effects, since they led to an increase in the EM and to a decrease in both E and TS. However, these fibers negatively affected the thermal stability and water barrier capacity of PHAs, which suggested a relatively poor adhesion between the fibers and the polymeric matrix due to the more hydrophilic nature of the fibers as compared to the hydrophobic matrix [112] achieved improved adhesion between PHBV and wood fibers in composite films by grafting PHBV with MA (PHBV-g-MA); this enhanced their mechanical and thermal properties, such as degradation temperature, glass transition temperature, melting temperature, and crystallinity. Moreover, the PHBV-g-MA composite showed decreased water absorption. Some other natural fibrous materials based on food-waste, such as keratin fibers from poultry feathers [113] and dried distillers grains with soluble produced as a coproduct during the dry-milling process of ethanol production [114] have also been successfully used as fillers in PHA-based films.

Different organic fillers in the nanometric range have been studied to obtain PHA-based bionanocomposites, since nanofillers have a higher specific surface area compared to microfillers and, therefore, a larger interface and stronger interaction with the polymer matrix. For this reason, lower nanofiller loadings are required to obtain comparable results to those achieved by common concentrations of microfillers [115]. In this sense, PHB biocomposites based on cellulose nanowhiskers (CNWs), at different concentrations, have exhibited improved functional properties as compared to the neat polymer [104] developed multilayer films containing bacterial CNWs (15wt%) in all layers, the inner and the outer sheets (PHB/TPS/PHB), which exhibited improved EM and TS, as well as reduced WVP and OP. Likewise, de O Patrício et al. [92] achieved enhanced polymer processability by using small amounts of wood pulp CNWs (<0.75wt%) dispersed on PEG, since the biocomposites had a larger processing window than the neat PHB. Nanofillers of differing geometries, such as cellulose nanofibers (CNFs) and cellulose nanocrystals (CNCs), have also been investigated for the purposes of obtaining PHBV nanocomposites with enhanced performance [116]. PHBV-CNC and PHBV-CNFs composites exhibited improved thermal stability. Both CNCs and CNFs acted as nucleating agents, thus decreasing the crystal size and increasing the T_m. Moreover, both nanofillers showed reinforcing effects in PHBV composites, which was greater for the CNCs at 1 wt%.

Table 3.7: PHA-based biocomposites for packaging materials.

Biocomposite	Origin	Polymer	Ratio	Process	Main results	Reference
Organic microfillers						
Lignocellulosic fibers	Wheat straw	PHBV	20 wt%	Extrusion and compression molding	– Reduced material cost – Decreased TS and E, but increased EM, which suggests a relatively poor adhesion between the fibers and the matrix due to the more hydrophilic nature of the fibers as compared to the hydrophobic matrix. – Further work: overcoming the lack of adhesion at the fiber–matrix interface (e.g., surface chemical modification)	[110]
	Wheat straw Olive mills	PHBV	20 wt%	Extrusion and compression molding	– Reduced material cost – Decreased PHBV thermal stability and crystallinity – Decreased TS and E due to the poor fiber/matrix adhesion – Composites based on wheat straw fibers showed higher WVP, while those based on olive mills have an improved water barrier capacity.	[109]
	Almond shell Rice husk Seagrass	PHB	10 wt% and 20 wt%	Melt blending and compression molding	– Reduced material cost – All the fibers showed a mechanical reinforcing effect in terms of increase in EM. – Decreased thermal stability and water barrier capacity of the fiber-based composites. – No effect on the PHB X_c or the disintegration rate of PHB. – Best performance with 10% of almond shell fiber.	[111]
	Wood	PHBV-g-MA	30 wt%	Extrusion and compression molding	– Better fiber/matrix adhesion in PHBV-g-MA composites – Improved thermal properties (T_{max}, T_g, T_m) and X_c – Improved mechanical properties – Decreased water absorption.	[112]

(continued)

Table 3.7 (continued)

Biocomposite	Origin	Polymer	Ratio	Process	Main results	Reference
Dried distillers grains with solubles	Coproduct during the dry-milling ethanol production	PHA	10 wt%	Extrusion and compression moulding	– Reduced material cost – Increased biodegradation rate – Enhanced mechanical properties	[114]
Keratin fiberes	Poultry feathers	PHBV containing 10 wt% of citric ester	0.5, 1, 3, 5, 10, 25, and 50 wt%	Melt blending and compression moulding	– Good fibere/matrix interaction – Composite containing 1 wt% of keratin showed reduced WVP, OP, and limonene permeability and exhibited optimum mechanical performance	[113]
Organic nanofillers						
CNWs	Bacterial cellulose	PHB/TPS/ PHB multilayer	2, 5, 10, 15, and 20 wt% of the TPS layer 15 wt% of the electrospun PHB layer	Melt blending and compression moulding	– TPS-CNWs by melt blending and compression moulding. Multilayer film by electrospinning of PHB-BCNW fiberes onto both sides of the TPS-CNWs and compression moulding. – CNWs in TPS films increased EM and TS, whereas they decreased E, WVP and OP – Best performance at 15 wt% CNW loading in TPS monolayer – The greatest reduction in WVP was achieved for multilayer structures incorporating BCNW in both the TPS inner layer and the PHB coating	[104]

				Observations	Ref.
Eucalyptus wood pulp	PHB	0.022, 0.045, 0.075, 0.15, 0.22, 0.45, and 0.75 wt% of the nanocomposite film	Solvent casing. Dispersion of CNWs in PEG added to the PHB polymer	– T_{max} attributed to PHB increased, whereas the T_m of the nanocomposites decreased as compared to that of neat PHB. – CNW concentrations up to 0.45 wt% improved the E without a significant loss in TS. – CNW concentration of 0.75 wt% decreased E compared to the other nanocomposites, due to a poor interface between the phases when the CNWs were not covered by PEG.	[92]
CNCs CNFs	Rice straw	PHBV	1, 2, 3, 4, 5, 6, and 7 wt%	Melt blending and compression molding	– Thermal stability of PHBV composites improved by both CNCs and CNFs. – Both CNCs and CNFs decreased the crystalline size of PHBV spherulites and CNCs acted as a better nucleating agent for PHBV than CNFs. – CNCs exhibited greater reinforcing effects than CNFs. – 1 wt% CNCs led to the best mechanical performance.	[116]
Chemically modified MWCNTs-COOH	Produced via chemical vapor deposition	PHA-g-MA	0.5, 1, 2, and 3 wt%	Melt blending and compression molding	Improved thermal and mechanical properties of the PHA-g-MA/ MWCNTs-COOH composites compared with PHA, because of the formation of ester carbonyl groups through the reaction between MA groups of PHA-g-MA and the carboxylic acid groups of the MWCNTs-COOH.	[117]

(continued)

Table 3.7 (continued)

Biocomposite	Origin	Polymer	Ratio	Process	Main results	Reference
Inorganic nanofillers						
ZnO nanoparticles	-	PHB	1, 2, 5, and 10 wt %	Solvent casting	– Increased T_c and X_c, thus increasing the polymer thermal stability. – Increased mechanical resistance due to the strong matrix–nanofiller interfacial adhesion. – Decreased WVP and OP as compared to neat PHB. – Decreased the migration of the PHB/ZnO composites as the nanoparticle content increased.	[42]

T_g: glass transition temperature; T_m: melting temperature; EM: elastic modulus; TS: tensile strength; E: elongation at break; T_c: crystallization temperature; X_c: degree of crystallinity; WVP: water vapor permeability; OP: oxygen permeability; T_{max}: maximum degradation temperature. TPS: thermoplastic starch; PHA-g-MA: maleic anhydride (MA)-grafted polyhydroxyalkanoate; CNW: cellulose nanowhiskers; CNCs: cellulose nanocrystals; CNFs: cellulose nanofibers; ZnO: zinc oxide; MWCNTs-COOH: multiple-walled carbon nanotubes.

Likewise, Díez-Pascual and Díez-Vicente [42] reported improved thermal stability and mechanical resistance for PHB nanocomposites based on inorganic nanofillers such as zinc oxide nanoparticles (ZnONP), due to their nucleating effect and the strong matrix-ZnONP interfacial adhesion attained via hydrogen bonding interactions. In addition, PHB-ZnONP nanocomposites lowered WVP and OP values as compared to those of pure PHB films.

Chemically modified nanocomposites have also been used in order to improve the compatibility with the polymer matrix. Thus chemically modified multiwalled carbon nanotubes (MWCNTs-COOH), have been used in PHA-g-MA matrices with improved compatibility and dispersibility of the MWCNTs within the matrix. In this case, the reaction between MA groups of PHA-g-MA and the carboxylic acid groups of the MWCNTs-COOH enhanced the thermal and mechanical properties of the material [117].

3.6 Conclusions

PHAs are very promising candidates to substitute petroleum-based plastics in a wide range of applications, being completely biobased and biodegradable, thus reducing both the dependency on fossil resources as well as the environmental problems related with their disposal. Despite the intrinsic brittleness of PHA-based materials, a great deal of progress has been made at improving this drawback. The use of additives, such as plasticizers or compatibilizers, their combination with other biodegradable polymers in blends or multilayer assemblies, as well as the development of biocomposites which incorporate micro or nanofillers, are some of the possible strategies which can improve the functional properties of PHAs, adapting them to the requirements of specific applications. However, more studies are necessary to reduce the high production cost compared to the synthetic plastics, such as those aimed at developing superior bacterial strains and transgenic plants by recombinant DNA technology to use waste as growth feedstock or optimizing the PHA extraction process.

References

[1]   Laycock, B., Halley, P., Pratt, S., Werker, A., and Lant, P. (2013). The chemomechanical properties of microbial polyhydroxyalkanoates. Prog. Polym. Sci. 38(3–4), 536–583.
[2]   Sudesh, K., Abe, H., and Doi, Y. (2000). Synthesis, structure and properties of polyhydroxyalkanoates: biological polyesters. Prog. Polym. Sci. 25(10), 1503–1555.
[3]   Khanna, S., and Srivastava, A. K. (2005). Recent advances in microbial polyhydroxyalkanoates. Process Biochem. 40(2), 607–619.

[4]   Muhammadi, S., Afzal, M., and Hameed, S. (2015). Bacterial polyhydroxyalkanoates-eco-friendly next generation plastic: production, biocompatibility, biodegradation, physical properties and applications. Green Chem. Lett. Rev. 8(3–4), 56–77.

[5]   Raza, Z. A., Abid, S., and Banat, I. M. (2018). Polyhydroxyalkanoates: characteristics, production, recent developments and applications. Int. Biodeterior. Biodegrad. 126, 45–56.

[6]   Piergiovanni, L., and Limbo, S. (2016). Food Packaging Materials. Springer.

[7]   Xu, J., and Guo, B. H. (2010). Plastics from Bacteria Natural Functions and Applications. Microbiology Monographs (Vol. 14). Springer-Verlag, Berlin, Heidelberg.

[8]   Snell, K. D., and Peoples, O. P. (2002). Polyhydroxyalkanoate polymers and their production in transgenic plants. Metab. Eng. 4(1), 29–40.

[9]   Hempel, F., Bozarth, A. S., Lindenkamp, N., Klingl, A., Zauner, S., Linne, U., and Maier, U. G. (2011). Microalgae as bioreactors for bioplastic production. Microb. Cell Fact. 10(1), 81.

[10]   Bohmert-Tatarev, K., McAvoy, S., Daughtry, S., Peoples, O. P., and Snell, K. D. (2011). High levels of bioplastic are produced in fertile transplastomic tobacco plants engineered with a synthetic operon for production of polyhydroxybutyrate. Plant Physiol. 110.

[11]   Bugnicourt, E., Cinelli, P., Lazzeri, A., and Alvarez, V. A. (2014). Polyhydroxyalkanoate (PHA): review of synthesis, characteristics, processing and potential applications in packaging. eXPRESS Polym. Lett. 8(11), 791–808.

[12]   Gumel, A. M., Annuar, M. S. M., and Chisti, Y. (2013). Recent advances in the production, recovery and applications of polyhydroxyalkanoates. J. Polym. Environ. 21(2), 580–605.

[13]   Jacquel, N., Lo, C. W., Wei, Y. H., Wu, H. S., and Wang, S. S. (2008). Isolation and purification of bacterial poly (3-hydroxyalkanoates). Biochem. Eng. J. 39(1), 15–27.

[14]   Kunasundari, B., and Sudesh, K. (2011). Isolation and recovery of microbial polyhydroxyalkanoates. eXPRESS Polym. Lett. 5(7).

[15]   Koller, M., Niebelschütz, H., and Braunegg, G. (2013). Strategies for recovery and purification of poly [(R)-3-hydroxyalkanoates](PHA) biopolyesters from surrounding biomass. Eng. Life Sci. 13(6), 549–562.

[16]   Koller, M., Sandholzer, D., Salerno, A., Braunegg, G., and Narodoslawsky, M. (2013). Biopolymer from industrial residues: life cycle assessment of poly (hydroxyalkanoates) from whey. Resour. Conserv. Recycl. 73, 64–71.

[17]   Albuquerque, P. B., and Malafaia, C. B. (2017). Perspectives on the production, structural characteristics and potential applications of bioplastics derived from polyhydroxyalkanoates. Int. J. Biol. Macromol. 107, 615–625.

[18]   Hazer, B., and Steinbüchel, A. (2007). Increased diversification of polyhydroxyalkanoates by modification reactions for industrial and medical applications. Appl. Microbiol. Biotechnol. 74(1), 1–12.

[19]   Keshavarz, T., and Roy, I. (2010). Polyhydroxyalkanoates: bioplastics with a green agenda. Curr. Opin. Microbiol. 13(3), 321–326.

[20]   Madison, L. L., and Huisman, G. W. (1999). Metabolic engineering of poly (3-hydroxyalkanoates): from DNA to plastic. Microbiol. Mol. Biol. Rev. 63(1), 21–53.

[21]   Yalpani, M. July 1993a. U.S. patent 5,225,227.

[22]   Yalpani, M. July 1993b. U.S. patent 5,229,158.

[23]   Steel, M. L., and Norton-Berry, P. July 1986. U.S. patent 4,603,070.

[24]   Marchessault, R. H., LePoutre, P. F., and Wrist, P. E. September 1995. U.S. patent 5,451,456.

[25]   Kauffman, T., Brady, F. X., Puletti, P. P., and Raykovitz, G. December 1992. U.S. patent 5,169,889.

[26]   Siracusa, V., Rocculi, P., Romani, S., and Dalla Rosa, M. (2008). Biodegradable polymers for food packaging: a review. Trends Food Sci. Technol. 19(12), 634–643.

[27]   Muizniece-Brasava, S., and Dukalska, L. (2006). Impact of biodegradable PHB packaging composite materials on dairy product quality. Proc. Latv. Univ. Agric. 16, 79–87.

[28] Bucci, D. Z., Tavares, L. B. B., and Sell, I. (2005). PHB packaging for the storage of food products. Polym. Test. 24(5), 564–571.

[29] Levkane, V., Muizniece-Brasava, S., and Dukalska, L. (2008). Pasteurization effect to quality of salad with meat in mayonnaise. Foodbalt, 6.

[30] Haugaard, V. K., Danielsen, B., and Bertelsen, G. (2003). Impact of polylactate and poly (hydroxybutyrate) on food quality. Eur. Food Res. Technol. 216(3), 233–240.

[31] Kantola, M., and Helén, H. (2001). Quality changes in organic tomatoes packaged in biodegradable plastic films. J. Food Qual. 24(2), 167–176.

[32] Modi, S., Koelling, K., and Vodovotz, Y. (2011). Assessment of PHB with varying hydroxyvalerate content for potential packaging applications. Eur. Polym. J. 47(2), 179–186.

[33] Cao, W., Wang, A., Jing, D., Gong, Y., Zhao, N., and Zhang, X. (2005). Novel biodegradable films and scaffolds of chitosan blended with poly (3-hydroxybutyrate). J. Biomater. Sci., Polym. Ed. 16(11), 1379–1394.

[34] Peelman, N., Ragaert, P., De Meulenaer, B., Adons, D., Peeters, R., Cardon, L., and Devlieghere, F. (2013). Application of bioplastics for food packaging. Trends Food Sci. Technol. 32(2), 128–141.

[35] Petersen, K., Nielsen, P. V., Bertelsen, G., Lawther, M., Olsen, M. B., Nilsson, N. H., and Mortensen, G. (1999). Potential of biobased materials for food packaging. Trends Food Sci. Technol. 10(2), 52–68.

[36] Zhang, M., and Thomas, N. L. (2011). Blending polylactic acid with polyhydroxybutyrate: the effect on thermal, mechanical, and biodegradation properties. Adv. Polym. Tech. 30(2), 67–79.

[37] Choi, J. S., and Park, W. H. (2004). Effect of biodegradable plasticizers on thermal and mechanical properties of poly (3-hydroxybutyrate). Polym. Test. 23(4), 455–460.

[38] Savenkova, L., Gercberga, Z., Bibers, I., and Kalnin, M. (2000). Effect of 3-hydroxy valerate content on some physical and mechanical properties of polyhydroxyalkanoates produced by Azotobacter chroococcum. Process Biochem. 36(5), 445–450.

[39] Sanchez-Garcia, M. D., Lopez-Rubio, A., and Lagaron, J. M. (2010). Natural micro and nanobiocomposites with enhanced barrier properties and novel functionalities for food biopackaging applications. Trends Food Sci. Technol. 21(11), 528–536.

[40] Castro-Mayorga, J. L., Fabra, M. J., and Lagaron, J. M. (2016). Stabilized nanosilver based antimicrobial poly (3-hydroxybutyrate-co-3-hydroxyvalerate) nanocomposites of interest in active food packaging. Innovative Food Sci. Emerg. Technol. 33, 524–533.

[41] Castro-Mayorga, J. L., Fabra, M. J., Pourrahimi, A. M., Olsson, R. T., and Lagaron, J. M. (2017). The impact of zinc oxide particle morphology as an antimicrobial and when incorporated in poly (3-hydroxybutyrate-co-3-hydroxyvalerate) films for food packaging and food contact surfaces applications. Food Bioprod. Process. 101, 32–44.

[42] Díez-Pascual, A. M., and Díez-Vicente, A. L. (2014). Poly (3-hydroxybutyrate)/ZnO bionanocomposites with improved mechanical, barrier and antibacterial properties. Int. J. Mol. Sci. 15(6), 10950–10973.

[43] Narayanan, A., and Ramana, K. V. (2013). Synergized antimicrobial activity of eugenol incorporated polyhydroxybutyrate films against food spoilage microorganisms in conjunction with pediocin. Appl. Biochem. Biotechnol. 170(6), 1379–1388.

[44] Xavier, J. R., Babusha, S. T., George, J., and Ramana, K. V. (2015). Material properties and antimicrobial activity of polyhydroxybutyrate (PHB) films incorporated with vanillin. Appl. Biochem. Biotechnol. 176(5), 1498–1510.

[45] Cerqueira, M. A., Fabra, M. J., Castro-Mayorga, J. L., Bourbon, A. I., Pastrana, L. M., Vicente, A. A., and Lagaron, J. M. (2016). Use of electrospinning to develop antimicrobial biodegradable multilayer systems: encapsulation of cinnamaldehyde and their physicochemical characterization. Food Bioprocess Technol. 9(11), 1874–1884.

[46] Fabra, M. J., Castro-Mayorga, J. L., Randazzo, W., Lagarón, J. M., López-Rubio, A., Aznar, R., and Sánchez, G. (2016). Efficacy of cinnamaldehyde against enteric viruses and its activity after incorporation into biodegradable multilayer systems of interest in food packaging. Food Environ. Virol. 8(2), 125–132.

[47] Arrieta, M. P., Castro-Lopez, M. D. M., Rayón, E., Barral-Losada, L. F., López-Vilariño, J. M., López, J., and González-Rodríguez, M. V. (2014). Plasticized poly (lactic acid)–poly (hydroxybutyrate)(PLA–PHB) blends incorporated with catechin intended for active food-packaging applications. J. Agric. Food Chem. 62(41), 10170–10180.

[48] Fan, X., Jiang, Q., Sun, Z., Li, G., Ren, X., Liang, J., and Huang, T. S. (2015). Preparation and characterization of electrospun antimicrobial fibrous membranes based on polyhydroxybutyrate (PHB). Fibers Polym. 16(8), 1751–1758.

[49] Correa, J. P., Molina, V., Sanchez, M., Kainz, C., Eisenberg, P., and Massani, M. B. (2017). Improving ham shelf life with a polyhydroxybutyrate/polycaprolactone biodegradable film activated with nisin. Food Packag. Shelf Life 11, 31–39.

[50] Hema, R., Ng, P. N., and Amirul, A. A. (2013). Green nanobiocomposite: reinforcement effect of montmorillonite clays on physical and biological advancement of various polyhydroxyalkanoates. Polym. Bull. 70(3), 755–771.

[51] Sevastianov, V. I., Perova, N. V., Shishatskaya, E. I., Kalacheva, G. S., and Volova, T. G. (2003). Production of purified polyhydroxyalkanoates (PHAs) for applications in contact with blood. J. Biomater. Sci., Polym. Ed. 14(10), 1029–1042.

[52] Chen, W., and Tong, Y. W. (2012). PHBV microspheres as neural tissue engineering scaffold support neuronal cell growth and axon–dendrite polarization. Acta Biomater. 8(2), 540–548.

[53] Porter, M. M., Lee, S., Tanadchangsaeng, N., Jaremko, M. J., Yu, J., Meyers, M., and McKittrick, J. (2013). Porous hydroxyapatite-polyhydroxybutyrate composites fabricated by a novel method via centrifugation. Mech. Biol. Syst. Mater. 5, 63–71.

[54] Zhou, J., Peng, S. W., Wang, Y. Y., Zheng, S. B., Wang, Y., and Chen, G. Q. (2010). The use of poly (3-hydroxybutyrate-co-3-hydroxyhexanoate) scaffolds for tarsal repair in eyelid reconstruction in the rat. Biomaterials 31(29), 7512–7518.

[55] Shishatskaya, E. I., Nikolaeva, E. D., Vinogradova, O. N., and Volova, T. G. (2016). Experimental wound dressings of degradable PHA for skin defect repair. J. Mater. Sci. Mater. Med. 27(11), 165.

[56] Volova, T. G. (2004). Polyhydroxyalkanoates-plastic materials of the twenty-first century: production, properties, applications. Nova Publishers.

[57] Valappil, S. P., Misra, S. K., Boccaccini, A. R., and Roy, I. (2006). Biomedical applications of polyhydroxyalkanoates, an overview of animal testing and in vivo responses. Expert Rev. Med. Devices 3(6), 853–868.

[58] Valappil, S. P., Boccaccini, A. R., Bucke, C., and Roy, I. (2007). Polyhydroxyalkanoates in Gram-positive bacteria: insights from the genera Bacillus and Streptomyces. Antonie Van Leeuwenhoek 91(1), 1–17.

[59] Kılıçay, E., Demirbilek, M., Türk, M., Güven, E., Hazer, B., and Denkbas, E. B. (2011). Preparation and characterization of poly (3-hydroxybutyrate-co-3-hydroxyhexanoate)(PHBHHX) based nanoparticles for targeted cancer therapy. Eur. J. Pharm. Sci. 44(3), 310–320.

[60] Wang, Z., Wu, H., Chen, J., Zhang, J., Yao, Y., and Chen, G. Q. (2008). A novel self-cleaving phasin tag for purification of recombinant proteins based on hydrophobic polyhydroxyalkanoate nanoparticles. Lab Chip 8(11), 1957–1962.

[61] Yao, Y. C., Zhan, X. Y., Zhang, J., Zou, X. H., Wang, Z. H., Xiong, Y. C., Chen, J., and Chen, G. Q. (2008). A specific drug targeting system based on polyhydroxyalkanoate granule binding protein PhaP fused with targeted cell ligands. Biomaterials 29(36), 4823–4830.

[62] Xiao, X. Q., Zhao, Y., and Chen, G. Q. (2007). The effect of 3-hydroxybutyrate and its derivatives on the growth of glial cells. Biomaterials 28(25), 3608–3616.

[63] Zou, X. H., Li, H. M., Wang, S., Leski, M., Yao, Y. C., Yang, X. D., and Chen, G. Q. (2009). The effect of 3-hydroxybutyrate methyl ester on learning and memory in mice. Biomaterials 30(8), 1532–1541.

[64] Gao, X., Chen, J. C., Wu, Q., and Chen, G. Q. (2011). Polyhydroxyalkanoates as a source of chemicals, polymers, and biofuels. Curr. Opin. Biotechnol. 22(6), 768–774.

[65] Wang, S. Y., Wang, Z., Liu, M. M., Xu, Y., Zhang, X. J., and Chen, G. Q. (2010). Properties of a new gasoline oxygenate blend component: 3-hydroxybutyrate methyl ester produced from bacterial poly-3-hydroxybutyrate. Biomass Bioenergy 34(8), 1216–1222.

[66] Chen, G. Q. (2009). A microbial polyhydroxyalkanoates (PHA) based bio-and materials industry. Chem. Soc. Rev. 38(8), 2434–2446.

[67] Grillo, R., Santo Pereira, A. D. E., de Melo, N. F. S., Porto, R. M., Feitosa, L. O., Tonello, P. S., Filho, N. L., Rosa, A. H., Lima, R., and Fraceto, L. F. (2011). Controlled release system for ametryn using polymer microspheres: preparation, characterization and release kinetics in water. J. Hazard. Mater. 186(2–3), 1645–1651.

[68] Zhang, X., Wei, C., He, Q., and Ren, Y. (2010). Enrichment of chlorobenzene and o-nitrochlorobenzene on biomimetic adsorbent prepared by poly-3-hydroxybutyrate (PHB). J. Hazard. Mater. 177(1–3), 508–515.

[69] Defoirdt, T., Sorgeloos, P., and Bossier, P. (2011). Alternatives to antibiotics for the control of bacterial disease in aquaculture. Curr. Opin. Microbiol. 14(3), 251–258.

[70] Wille, M., De Schryver, P., Defoirdt, T., Bossier, P., and Sorgeloos, P. (2010). The effect of poly β-hydroxybutyrate on larviculture of the giant freshwater prawn Macrobrachium rosenbergii. Aquaculture 302(1–2), 76–81.

[71] Fabra, M. J., Lopez-Rubio, A., and Lagaron, J. M. (2013). High barrier polyhydroxyalcanoate food packaging film by means of nanostructured electrospun interlayers of zein. Food Hydrocolloids 32(1), 106–114.

[72] Holmes, P. A. (1988). Biologically Produced (R)-3-hydroxy-alkanoate Polymers and Copolymers. Developments in Crystalline Polymers. Springer, Dordrecht, 1–35.

[73] Chuah, J. A., Yamada, M., Taguchi, S., Sudesh, K., Doi, Y., and Numata, K. (2013). Biosynthesis and characterization of polyhydroxyalkanoate containing 5-hydroxyvalerate units: effects of 5HV units on biodegradability, cytotoxicity, mechanical and thermal properties. Polym. Degrad. Stab. 98(1), 331–338.

[74] Dhar, P., Bhardwaj, U., Kumar, A., and Katiyar, V. (2015). Poly (3-hydroxybutyrate)/cellulose nanocrystal films for food packaging applications: barrier and migration studies. Polym. Eng. Sci. 55(10), 2388–2395.

[75] Erkske, D., Viskere, I., Dzene, A., Tupureina, V., and Savenkova, L. (2006). Biobased polymer composites for films and coatings. Proc. Est. Acad. Sci. Chem. 55(2), 70–77.

[76] Zhijiang, C., and Guang, Y. (2011). Optical nanocomposites prepared by incorporating bacterial cellulose nanofibrils into poly (3-hydroxybutyrate). Mater. Lett. 65(2), 182–184.

[77] Godbole, S., Gote, S., Latkar, M., and Chakrabarti, T. (2003). Preparation and characterization of biodegradable poly-3-hydroxybutyrate–starch blend films. Bioresour. Technol. 86(1), 33–37.

[78] Arrieta, M. P., López, J., Hernández, A., and Rayón, E. (2014c). Ternary PLA–PHB–Limonene blends intended for biodegradable food packaging applications. Eur. Polym. J. 50, 255–270.

[79] Fabra, M. J., Lopez-Rubio, A., and Lagaron, J. M. (2014). Nanostructured interlayers of zein to improve the barrier properties of high barrier polyhydroxyalkanoates and other polyesters. J. Food Eng. 127, 1–9.

[80] Sanchez-Garcia, M. D., Lagaron, J. M., and Hoa, S. V. (2010). Effect of addition of carbon nanofibers and carbon nanotubes on properties of thermoplastic biopolymers. Compos. Sci. Technol. 70(7), 1095–1105.

[81] Orts, W. J., Marchessault, R. H., Bluhm, T. L., and Hamer, G. K. (1990). Observation of strain-induce β form in poly (β-hydroxyalkanoates). Macromolecules 23(26), 5368–5370.

[82] Boyandin, A. N., Prudnikova, S. V., Karpov, V. A., Ivonin, V. N., Đỗ, N. L., Nguyễn, T. H., HiệpLê, T. H., Filichev, N. L., Levin, A. L., Filipenko, M. L., Volova, T. G., and Gitelson, I. I. (2013). Microbial degradation of polyhydroxyalkanoates in tropical soils. Int. Biodeterior. Biodegrad. 83, 77–84.

[83] Volova, T. G., Prudnikova, S. V., Vinogradova, O. N., Syrvacheva, D. A., and Shishatskaya, E. I. (2017). Microbial degradation of polyhydroxyalkanoates with different chemical compositions and their biodegradability. Microb. Ecol. 73(2), 353–367.

[84] Lee, S. Y. (1996). Bacterial polyhydroxyalkanoates. Biotechnol. Bioeng. 49(1), 1–14.

[85] Thellen, C., Coyne, M., Froio, D., Auerbach, M., Wirsen, C., and Ratto, J. A. (2008). A processing, characterization and marine biodegradation study of melt-extruded polyhydroxyalkanoate (PHA) films. J. Polym. Environ. 16(1), 1–11.

[86] Wang, S., Ma, P., Wang, R., Wang, S., Zhang, Y., and Zhang, Y. (2008). Mechanical, thermal and degradation properties of poly (d, l-lactide)/poly (hydroxybutyrate-co-hydroxyvalerate) /poly (ethylene glycol) blend. Polym. Degrad. Stab. 93(7), 1364–1369.

[87] Vieira, M. G. A., da Silva, M. A., dos Santos, L. O., and Beppu, M. M. (2011). Natural-based plasticizers and biopolymer films: a review. Eur. Polym. J. 47(3), 254–263.

[88] Jost, V., and Langowski, H. C. (2015). Effect of different plasticisers on the mechanical and barrier properties of extruded cast PHBV films. Eur. Polym. J. 68, 302–312.

[89] Branciforti, M. C., Corrêa, M. C. S., Pollet, E., Agnelli, J. A. M., de Paula Nascente, P. A., and Avérous, L. (2013). Crystallinity study of nano-biocomposites based on plasticized poly (hydroxybutyrate-co-hydroxyvalerate) with organo-modified montmorillonite. Polym. Test. 32(7), 1253–1260.

[90] Solaiman, D. K., Ashby, R. D., Zerkowski, J. A., Krishnama, A., and Vasanthan, N. (2015). Control-release of antimicrobial sophorolipid employing different biopolymer matrices. Biocatal. Agric. Biotechnol. 4(3), 342–348.

[91] Yoshie, N., Nakasato, K., Fujiwara, M., Kasuya, K., Abe, H., Doi, Y., and Inoue, Y. (2000). Effect of low molecular weight additives on enzymatic degradation of poly (3-hydroxybutyrate). Polymer 41(9), 3227–3234.

[92] de O Patrício, P. S., Pereira, F. V., dos Santos, M. C., de Souza, P. P., Roa, J. P., and Orefice, R. L. (2013). Increasing the elongation at break of polyhydroxybutyrate biopolymer: effect of cellulose nanowhiskers on mechanical and thermal properties. J. Appl. Polym. Sci. 127(5), 3613–3621.

[93] Armentano, I., Fortunati, E., Burgos, N., Dominici, F., Luzi, F., Fiori, S., Jiménez, A., Yoon, K., Ahn, J., Kang, S., and Kenny, J. M. (2015). Processing and characterization of plasticized PLA/PHB blends for biodegradable multiphase systems. eXPRESS Polym. Lett. 9(7).

[94] Arrieta, M. P., Samper, M. D., López, J., and Jiménez, A. (2014). Combined effect of poly (hydroxybutyrate) and plasticizers on polylactic acid properties for film intended for food packaging. J. Polym. Environ. 22(4), 460–470.

[95] Ma, X., Wang, Y., Wang, J., and Xu, Y. (2017). Effect of PBAT on property of PLA/PHB film used for fruits and vegetables. MATEC Web Conf. 88, 02009. EDP Sciences.

[96] Ferreira, B. M. P., Zavaglia, C. A. C., and Duek, E. A. R. (2001). Films of poly (L-lactic acid)/poly (hydroxybutyrate-co-hydroxyvalerate) blends: in vitro degradation. Mater. Res. 4(1), 34–42.

[97] Jost, V., and Kopitzky, R. (2015). Blending of polyhydroxybutyrate-co-valerate with polylactic acid for packaging applications–reflections on miscibility and effects on the mechanical and barrier properties. Chem. Biochem. Eng. Q. 29(2), 221–246.

[98] Parulekar, Y., and Mohanty, A. K. (2007). Extruded biodegradable cast films from polyhydroxyalkanoate and thermoplastic starch blends: fabrication and characterization. Macromol. Mater. Eng. 292(12), 1218–1228.

[99] Sun, S., Liu, P., Ji, N., Hou, H., and Dong, H. (2017). Effects of low polyhydroxyalkanoate content on the properties of films based on modified starch acquired by extrusion blowing. Food Hydrocolloids 72, 81–89.

[100] Sun, S., Liu, P., Ji, N., Hou, H., and Dong, H. (2017). Effects of various cross-linking agents on the physicochemical properties of starch/PHA composite films produced by extrusion blowing. Food Hydrocolloids 77, 964–975.

[101] Muller, J., González-Martínez, C., and Chiralt, A. (2017). Combination of poly (lactic) acid and starch for biodegradable food packaging. Materials 10(8), 952.

[102] Cherpinski, A., Torres-Giner, S., Vartiainen, J., Peresin, M. S., Lahtinen, P., and Lagaron, J. M. (2018). Improving the water resistance of nanocellulose-based films with polyhydroxyalkanoates processed by the electrospinning coating technique. Cellulose, 25(2), 1291–1307.

[103] Fabra, M. J., López-Rubio, A., and Lagaron, J. M. (2015). Three-layer films based on wheat gluten and electrospun PHA. Food Bioprocess Technol. 8(11), 2330–2340.

[104] Fabra, M. J., López-Rubio, A., Ambrosio-Martín, J., and Lagaron, J. M. (2016). Improving the barrier properties of thermoplastic corn starch-based films containing bacterial cellulose nanowhiskers by means of PHA electrospun coatings of interest in food packaging. Food Hydrocolloids 61, 261–268.

[105] Thellen, C., Cheney, S., and Ratto, J. A. (2013). Melt processing and characterization of polyvinyl alcohol and polyhydroxyalkanoate multilayer films. J. Appl. Polym. Sci. 127(3), 2314–2324.

[106] de Andrade, C. S., Fonseca, G. G., Mei, I., Helena, L., and Fakhouri, F. M. (2017). Development and characterization of multilayer films based on polyhydroxyalkanoates and hydrocolloids. J. Appl. Polym. Sci. 134(6).

[107] Boufarguine, M., Guinault, A., Miquelard-Garnier, G., and Sollogoub, C. (2013). PLA/PHBV films with improved mechanical and gas barrier properties. Macromol. Mater. Eng. 298(10), 1065–1073.

[108] Díez-Pascual, A. M. (2017). Biodegradable food packaging nanocomposites based on ZnO-reinforced polyhydroxyalkanoates. Food Packag. 185–221.

[109] Berthet, M. A., Angellier-Coussy, H., Machado, D., Hilliou, L., Staebler, A., Vicente, A., and Gontard, N. (2015). Exploring the potentialities of using lignocellulosic fibres derived from three food by-products as constituents of biocomposites for food packaging. Ind. Crops Prod. 69, 110–122.

[110] Martino, L., Berthet, M. A., Angellier-Coussy, H., and Gontard, N. (2015). Understanding external plasticization of melt extruded PHBV–wheat straw fibers biodegradable composites for food packaging. J. Appl. Polym. Sci. 132(10).

[111] Sánchez-Safont, E. L., Aldureid, A., Lagarón, J. M., Gámez-Pérez, J., and Cabedo, L. (2018). Biocomposites of different lignocellulosic wastes for sustainable food packaging applications. Composites Part B, 145, 215–225.

[112] Yatigala, N. S., Bajwa, D. S., and Bajwa, S. G. (2018). Compatibilization improves physico-mechanical properties of biodegradable biobased polymer composites. Composites Part A 107, 315–325.

[113] Pardo-Ibáñez, P., Lopez-Rubio, A., Martínez-Sanz, M., Cabedo, L., and Lagaron, J. M. (2014). Keratin–polyhydroxyalkanoate melt-compounded composites with improved barrier properties of interest in food packaging applications. J. Appl. Polym. Sci. 131(4).

[114] Madbouly, S. A., Schrader, J. A., Srinivasan, G., Liu, K., McCabe, K. G., Grewell, D., and Kessler, M. R. (2014). Biodegradation behavior of bacterial-based polyhydroxyalkanoate (PHA) and DDGS composites. Green Chem. 16(4), 1911–1920.

[115] Naffakh, M., Díez-Pascual, A. M., Marco, C., Ellis, G. J., and Gómez-Fatou, M. A. (2013). Opportunities and challenges in the use of inorganic fullerene-like nanoparticles to produce advanced polymer nanocomposites. Prog. Polym. Sci. 38(8), 1163–1231.

[116] Jun, D., Guomin, Z., Mingzhu, P., Leilei, Z., Dagang, L., and Rui, Z. (2017). Crystallization and mechanical properties of reinforced PHBV composites using melt compounding: effect of CNCs and CNFs. Carbohydr. Polym. 168, 255–262.

[117] Wu, C. S., and Liao, H. T. (2017). Interface design of environmentally friendly carbon nanotube-filled polyester composites: fabrication, characterisation, functionality and application. eXPRESS Polym. Lett. 11(3), 187.

[118] Abe, H., Doi, Y., Aoki, H., Akehata, T., Hori, Y., & Yamaguchi, A. (1995). Physical properties and enzymic degradability of copolymers of (R)-3-hydroxybutyric and 6-hydroxyhexanoic acids. Macromolecules, 28(23), 7630–7637.

Aleksandra Nesic, Arash Moeini, Gabriella Santagata

4 Marine biopolymers: alginate and chitosan

Abstract: Marine environment has been recognized as a valuable source of bioactive polymers with industrial potential. Some families of polysaccharides, such as chitosan and alginate, are highly abundant renewable biomasses that might be obtained from microorganisms, shrimps, clams, or algae. These polysaccharides exhibit various biological activities, including antioxidant, anti-infection, anticoagulant, anti-inflammatory, and antidiabetic effects. Besides their natural biological and structural functions, the marine biopolymers can be tailored to be new biomaterials with novel functionalities. Hence, this chapter will summarize the most important novel concepts and inventions along with marine biopolymer engineering.

Keywords: marine biopolymers, chitosan, alginate

4.1 Introduction

Nowadays, due to heavy consumption of synthetic polymers, the world is facing serious problems and massive environmental contamination, urging the use of natural resources and to obtain biodegradable polymers as a replacement for the petroleum-derived ones. Oceans cover more than 70% of the earth's surface, representing an enormous resource for the discovery of potential bioactive compounds. Marine biopolymers have been found as the main component in cell wall structures of seaweeds (alginate) or exoskeletons of crustaceans, such as shrimps and clams (chitosan). Due to their diverse and versatile properties, they could both directly replace synthetically derived materials in traditional applications and open up a range of new commercial opportunities. Up to date, these biopolymers have been widely investigated for potential application in biomedicine, drug delivery sector, food packaging, cosmetics, food additives, textile industry, and water treatment. In this chapter, the chemical nature and various biological potentials of main marine biopolymers are presented, with the main focus on their properties and potential applications.

Aleksandra Nesic, Vinca Institute of Nuclear Sciences, University of Belgrade, Belgrade, Serbia
Arash Moeini, Department of Chemical Sciences, University of Naples Federico II, Naples, Italy
Gabriella Santagata, Institute for Polymers, Composites and Biomaterials (IPCB-CNR), Pozzuoli, Italy

https://doi.org/10.1515/9783110590586-004

4.2 Alginate

Alginate is a natural polysaccharide found in the cell wall of brown seaweeds (Phaeophyceae) and soil bacteria capsules (*Pseudomonas* and *Azotobacter*) [1]. The primary role of alginate in these organisms is protection: it provides mechanical strength and flexibility to algae and prevents their dehydration, while the bacteria enter into the composition of protective cysts and biofilms.

Commercially, alginate is found in the form of sodium alginate, that is, the sodium salt of alginic acid. The molecular weight of the commercially available sodium alginate varies in a wide range of 32 000–400 000 g/mol. The viscosity of the solution is dependent on the pH value and reaches the highest value at a pH of about 3–3.5. According to the structure, sodium alginates are linear polysaccharides composed of two-component copolymers of uronic residues: β-D-mannuronate residues (M) and α-L-guluronate residues (G) (Figure 4.1). The β-D-mannuronate and α-L-guluronate residues are connected through 1,4-glycosidic bonds and can be organized into homopolymer (MM and GG blocks) and heteropolymer (MG block) structure [2]. The relationship and order of M and G units within the polymer vary depending on the type of algae from which the alginate is isolated. Moreover, the physicochemical properties of alginate depend on M/G ratio and their distribution [3].

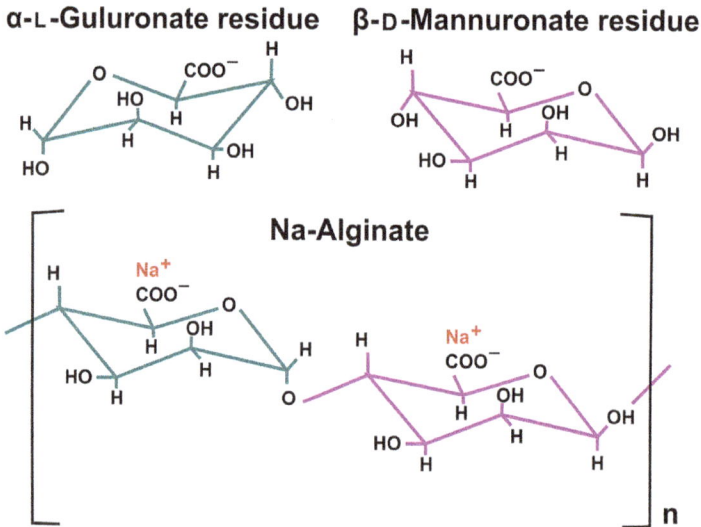

Figure 4.1: Molecular structure of uronic residues and sodium alginate.

Alginates, which are commercially available, are almost without exception obtained by extraction procedure from brown seaweeds of the Phaeophyceae family. Alginates are present in cell walls of brown algae in the form of insoluble salts of alginic acid (mostly calcium, magnesium, barium, and strontium salts), where the concentration of these ions is determined by their equilibrium with seawater [4]. In principle, the extraction process of alginate includes several stages. First, wet chopped seaweeds are subjected to blanching in presence of sodium hypochlorite and purification, in terms of removing the extra content of polyphenols and dye pigments by the use of formaldehyde. The next step includes preextraction with hydrochloric acid, causing conversion of salt to alginic acid. After washing and filtration, the pretreated seaweeds are immersed into alkaline solution (in the presence of NaOH or Na_2CO_3). Lastly, sodium alginate is precipitated from the solution by alcohol and reprecipitated in the same way. The yield, molecular weight, viscosity, and M/G unit distribution of alginate strongly depend on the origin of algae, time of extraction process, temperature, and concentration of extractive solvents during the preextraction (HCl) and extraction step (NaOH).

4.2.1 Properties of alginate

Sodium alginate is soluble in water and results in solutions of high viscosity. The viscosity of an alginate solution depends on the molecular weight, concentration of the polymer, and distribution of M and G blocks. Aqueous solutions of alginates have shear-thinning characteristics, which means that viscosity decreases as the shear rate, or stripped speed, increases. This property is known as pseudoplasticity or non-Newtonian flow. The pH of the solvent is an essential parameter determining the solubility of alginates in water, as it influences the presence of electrostatic charges on the uronic acid residues. In the case of a slight decrease of pH value of the solution below the pKa of guluronic acid (pKa = 3.65) and mannuronic acid (pKa = 3.38), alginate turns into a gel state, forming the so-called acidic gel, stabilized by intermolecular hydrogen bonds. Alginates containing MG-block structure precipitate at lower pH compared to the homogeneous polymeric G and M blocks. In addition to the pH, being important for alginate solubility in water, the ionic strength of the solution and the presence of divalent ions play important roles due to salting out effects and gelling, respectively. Sodium alginate is insoluble in alcohol and organic solvents, while calcium alginate is insoluble in water.

Alginate can degrade by different mechanism, which includes oxidative–reductive free radical depolymerization (ORD), acid and alkaline degradation, and enzymatic-catalyzed degradation. During degradation, cleavage of the glycoside bonds occurs, to obtain monomers [5] and there is also a subsequent decrease in viscosity. Concentrations of reactants and temperature are the main factors affecting degradation rates. Acid hydrolysis (decreasing pH values: pH 6.6 → 4.3) and β-elimination (caused by alkaline conditions) are the primary mechanisms responsible for the depolymerization. Depolymerization process is accelerated at increased temperature. Transition metals may catalyze an ORD degradation mechanism.

4.2.2 Gelation of alginate

The main property of the alginate is its ability to form hydrogels in the presence of divalent cations. The mechanism of alginate gelation is explained by the "egg-box" model [6] where the cross-links are formed by calcium ions occupying electronegative cavities in the two-fold buckled ribbon structure of the carboxylic groups (Figure 4.2). The cavities where the interaction between the negative and positive charges takes place are called junction zones. The higher the concentration of available divalent cations, the more efficient the gelation will be, leading to the release of a larger amount of structural water and reduced size and mass of the resulting hydrogels. Gelation takes place spontaneously, because the process is exothermic and is characterized by a positive change in entropy due to the release of a large amount of water molecules.

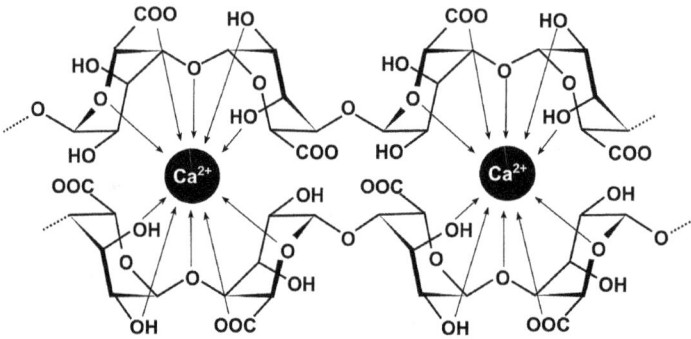

Figure 4.2: Egg-box model cross-linking of alginate by calcium ions.

There are two fundamental methods for preparing alginate gels by ionic cross-linking: (a) the diffusion method and (b) the internal gelation method. The methods differ in the way the cross-linking ions are introduced and hence the gelling kinetics of the two methods is very different. The diffusion method is based on diffusion of calcium ions (or other divalent ions) from an outer reservoir into the alginate solution (Figure 4.3). This method is very rapid and it is commonly used in food industry, as well as in drug delivery. The internal gelation method assumes mixing of an inactive form of the cross-linking ion with alginate solution (Figure 4.3). As an inactive cross-linking ions, usually insoluble $CaCO_3$ or Ca^{2+} ions complexed with chelating agents like EDTA or citrate, are used. The Ca^{2+} ions are released by lowering the pH, adding organic acids, or hydrolyzing lactones. This reaction is slow and allows controlled formation of homogenous hydrogels. Using a system of alginate mixed with $CaCO_3$ and D-glucono-δ-lactone, the gelling kinetics can be controlled by the calcium carbonate particle size.

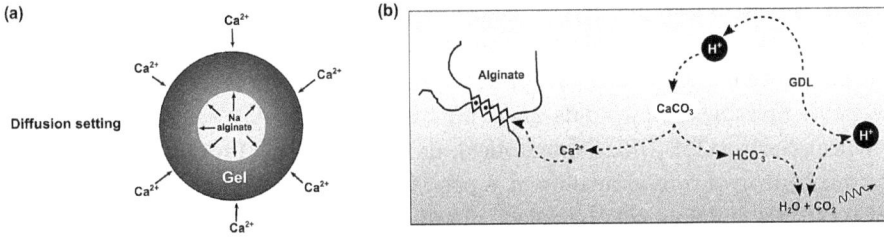

Figure 4.3: Different types of gelation of alginate: (a) external gelation and (b) internal gelation.

The affinity of the alginate to bind divalent cations is not the same for all metal ions and is growing in the following order [2]:

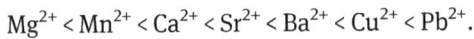

$$Mg^{2+} < Mn^{2+} < Ca^{2+} < Sr^{2+} < Ba^{2+} < Cu^{2+} < Pb^{2+}.$$

Affinity of divalent cations strongly depends on the alginate composition. The selectivity for alkaline earth metals increases with increasing G-unit content. Moreover, GG blocks have stronger affinity for divalent cations than MM and MG blocks. The physical properties of the hydrogel are also dependent on the formulation composition, that is, the proportion between G and M residues, in the order of sequence they are, as well as on molecular weight of alginate. Higher content of GG blocks leads to formation of more stiff, brittle, and porous hydrogels, whereas higher content of MM blocks leads to more elastic, less porous hydrogels, which can easily be disintegrated over time.

4.2.3 Application of alginate

As alginate is readily available, relatively inexpensive, and nontoxic, this biopolymer has several application areas. Alginate is commercially used in food industry as a thickening, gelling, and stabilizing agent. In dentistry, it is used as an impression material in orthodontic models, sports mouth guards, and bleaching trays. Nowadays, extensive research has been undertaken to investigate potentials of alginate hydrogels in biomedicine. Alginate hydrogels, most often in the form of Ca-alginate, possess many desirable characteristics that make them suitable for various biomedical applications such as carriers for cell tissue engineering, controlled drug release matrices, or coatings for wounds. They are hydrophilic, biocompatible, nontoxic, and hypoallergenic and have high capacity of absorption of liquids. Alginate hydrogels are formed under mild conditions such as room or body temperature and at neutral pH [1, 7]. Hence, the following section summarizes few of most important future key roles of alginate in biomedicine.

4.2.3.1 Wound dressing

The most severe cases of clinical application of alginate hydrogels are commercially available dressings for wounds and burns based on alginate: AlgicellTM (Derma Science), SorbsanTM (UDL Laboratories), or Comfeel PlusTM (Coloplast). Knowing the mechanism of wound healing, it is possible to develop functional wound dressings, which simultaneously represent a barrier that prevents the growth of microorganisms and protects the tissue, and it allows smooth transfer of oxygen, ensures adequate moisture with simultaneous absorption of excess exudate, provides an optimal environment in which regenerative processes are promoted and accelerates healing [8, 9]. Alginate intended for use in wound treatment is most often found in an anhydrous form, which, after contact with the wound, swell and recovers again into the hydrogel state and adapts the shape of the wound (Figure 4.4). Additional advantage of application of alginate in wound treatment is also the fact that the process of covering the wound is painless and comfortable for patients because the alginate is easily removed by dissolution [10].

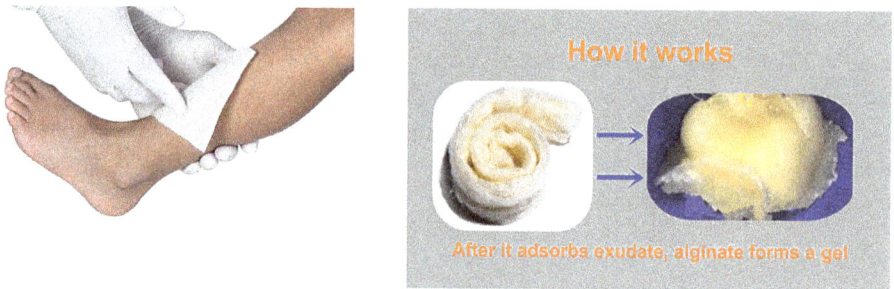

Figure 4.4: Alginate wound dressing.

Numerous studies have shown that the alginate is involved in several aspects in the process of wound healing. It has been found that released calcium from Ca-alginate acts homeostatically, that is stops bleeding [8, 11], activates macrophages [12], and induces proliferation of fibroblasts, which are responsible for remodeling of extracellular matrix and wound healing. In order to add more functionality to alginate hydrogel wound dressings, they can be cross-linked by zinc ions [13, 14] or incorporated with bioactive compounds [15], providing antimicrobial activity, as well.

4.2.3.2 Drug delivery

Due to porous structure and biocompatibility, alginate hydrogel might be attractive as a matrix for controlled and localized delivery and release of drugs or physiologically active agents [16]. These systems are especially suitable for proteins (growth factors, monoclonal antibodies, etc.), because they can be immobilized in hydrogel at mild conditions (physiological pH and temperature) and thus preserve their activity [17, 18]. Taking into account that the alginate can be used both for oral administration and for tissue-specific routes, in these systems they not only act as a carrier but also protects the drug from external factors and provides "identification" of the specific environment in which the medicine will be released. Theoretically, alginate hydrogels shrink at low pH values (i.e., in the stomach), so there is no release of the encapsulated active component [19]. In gastric fluid, the hydrated sodium alginate is converted into a porous, insoluble alginic acid. After passing through the intestinal part of the tract, characterized by higher pH value, the alginic acid is converted into a soluble, viscous compound. In fact, the pH dependent behavior of alginate can be used for designing the desired release profile of the active substance [20]. For example, when developing an oral administration system for insulin, the alginate/chitosan blend hydrogel shows to be effective preventing the release and inactivation of proteins at low pH in the stomach and enables delivery and release of insulin at neutral pH of small intestine [21]. The release kinetics of the drug depends on the degree of cross-linking of the hydrogel, the nature of the drug, and its interactions with alginate. It has been observed that covalent alginate networks lead to prolonged release of the incorporated drug due to increased stability and decreased swelling of hydrogel [22]. Also, inert drugs that do not react with alginate are released faster in relation to those that are, for example, bound to alginate chains by covalent bonds [23].

However, alginate hydrogels have some shortcoming as a carriers for drug delivery. One of them is the reduced chemical stability of alginate in the presence of compounds that might lead to the formation of chelating complexes. For example, Ca-alginate is sensitive and chemically unstable in the presence of phosphate, citrate, and anions, as well as in the presence of Na^+ and Mg^{2+} cations. Further, alginate hydrogels disintegrate fast at higher pH, which can lead to fast initial release of active component. Moreover, the leakage of the active component during the process of alginate hydrogel formation might occur. These undesired characteristics of alginate hydrogels are mostly caused by large pore size. The stabilization of the alginate hydrogel and the reduction of its porosity are achieved by coating/blending with another polymer; for these purposes, synthetic polymers [24], polypeptides [25], or organic polycations, such as chitosan [26], can be used.

4.2.3.3 Tissue engineering and regeneration medicine

A special area of biomedical application of alginate hydrogels is tissue engineering and regenerative medicine. Thanks to biocompatibility, mild conditions gelation, and modification capabilities, alginate hydrogels are the ideal cell carriers at in vitro and in vivo conditions. In vitro systems represent the 2D and 3D cultures of cells that can serve as tissue constructs or model systems for the study of normal and pathological process [27]. Depending on the type of cell that is immobilized, as well as on conditions under which are cultivated (presence of chemical, biological, and biophysical stimulus), it is possible to induce and maintain a specific cell phenotype to stimulate the extracellular synthesis matrix and get a tissue construct, which can be implanted in the body. Study with immobilized primary chondrocytes has shown that copper-alginate hydrogel has a positive effect on the proliferation and synthetic cell activity and is effective in induction of chondrogenic phenotype in 3D culture [26]. Taking into account that the primary goal of tissue engineering is the cultivation of equivalent tissues and organs in vitro [28], so far many cell types have been successfully immobilized in alginate, such as osteoblasts [27, 29], chondrocytes [30, 31], or stem cells [32, 33].

In contrast to tissue engineering, regenerative medicine involves activation of regenerative potential of the organism itself, in order to repair the damaged structure and functions. In this field, alginate has been widely investigated. Namely, in vivo studies have shown that in ischemic regions of the tissue, it is possible to successfully induce angiogenesis through controlled delivery and continual release of growth factors [34] and/or cell [35] using alginate matrix. In addition to blood vessels, there are numerous literature reports related to bone or cartilage regeneration [36–38] by using alginate as a matrix (Figure 4.5).

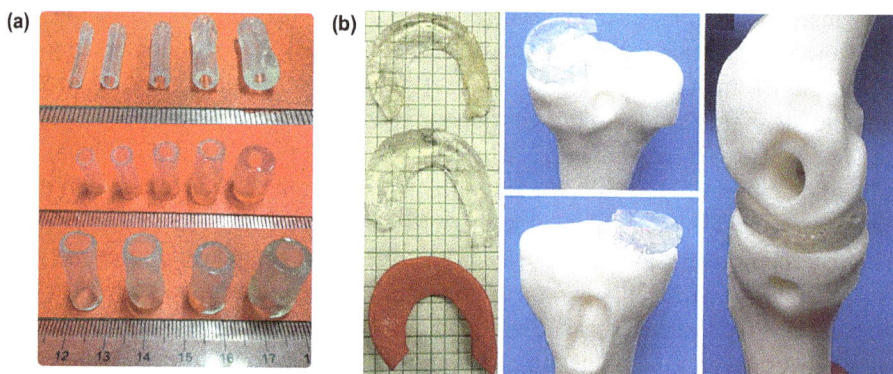

Figure 4.5: (a) Three-dimensional printing of alginate-based artificial blood vessel (Copyright from Huanbao Liu, Huixing Zhou, Haiming Lan, Tianyu Liu, Xiaolong Liu and Hejie Yu, Micromachines 2017, 8, 237; doi:10.3390/mi8080237) and (b) alginate-based artificial cartilage (Copyright from Journal of Materials Science, doi:10.1007/s10853-015-9091-0).

However, there are some limitations in the biomedical and therapeutic use of the alginate. The main disadvantage refers to their passive role in the organism, that is, the lack of bioactivity and antimicrobial potentiality. By bioactivity, it means the ability of biomaterial to induce a specific biological response in the physiological environment by supporting the process of cellular communication, migration, proliferation, and differentiation. Also, the integrity and mechanical properties of alginate hydrogels are, to a large extent, depending on pH environment [39]. Hence, great effort has been invested in the development of methods for improving structural and functional characteristics of alginate-based biomaterials such as chemical modifications [40], composite synthesis [41], or the incorporation of active compounds [19].

4.3 Chitin

Chitin or poly-β-(1→4)-N-acetyl-D-glucosamine (Figure 4.6) is a colorless polysaccharide and one of the most abundant natural marine biopolymer. Chitin has similar molecular structure of cellulose; the only difference is that the 2-hydroxy group of the cellulose has been replaced with an acetamide in chitin or chitosan. Therefore, the physicochemical properties of chitin and cellulose are presumably similar. Chitin can be mainly found in the shells of crustaceans and mollusks, in the backbone of squids and in the cuticle of insects. It is also present naturally in the cell wall of spores and hyphae of few species of fungi. Chitin is biodegradable and has low toxicity, so it is inert in the gastrointestinal tract of mammals. The biodegradability of chitin is due to the presence of the chitinase enzyme. In general, chitin has three different forms: α-, β-, and γ-chitin. Chitin forms differ in the chain packing and the polarities of adjacent chains. For instance, in α-chitin, all the chains are aligned in parallel way and in β-chitin, chains are organized in antiparallel way, while γ-chitin contain two parallel and one antiparallel chain. Chitin represents 14–27% of the dry weight of shrimp and 13–15% of crab processing wastes.

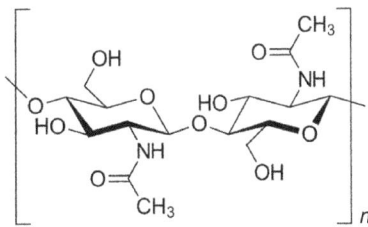

Figure 4.6: Molecular structure of chitin.

Industrial production of chitin from the shells of crabs, shrimps, prawns, and lobsters includes three steps: (a) demineralization process in acidic solution (2 M HCl) to remove calcium carbonate, (b) deproteinization process in alkaline solution (1 M NaOH) to remove proteins, and (c) discoloration by use of ethanol, acetone, and hydrogen peroxide to remove pigments. Extraction process from fungal sources is similar to conventional industrial process, except that no demineralization treatment is required due to low mineral content in fungal mycelia. Generally, the extraction procedure of chitin from fungi consists of two steps: (a) alkaline treatment to remove protein and alkali soluble polysaccharides and (b) acid reflux to separate chitin and chitosan.

4.3.1 Properties of chitin

Chitin has lower solubility than cellulose because of the hydrogen bonds through the acetamido group of chitin. That is why chitin is insoluble in water and in most of the organic solvents. However, it is soluble in dimethylacetamide containing 5% lithium chloride and in fluorinated solvents such as hexafluoroisopropanol and hexafluoroacetone. The molecular weight of native chitin is usually >1,000,000 Da and its degree of deacetylation is usually in the range of 5% to 15%. The most important derivative of chitin is chitosan.

4.4 Chitosan

Chitosan is a pale yellow polysaccharide, which includes β-(1–4)-2-acetamido-2-deoxy-β-D-glucopyranose-N-acetylglucosamine) and 2-amino-2-deoxy-β-D-glucopyranose (D-glucosamine) units (Figure 4.7). It is produced by N-deacetylation of α-chitin. Chitosan is generally obtained from chitin by use of concentrated NaOH. In order to obtain highly deacetylated chitosan and avoid the use of an excess amount of NaOH, chitin and solid NaOH are mixed (weight ratio 1:5) in extruder at 180 °C.

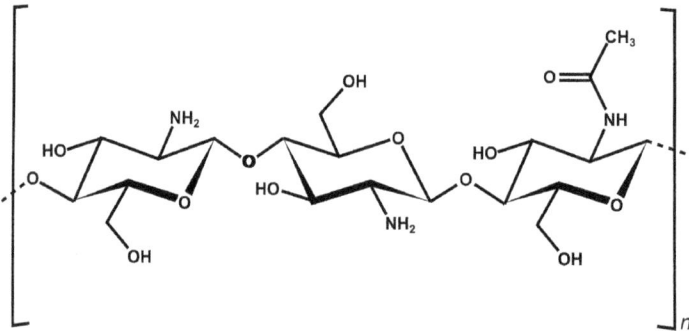

Figure 4.7: Molecular structure of chitosan.

4.4.1 Properties of chitosan

Since chitosan contain primary aliphatic amine groups, it can be protonated at acidic conditions and easily dissolved. Chitosan is mainly soluble in diluted organic acidic solutions like formic, acetic, pyruvic, citric, and lactic acid, where the pH is lower than 6.5. In general, thanks to the crystalline structure of chitosan, which is attributed to extensive intermolecular hydrogen bonding between chains, it is not soluble in water. The interaction between chitosan and solvent results in the conversion of 2-amino-2-deoxyglucose units to salt. In comparison with cellulose, chitosan can absorb more water. In fact, the high content of primary amino groups is mainly involved in interactions with molecules of water, thus causing higher water absorption. Moreover, due to the presence of free $-NH_2$ groups, chitosan has a pronounced ability to form complex with metal ions. The ability to form complex depends on the physical state of the chitosan, the degree of deacetylation, the arrangement of free $-NH_2$ group, pH of the surrounding medium, the species, and content of cations [42]. Chitosan affinity to bind two- and three-valence metal ions is in the following order: $Cu^{2+} > Hg^{2+} > Zn^{2+} > Cd^{2+} > Ni^{2+} > Co^{2+} \sim Ca^{2+} > Nd^{3+} > Cr^{3+} \sim Pr^{3+}$ [43].

The physicochemical properties of chitosan like viscosity, solubility, elasticity, and tear strength depend on the molecular weight and degree of deacetylation. Chitosan has four different average molecular weight ranges based on its origin: the range from 50 to 1,000 kDa, 3.8 to 2,000 kDa, or 50 to 2,000 kDa. The average molecular weight is effective on the chitosan's antimicrobial properties. The inhibitory effect toward phytopathogens is higher by use of chitosan of low molecular weight, than of high molecular weight [44]. Regardless of its molecular weight, chitosan is a strong antibiotic against gram-positive bacteria [45].

4.4.2 Application of chitin and its derivatives

Due to diverse properties of chitin and chitosan, such as biocompatibility, biodegradability, hemostatic, fungistatic, and antitumor property, they have huge potential for variety of different applications. Chitin and chitosan are commercially used in food industry as a preservative and biologically active additive, emulsifier, and viscosity regulator. Moreover, due to its ability to from complexes in presence of metal ions, chitosan is used as a coagulant in wastewater treatments and in processes where recovery of solid materials from food processing is required. Chitin and its derivatives are also used in paper industry to improve paper's mechanical properties, hydrophobicity, and air permeability. In order to improve the strength of paper sheets, chitin and its derivatives can be either added as ingredient in paper furnish [46, 47] or used as coating [48]. One of the most prominent commercial applications of chitosan is in the field of wound dressings [49]. Several chitosan based wound dressings are available on the market for clinical use, including HemCon® Bandage and ChitoFlex wound dressing (West Yorkshire, UK) and Celox (Crewe, England), both claimed to be FDA approved. Due to the wide scope of applications (pharmaceutical industry, paper industry, food packaging, and agriculture), only recent trends will be further discussed in this section with the main focus on new and innovative ecosustainable technological concepts.

4.4.2.1 Food packaging

The global consumption of synthetic plastics is estimated to be more than 200 million tons per year, with an annual growth of approximately 5%. Since plastic packaging materials are nondegradable and have tendency to stay intact in nature for more than 100 years, they pose serious problems to environment. The use of edible films and coatings represents an innovative approach in food packaging sector, as a green alternative to the plastic packages. The main role of edible films and coatings is to extend shelf life and improve the quality of fresh, frozen, and processed foods. The mechanism of extending the shelf life of food products by film coating includes controlling many parameters of packaging system, such as water or moisture permeability, antimicrobial and antioxidants properties, package temperature, pressure of oxygen inside the package, respiration rate, impermeability to certain substances like fats and oils. Due to nontoxicity and antimicrobial property, chitosan has been widely investigated as edible matrix to protect food products (Figure 4.8). It can provide some supplementary and essential properties in the packaging system, by means of controlling physiological, morphological, and physicochemical changes in food products [50].

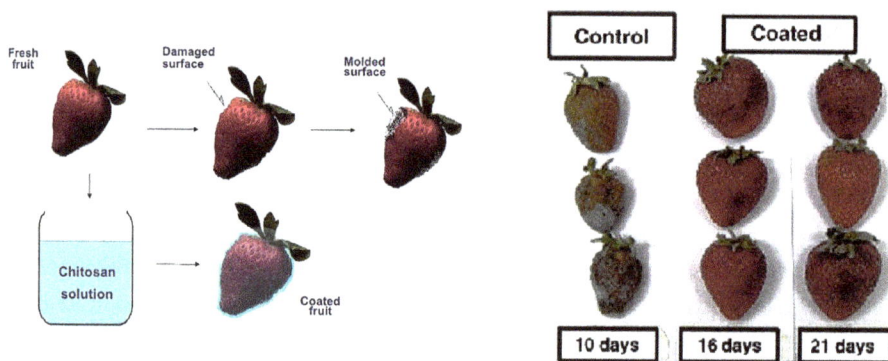

Figure 4.8: The influence of chitosan coating on the shelf life of strawberries. Copyright from: Materials Science and Engineering C, doi: 10.1016/j.msec.2013.01.010.

Encapsulation is another technological approach, where chitosan can be used as encapsulating agent for bioactive components in a form of cross-linked microbeads/microcapsules. Encapsulation is a beneficial technology to carry natural active substances, preventing their loss and maintaining their stability until use. Chitosan microcapsules can be used in intelligent food packaging, providing long-term microbiological quality of food products. For example, chitosan microbeads encapsulating with ungeremine showed high antifungal activity toward *Penicillium roqueforti*, a filamentous fungus responsible for the bakery products deterioration [51–54].

4.4.2.2 Dentistry

Chitosan has a wide range of application in dentistry. Indeed, chitosan has a potential to be used as an additive in toothpaste, because it shows reduced erosion/abrasion effect of dentine [55]. Also, it can be used as a bioactive component against dental microorganisms. Namely, the growth of some microorganisms such as *Streptococcus mutans* and *Candida albicans* are the main reason of secondary caries, where secondary caries is a disease that occurs on the tooth after the filling has been used for a period of time [56]. Hence, chitosan has been investigated as an antibacterial agent that can be incorporated into restorative material-composite resin as filler cement or adhesive, in order to prevent occurrence of caries [57, 58]. From drug delivery point of view, it is worth to mention that chitosan can be used in a form of hydrogel for rapid mineralization of enamel repair. In fact, tooth enamel is the hardest, nonvascular tissue of body representing one of the biggest challenges to repair [59]. In addition, chitosan can be used in oral surgery for guided bone regeneration or repair temporomandibular joint disc-guided periodontal tissue regeneration [60, 61].

4.4.2.3 Agriculture

Treatment of seeds of agricultural crops is a permanent and necessary task, in order to suppress harmful factors that are transmitted by seeds or are in the soil. For this aim, fungicides have been used by spraying in the soil and seeds, to suppress or control fungal diseases of crops. The usage of highly toxic chemicals for control of plant diseases has increased exponentially, in order to achieve higher productivity of crops. The use of toxic chemicals has led to their deposition in soil and accumulation in plants, which can be transferred to humans' trough food chains and cause serious health problems. Moreover, the extent of use of fungicide may induce resistance in target pathogens within the time. Hence, there is an urgent need to explore bioactive and eco-friendly compounds for control of fungal disease. Due to high antifungal activity, strong affinity for biological cells adhesion property to biological membrane, chitin and chitosan show to be good candidate as a biofungicide [62] chitosan. Moreover, as a potent elicitor of plant defense responses, produces activate the expression of plant defensive genes and induce the production of pathogen related proteins, such a chitinases and other hydrolytic enzymes [63]. These enzymes cause hydrolysis, chitin and chitosan in fungal cell walls, consequently leading to growth inhibition and/or death. Chitosan is found to be effective in reducing the amount of polygalacturonases produced by *Botrytis cinerea* and cause severe cytological damage to the invading hyphae in bell pepper fruit [64]. Chitosan may induce the synthesis of phytoalexin in rice leaves, which is a potent suppressor of fungal growth [65] (Figure 4.9). In the cucumber plants sprayed with chitosan or chitin before

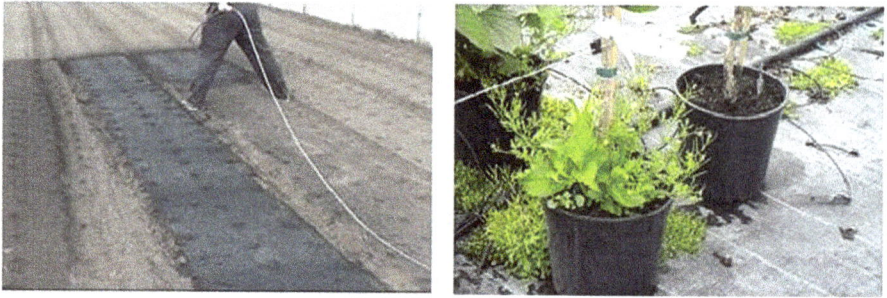

Figure 4.9: Treatment of plants with chitosan solution.

the inoculating of *Botrytis cinerea*, chitosanase and peroxidase activity increased and growth of *Botrytis cinerea* was successfully inhibited [66].

In addition, chitosan can be used as antitranspirant, in order to reduce the water losses during the vegetative growth period either before or after fruits harvesting. Generally, antitranspirants are used to enhance the yield of crops exposed to water stress during growth and are applied by foliar spraying. For example, foliar application of chitosan on pepper plant decrease transpiration rate and reduce additional irrigation by 26–43% [67]. Moreover, foliar application of chitosan may have significant effect on plant growth, since it can increase key enzymes activities of nitrogen metabolism (nitrate reductase, glutamine synthetase, and rotease), thus improving the transportation of nitrogen in the functional leaves, which as a result enhances plant growth and development [68, 69]. Chitosan-based plant growth stimulators are commercially available (e.g., ChitoPlant® and SilioPlant®; ChiPro GmbH, Germany). They presumably stimulate the plant immune response against pathogens and have a growth promoting activity.

4.5 Conclusions

Unlike other polymers, marine biopolymers, such as chitosan and alginate, show various and specific physical and chemical properties related to their complex structure, which is the reason for their very wide spectra of application, as shown in this chapter. Biocompatibility, biodegradability, and nontoxicity are properties that provide the possibility of their application in food and food packaging technology, cosmetics, and biomedicine. Therefore, a large number of recent scientific publications refer to hydrogels and films obtained from these marine biopolymers. The review of properties and applications of these unique polysaccharide polymers offers a significant contribution to further research in mentioned applicative areas.

Acknowledgments: This work was financially supported by the Ministry of Education, Science and Technological Development of the Republic of Serbia (research projects 43009), and the author acknowledge of the project "SOSTENIBILITA' DELLAFILIERA AGROALIMENTARE (SO.FI.A)" in the frame of National Technological Cluster of Ministry of University and Research (MIUR), for the financial support.

References

[1]   Donati, I., Asaro, F., and Paoletti, S. (2009). Experimental evidence of counterion affinity in alginates: the case of nongelling ion Mg 2+. J. Phys. Chem. B 113(39), 12877–12886. https://doi.org/10.1021/jp902912m.

[2]   Haug, A., Larsen, B., Fykse, O., Block-Bolten, A., Toguri, J. M., and Flood, H. (1962). Quantitative determination of the uronic acid composition of alginates. Acta Chem. Scand. 16, 1908–1918. https://doi.org/10.3891/acta.chem.scand.16-1908.

[3]   George, M., and Abraham, T. E. (2006). Polyionic hydrocolloids for the intestinal delivery of protein drugs: alginate and chitosan – a review. J. Control. Release 114(1), 1–14. https://doi.org/10.1016/j.jconrel.2006.04.017.

[4]   Pawar, S. N., and Edgar, K. J. (2012). Alginate derivatization: a review of chemistry, properties and applications. Biomaterials 33(11), 3279–3305. https://doi.org/10.1016/j.biomaterials. 2012.01.007.

[5]   Holme, H. K., Lindmo, K., Kristiansen, A., and Smidsrød, O. (2003). Thermal depolymerization of alginate in the solid state. Carbohydr. Polym. 54(4), 431–438. https://doi.org/10.1016/ S0144-8617(03)00134-6.

[6]   Grant, G. T., Morris, E. R., Rees, D. A., Smith, P. J. C., and Thom, D. (1973). Biological interactions between polysaccharides and divalent cations: the egg-box model. FEBS Lett. 32(1),195–198. https://doi.org/10.1016/0014-5793(73)80770-7.

[7]   Lee, K. Y., and Mooney, D. J. (2012). Alginate: properties and biomedical applications. Prog. Polym. Sci. 37(1),106–126. https://doi.org/10.1016/j.progpolymsci.2011.06.003.

[8]   Lloyd, L. L., Kennedy, J. F., Methacanon, P., Paterson, M., and Knill, C. J. 1998. Carbohydrate polymers as wound management aids. Carbohydr. Polym. 37(3), 315–322. https://doi.org/10.1016/S0144-8617(98)00077-0.

[9]   Queen, D., Orsted, H., Sanada, H., and Sussman, G. (2004). A dressing history. Int. Wound J. 1(1),59–77. https://doi.org/10.1111/j.1742-4801.2004.0009.x.

[10]   Gilchrist, T., and Martin, A. (1983). Wound treatment with sorbsan – an alginate fibre dressing. Biomaterials 4(4),317–320. https://doi.org/10.1016/0142-9612(83)90036-4.

[11]   Boateng, J. S., Matthews, K. H., Stevens, H. N. E., and Eccleston, G. M. (2008). Wound healing dressings and drug delivery systems: a review. J. Pharm. Sci. 97(8), 2892–2923. https://doi.org/10.1002/jps.21210.

[12]   Thomas, S. (2000). Alginate dressings in surgery and wound management – part 1. J. Wound Care 9(2),56–60. https://doi.org/10.12968/jowc.2000.9.2.26338.

[13]   Straccia, M. C., D'Ayala, G. G., Romano, I., and Laurienzo, P. (2015). Novel zinc alginate hydrogels prepared by internal setting method with intrinsic antibacterial activity. Carbohydr. Polym. 125(July), 103–112. https://doi.org/10.1016/j.carbpol.2015.03.010.

[14]   Straccia, M. C., Romano, I., Oliva, A., Santagata, G., and Laurienzo, P. (2014). Crosslinker effects on functional properties of alginate/N-succinylchitosan based hydrogels. Carbohydr. Polym. 108(1), 321–330. https://doi.org/10.1016/j.carbpol.2014.02.054.

[15] Catanzano, O., Straccia, M. C., Miro, A., Ungaro, F., Romano, I., Mazzarella, G., Santagata, G., Quaglia, F., Laurienzo, P., and Malinconico, M. (2015). Spray-by-spray in situ cross-linking alginate hydrogels delivering a tea tree oil microemulsion. Eur. J. Pharm. Sci. 66, 20–28. https://doi.org/10.1016/j.ejps.2014.09.018.

[16] Boontheekul, T., Kong, H.-J., and Mooney, D. J. (2005). Controlling alginate gel degradation utilizing partial oxidation and bimodal molecular weight distribution. Biomaterials 26(15), 2455–2465. https://doi.org/10.1016/j.biomaterials.2004.06.044.

[17] Chan, A. W., and Neufeld, R. J. (2010). Tuneable semi-synthetic network alginate for absorptive encapsulation and controlled release of protein therapeutics. Biomaterials 31(34), 9040–9047. https://doi.org/10.1016/j.biomaterials.2010.07.111.

[18] Wells, L. A., and Sheardown, H. (2007). Extended release of high PI proteins from alginate microspheres via a novel encapsulation technique. Eur. J. Pharm. Biopharm. 65(3), 329–335. https://doi.org/10.1016/j.ejpb.2006.10.018.

[19] Chen, S.-C., Wu, Y.-C., Mi, F.-L., Lin, Y.-H., Yu, L.-C., and Sung, H.-W. (2004). A novel PH-sensitive hydrogel composed of N,O-carboxymethyl chitosan and alginate cross-linked by genipin for protein drug delivery. J. Control. Release 96(2),285–300. https://doi.org/10.1016/j.jconrel.2004.02.002.

[20] Silva, C. M., Ribeiro, A. J., Figueiredo, I. V., Gonçalves, A. R., and Veiga, F. (2006). Alginate microspheres prepared by internal gelation: development and effect on insulin stability. Int. J. Pharm. 311(1–2), 1–10. https://doi.org/10.1016/j.ijpharm.2005.10.050.

[21] Maiti, S., Singha, K., Ray, S., Dey, P., and Sa, B. (2009). Adipic acid dihydrazide treated partially oxidized alginate beads for sustained oral delivery of flurbiprofen. Pharm. Dev. Technol. 14(5), 461–470. https://doi.org/10.1080/10837450802712658.

[22] Bouhadir, K. H., Lee, K. Y., Alsberg, E., Damm, K. L., Anderson, K. W., and Mooney, D. J. (2001). Degradation of partially oxidized alginate and its potential application for tissue engineering. Biotechnol. Progr. 17(5), 945–950. https://doi.org/10.1021/bp010070p.

[23] Martínez-Gómez, F., Guerrero, J., Matsuhiro, B., and Pavez, J. (2017). In vitro release of metformin hydrochloride from sodium alginate/polyvinyl alcohol hydrogels. Carbohydr. Polym. 155(January), 182–191. https://doi.org/10.1016/j.carbpol.2016.08.079.

[24] Severino, P., Chaud, M. V., Shimojo, A., Antonini, D., Lancelloti, M., Santana, M. H. A., and Souto, E. B. (2015). Sodium alginate-cross-linked polymyxin B sulphate-loaded solid lipid nanoparticles: antibiotic resistance tests and HaCat and NIH/3T3 cell viability studies. Colloids Surf., B 129(May), 191–197. https://doi.org/10.1016/j.colsurfb.2015.03.049.

[25] Zhang, Y., Wei, W., Lv, P., Wang, L., and Ma, G. (2011). Preparation and evaluation of alginate–chitosan microspheres for oral delivery of insulin. Eur. J. Pharm. Biopharm. 77(1), 11–19. https://doi.org/10.1016/j.ejpb.2010.09.016.

[26] Meyer, U., Meyer, Th., Handschel, J., and Wiesmann,. (2009). Fundamentals of Tissue Engineering and Regenerative Medicine (Meyer, U., Handschel, J., Wiesmann, H. P., and Meyer, T., Eds.). Springer, Berlin, Heidelberg. https://doi.org/10.1007/978-3-540-77755-7.

[27] Madzovska-Malagurski, I., Vukasinovic-Sekulic, M., Kostic, D., and Levic, S. 2016. Towards antimicrobial yet bioactive Cu-alginate hydrogels. Biomed. Mater. 11(3), 035015. https://doi.org/10.1088/1748-6041/11/3/035015.

[28] Valente, J. F. A., Valente, T. A. M., Alves, P., Ferreira, P., Silva, A., and Correia, I. J. (2012). Alginate based scaffolds for bone tissue engineering. Mater. Sci. Eng. C 32(8), 2596–2603. https://doi.org/10.1016/j.msec.2012.08.001.

[29] Degala, S., Williams, R., Zipfel, W., and Bonassar, L. J. (2012). Calcium signaling in response to fluid flow by chondrocytes in 3D alginate culture. J. Orthop. Res. 30(5), 793–799. https://doi.org/10.1002/jor.21571.

[30] Hsiong, S. X., Carampin, P., Kong, H.-J., Lee, K.-Y., and Mooney, D. J. (2008)."Differentiation stage alters matrix control of stem cells. J. Biomed. Mater. Res. Part A 85A(1), 145–156. https://doi.org/10.1002/jbm.a.31521.

[31] Wang, L., Shelton, R. M., Cooper, P. R., Lawson, M., Triffitt, J. T., and Barralet, J. E. (2003). Evaluation of sodium alginate for bone marrow cell tissue engineering. Biomaterials 24(20),3475–3481. https://doi.org/10.1016/S0142-9612(03)00167-4.

[32] Jay, S. M., and Saltzman, W. M. (2009). Controlled delivery of VEGF via modulation of alginate microparticle ionic crosslinking. J. Control. Release 134 (1), 26–34. https://doi.org/10.1016/j.jconrel.2008.10.019.

[33] Sun, Q., Silva, E. A., Wang, A., Fritton, J. C., Mooney, D. J., Schaffler, M. B., Grossman, P. M., and Rajagopalan, S. (2010). Sustained release of multiple growth factors from injectable polymeric system as a novel therapeutic approach towards angiogenesis. Pharm. Res. 27(2), 264–271. https://doi.org/10.1007/s11095-009-0014-0.

[34] Peters, M. C., Polverini, P. J., and Mooney, D. J. (2002). Engineering vascular networks in porous polymer matrices. J. Biomed. Mater. Res. 60(4), 668–678. https://doi.org/10.1002/jbm.10134.

[35] Kolambkar, Y. M., Dupont, K. M., Boerckel, J. D., Huebsch, N., Mooney, D. J., Hutmacher, D. W., and Guldberg, R. E. 2011. An alginate-based hybrid system for growth factor delivery in the functional repair of large bone defects. Biomaterials 32(1), 65–74. https://doi.org/10.1016/j.biomaterials.2010.08.074.

[36] Igarashi, T., Iwasaki, N., Kasahara, Y., and Minami, A. 2010. A cellular implantation system using an injectable ultra-purified alginate gel for repair of osteochondral defects in a rabbit model. J. Biomed. Mater. Res. Pt A 9999A, NA-NA. https://doi.org/10.1002/jbm.a.32762.

[37] Ma, H.-L., Hung, S.-C., Lin, S.-Y., Chen, Y.-L., and Lo, W.-H. (2003). Chondrogenesis of human mesenchymal stem cells encapsulated in alginate beads. J. Biomed. Mater. Res. 64A(2), 273–281. https://doi.org/10.1002/jbm.a.10370.

[38] Bajpai, S. K., and Sharma, S. (2004). Investigation of swelling/degradation behaviour of alginate beads crosslinked with Ca2+ and Ba2+ ions. React. Funct. Polym. 59(2), 129–140. https://doi.org/10.1016/j.reactfunctpolym.2004.01.002.

[39] Priddy, L. B., Chaudhuri, O., Stevens, H. Y., Krishnan, L., Uhrig, B. A., Willett, N. J., and Guldberg, R. E. (2014). Oxidized alginate hydrogels for bone morphogenetic protein-2 delivery in long bone defects. Acta Biomater. 10(10), 4390–4399. https://doi.org/10.1016/j.actbio.2014.06.015.

[40] Venkatesan, J., Bhatnagar, I., Manivasagan, P., Kang, K.-H., and Kim, S.-K. (2015). Alginate composites for bone tissue engineering: a review. Int. J. Biol. Macromol. 72 (January), 269–281. https://doi.org/10.1016/j.ijbiomac.2014.07.008.

[41] Neves, N., Campos, B. B., Almeida, I. F., Costa, P. C., Cabral, A. T., Barbosa, M. A., and Ribeiro, C. C. (2016). Strontium-rich injectable hybrid system for bone regeneration. Mater. Sci. Eng. C 59(February), 818–827. https://doi.org/10.1016/j.msec.2015.10.038.

[42] Wong, T. (2009). Chitosan and its use in design of insulin delivery system. Recent Pat. Drug Delivery Formulation 3(1), 8–25. https://doi.org/10.2174/187221109787158346.

[43] Rinaudo, M. (2006). Chitin and chitosan: properties and applications. Prog. Polym. Sci. (Oxford) 31(7),603–632. https://doi.org/10.1016/j.progpolymsci.2006.06.001.

[44] Hirano, S., and Nagao, N. (1989). Effects of chitosan, pectic acid, lysozyme, and chitinase on the growth of several phytopathogens. Agric. Biol. Chem. 53(11), 3065–3066. https://doi.org/10.1080/00021369.1989.10869777.

[45] No, H. K., Kim, S. D., Kim, D. S., Kim, S. J., and Meyers, S. P. (1999). Effect of physical and chemical treatments on chitosan viscosity TT – effect of physical and chemical treatments on

chitosan viscosity. 한국키틴키토산학회지 4(4), 177–183. http://kiss.kstudy.com/search/detail_page.asp?key=1808012.

[46] Agusnar, H., Nainggolan, I., and Sukirman,. (2013). Mechanical properties of paper from oil palm pulp treated with chitosan from horseshoe crab. Adv. Environ. Biol. 7(SPEC. ISSUE 12), 3857–3860.

[47] Vikele, L., Laka, M., Sable, I., Rozenberga, L., Grinfelds, U., Zoldners, J., Passas, R., and Mauret, E., & Lamar@edi Lv. (2017). Effect of chitosan on properties of paper for packaging. Cellulose Chem. Technol. 51(12), 67–73. https://doi.org/10.5897/JCEMS2015.0235.

[48] Vellingiri, K., Ramachandran, T., and Senthilkumar, M. (2013). Eco-friendly application of nano chitosan in antimicrobial coatings in the textile industry. Nanosci. Nanotechnol. Lett. 5(5), 519–529. https://doi.org/10.1166/nnl.2013.1575.

[49] Moeini, Arash, Parisa Pedram, Pooyan Makvandi, Mario Malinconico, and Giovanna Gomez d'Ayala. 2020. "Wound Healing and Antimicrobial Effect of Active Secondary Metabolites in Chitosan-Based Wound Dressings: A Review." *Carbohydrate Polymers*, 115839. https://doi.org/10.1016/j.carbpol.2020.115839.

[50] Elsabee, M. Z., and Abdou, E. S. (2013). Chitosan based edible films and coatings: a review. Mater. Sci. Eng. C 33(4), 1819–1841. https://doi.org/10.1016/j.msec.2013.01.010.

[51] Moeini, A., Cimmino, A., Poggetto, G. D., Biase, M. D., Evidente, A., Masi, M., Lavermicocca, P., et al. (2018). Effect of PH and TPP concentration on chemico-physical properties, release kinetics and antifungal activity of chitosan-TPP-ungeremine microbeads. Carbohydr. Polym. 195(September), 631–641. https://doi.org/10.1016/j.carbpol.2018.05.005.

[52] Valerio, F., Masi, M., Cimmino, A., Moeini, S. A., Lavermicocca, P., and Evidente, A. (2017). Antimould microbial and plant metabolites with potential use in intelligent food packaging. Nat. Prod. Res. 6419(October), 1–6. https://doi.org/10.1080/14786419.2017.1385018.

[53] Moeini, Arash, Alessio Cimmino, Marco Masi, Antonio Evidente, and Albert Van Reenen. 2020. "The Incorporation and Release of Ungeremine, an Antifungal Amaryllidaceae Alkaloid, in Poly(Lactic Acid)/Poly(Ethylene Glycol) Nanofibers." *Journal of Applied Polymer Science* n/a (n/a): 49098. https://doi.org/10.1002/app.49098.

[54] Moeini, Arash, Salvatore Mallardo, Alessio Cimmino, Giovanni Dal Poggetto, Marco Masi, Mariaelena Di Biase, Albert van Reenen, et al. 2020. "Thermoplastic Starch and Bioactive Chitosan Sub-Microparticle Biocomposites: Antifungal and Chemico-Physical Properties of the Films." *Carbohydrate Polymers* 230: 115627. https://doi.org/10.1016/j.carbpol.2019.115627.

[55] West, N. X., Hooper, S. M., O'Sullivan, D., Hughes, N., North, M., Macdonald, E. L., Davies, M., and Claydon, N. C. A. (2012). In situ randomised trial investigating abrasive effects of two desensitising toothpastes on dentine with acidic challenge prior to brushing. J. Dent. 40(1), 77–85. https://doi.org/10.1016/j.jdent.2011.10.010.

[56] Makvandi, P., Jamaledin, R., Jabbari, M., Nikfarjam, N., and Borzacchiello, A. (2018). Antibacterial quaternary ammonium compounds in dental materials: a systematic review. Dent. Mater. 34(6), 851–867. https://doi.org/10.1016/j.dental.2018.03.014.

[57] D'Ayala, G., Rosa, A., Laurienzo, P., and Malinconico, M. (2007). Development of a new calcium sulphate-based composite using alginate and chemically modified chitosan for bone regeneration. J. Biomed. Mater. Res. Pt A. 81. https://doi.org/10.1002/jbm.a.31009.

[58] Saker, S., Alnazzawi, A., and Özcan, M. (2014). Adhesive Strength of Self-Adhesive Resins to Lithium Disilicate Ceramic and Dentin: Effect of Dentin Chelating Agents. Odontology / the Society of the Nippon Dental University (Vol. 104). https://doi.org/10.1007/s10266-014-0180-3.

[59] Ruan, Q., Siddiqah, N., Li, X., Nutt, S., and Moradian-Oldak, J. (2014). Amelogenin–chitosan matrix for human enamel regrowth: effects of viscosity and supersaturation degree. Connective Tissue Res. 55(01), 150–154. https://doi.org/10.3109/03008207.2014.923856.

[60] Shin, S.-Y., Park, H.-N., Kim, K.-H., Lee, M.-H., Choi, Y. S., Park, Y.-J., Lee, Y.-M., et al. (2005). Biological evaluation of chitosan nanofiber membrane for guided bone regeneration. J. Periodontology 76(10), 1778–1784. https://doi.org/10.1902/jop.2005.76.10.1778.

[61] Wu, Y., Gong, Z., Li, J., Meng, Q., Fang, W., and Long, X. (2014). The pilot study of fibrin with temporomandibular joint derived synovial stem cells in repairing TMJ disc perforation. BioMed Res. Int. 2014, 454021. https://doi.org/10.1155/2014/454021.

[62] Falcón-Rodríguez, A. B., Costales, D., Cabrera, J. C., and Martínez-Téllez, M. Á. (2011). Chitosan physico–chemical properties modulate defense responses and resistance in tobacco plants against the oomycete phytophthora nicotianae. Pestic. Biochem. Physiol. 100(3), 221–228. https://doi.org/10.1016/j.pestbp.2011.04.005.

[63] Doares, S. H., Syrovets, T., Weiler, E. W., and Ryan, C. A. (1995). Oligogalacturonides and chitosan activate plant defensive genes through the octadecanoid pathway. Proc. Natl. Acad. Sci. U.S.A. 92(10), 4095–4098. http://www.ncbi.nlm.nih.gov/pmc/articles/PMC41892/.

[64] Ghaouth, A., El Arul, J, Wilson, C., and Benhamou, N. (1997). Biochemical and cytochemical aspects of the interactions of chitosan and botrytis cinerea in bell pepper fruit. Postharvest Biol. Technol. 12(2), 183–194. https://doi.org/10.1016/S0925-5214(97)00056-2.

[65] Agrawal, G. K., Rakwal, R., Tamogami, S., Yonekura, M., Kubo, A., and Saji, H. (2002). Chitosan activates defense/stress response(s) in the leaves of Oryza sativa seedlings. Plant Physiol. Biochem. 40(12), 1061–1069. https://doi.org/10.1016/S0981-9428(02)01471-7.

[66] Ben-Shalom, N., Ardi, R., Pinto, R., Aki, C., and Fallik, E. (2003). Controlling gray mould caused by Botrytis cinerea in cucumber plants by means of chitosan. Crop Prot. 22(2), 285–290. https://doi.org/10.1016/S0261-2194(02)00149-7.

[67] Bittelli, M., Flury, M., Campbell, G. S., and Nichols, E. J. (2001). Reduction of transpiration through foliar application of chitosan. Agric. For. Meteorol. 107(3), 167–175. https://doi.org/ https://doi.org/10.1016/S0168-1923(00)00242-2.

[68] Górnik, K., Mieczyslaw, G., and Duda, B. R. (2008). The effect of chitosan on rooting of grapevine cuttings and on subsequent plant growth under drought and temperature stress. J. Fruit Ornam. Plant Res. 16.

[69] Mondal, M., Malek, M. A., Puteh, A, Ismail, M. R., Ashrafuzzaman, Md., and Naher, L. (2012). Effect of foliar application of chitosan on growth and yield in okra. Aust. J. Crop Sci. 6.

Gaetano Zuccaro, Jean-Philippe Steyer

5 Terrestrial biopolymers: cellulose, starch, lignin

Abstract: The potential of terrestrial biodegradable polymers intercepts the interest of industrial application since it represents a low cost source and an environmentally-conscious alternative. The main substrates from which to obtain such biopolymers are cellulose, starch and lignin. This chapter aims to describe the basic chemical structures and the associated reactions that lead to the various biomaterials, some of which are already marketed and others that represent a strong potential in terms of environmental impact and beyond.

Keywords: Cellulose, starch, lignin, biodegradability, mechanical properties, trade suppliers

5.1 Introduction

Since their introduction on the market, earlier than 1980s, a large number of biodegradable polymers have been synthesized [1] with the major objective of developing polymeric systems with stable electrical and mechanical properties during their programmed service life. In particular, biodegradable polymers are of great interest since these biomaterials are able to break down and be catabolized after use, eventually in carbon dioxide and water, by microorganisms under natural environment. Biodegradable polymers can be classified into different categories depending on the synthesis and on the sources, such as agro- or natural polymers that consist of lignocellulosic products, polysaccharides, and others [2]. Biopolymers fall into the natural polymers category, and represent one of the hot topics of polymer research. They are useful in medicine, agriculture, drug release, and packaging fields. In this chapter, the most important terrestrial biopolymers are reviewed in terms of source, properties, blending with other biodegradable polymers, and applications.

5.2 Biodegradable polymers from renewable resources

This group consists of naturally occurring polymers, which are typically formed by complex metabolic processes and synthetized by enzyme-catalyzed reactions.

https://doi.org/10.1515/9783110590586-005

5.2.1 Cellulose

Cellulose is a complex polymer consisting of glucose unit linked by β-1,4 glycosidic bonds organized in a linear chain (Figure 5.1) and generally synthesized by plants through several processes to separate it from lignin and hemicellulose and generally classified into physical, chemical, physical–chemical, biological, and their combinations.

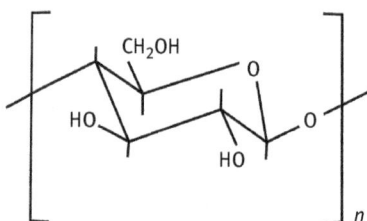

Figure 5.1: Structure of repeating units for cellulose.

It is in a very high crystalline state, with a high molecular weight, infusible, insoluble in all organic solvents [3], with a high tensile strength of 60–500 MPa. As a result of this, cellulose is usually transformed in processable compounds produced by chemical modification of hydroxyl groups present in the repeating units. There are three main groups that are produced using these methodologies: ethers, esters, and regenerated cellulose. Cellulose ethers are processed with chemicals and etherification reactions, which are carried out using agents such as chlorinated ethylene, chlorinated propylene, and oxidized ethylene. They have several applications, from construction products, ceramics, paint to foods, cosmetics, and pharmaceuticals [4], and they are also used as surfactants, lubricants, and emulsifiers. Cellulose ethers are marketed as carboxymethylcellulose, methylcellulose and derivatives, hydroxyethylcellulose and derivatives, hydroxylpropylcellulose, and ethylcellulose. There are different trade names of these cellulose-based polymers as Kimicell®, Akucell®, Bermocoll®, Bermocell®, Depramin®, Gabroil®, Peridur®, Staflo®, Gabroil™, and Gabrosa™. Table 5.1 shows the international suppliers of different cellulose ethers.

Cellulose esters are obtained by reaction of cellulose with acetic anhydride and acetic acid in the presence of sulfuric acid, which is subsequently neutralized during processing. The reason for the conversion of cellulose in its organic derivatives is related to its poor solubility in common solvents and to a poor processability, since it decomposes before the melt flow. On the other hand, its derivatives are processable in several useful forms and their modified physical properties give entry into a range of applications more expanded from those available to the parent polysaccharides. Esters, for example, cellulose acetate and cellulose xanthate are mainly developed for film and fiber applications.

Cellulose acetate is commercially available and currently used for its tensile strength comparable to polystyrene [5]. Agricultural feedstocks based on lignocellulose

Table 5.1: International suppliers of cellulose ethers.

Company	Country
Kingstone Chemical Co. Ltd.	China
SE Tylose GmbH & Co. KG	England
The Dow Chemical Company	England
Ashland Aqualon Functional Ingredients	England
Dow Construction Chemicals (former Wolff Walsrode)	England; Germany
Akzo Nobel Chemicals AG	Netherlands
Lotte Fine Chemical (former Samsung FC)	England
Henan Botai Chemical Building Materials, Co. Ltd.	China; England
Henan Tiangsheng	China; England
Cellotech	China; England
Zhejiang Kehong Chemical Co. Ltd.	China
Demacasa, Derivados Macroquimicos S.A. de C.V.	Mexico
Celanese Mexicana S.A. de C.V.	Mexico
Shin-Etsu Chemical Co. Ltd.	China
Prochem AG	Switzerland
Nordmann, Rassmann GmbH (NRC)	Germany

are commonly used for biofuels, but they also produce cellulose as residue and it could be used, for example, for cellulose acetate production. It was one of the first materials used in membrane separation technology and filtration applications, because of the reasonable resistance to degradation by chlorine and less fouling. Cellulose acetate is used in seawater desalination, drinking water purification, wastewater treatment, concentration of fruit juices, and as artificial kidneys. An application that has increased strongly during the last years is the use of cellulose acetate films in the production of liquid crystal display screens.

In the market, it is known as Acetow®, Auracell™, Celaire®, Clarifoil®, and Estron. The international suppliers of cellulose acetate are reported in Table 5.2.

Table 5.2: International suppliers of cellulose acetate.

Company	Country
Celanese	USA
Daicel Corporation	Japan
Eastman	USA
Nagase, Co, Ltd.	Japan
Rotuba	USA
Solvay	Belgium
Mazzucchelli	Italy
IFA	Austria
UCB	Germany

Finally, regenerated cellulose is commonly the product of cellulose regeneration by a physical process, with no accompanying chemical reactions. The regeneration is an important pathway since it allows gaining unsurpassed physical and chemical properties, inherent renewability, sustainability, and biocompatibility. Farther, it is a pathway to transform native cellulose to useful materials, such as filters, fibers, films, membranes beads, microspheres, hydrogels, and aerogels [6], but they also include apparel, upholstery, tire cords, hoses, surgical materials, and feminine hygiene products. The manufacturers and distributors are reported in Table 5.3.

Table 5.3: Manufactures and distributors of regenerated cellulose.

Company	Country
Mitsubishi Rayon	Japan
Hebei JiHao Chemical Fiber	China
Lenzing	Austria
Avtex Fibers	USA
TOYOBO	Japan
Sateri	Japan
Shanghai Development of LiAo	China
Akzo Nobel Chemicals AG	Netherlands
FUJIBO	Japan
Courtaulds	England
Dandong Chemical Fiber Group	China
Fortress paper	Canada
Asahi Kasei Corporation	Japan
Invista	USA
Sniace	Spain
Trevira GmbH	Germany

However, the biodegradability of cellulose derivatives remains unsolved. It is known that ether linkages are resistant to microbial attack. By contrast, conflicting reports are referred to the potential biodegradation of cellulose esters, maybe due to the synergistic action of esterase or simple chemical hydrolysis and cellulase-producing microorganisms [3].

5.2.2 Starch

Starch is a combination of linear and branched polysaccharides, amylose (soluble in boiling water), and amylopectin (insoluble), respectively (Figure 5.2). It contains only a single type of carbohydrate, glucose. Starch differs from cellulose since

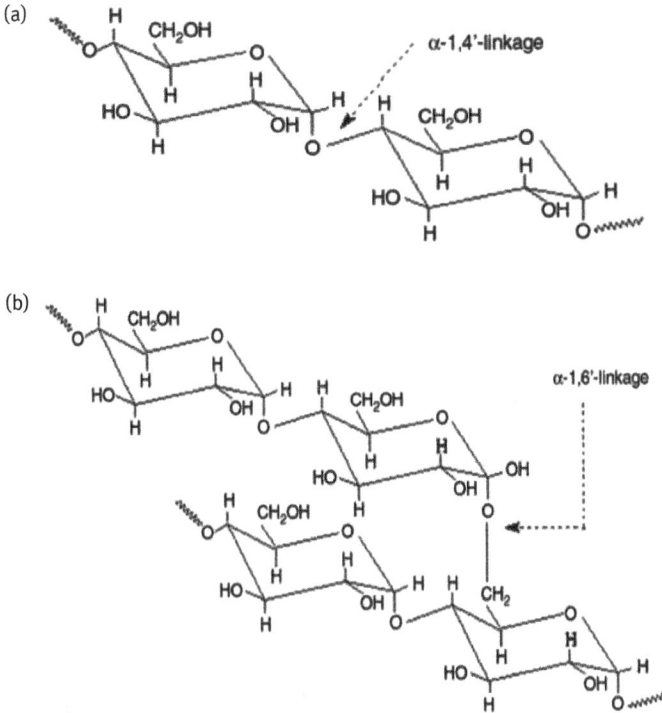

Figure 5.2: Molecular structure of starch: amylose (a) and amylopectin (b).

glucose units (or glucopyranoside rings) are linked by α-1,4-glycosidic bonds, whereas the bonds in cellulose are in β-1,4 form.

Essentially, starch is a hydrocolloid biopolymer and the end product of photosynthesis in plants [7]. The various sources used for its production include wheat, corn, rice, and potatoes, where it is produced in the form of hydrophilic granules. The thermal and mechanical properties and its biodegradability are strictly related to the relative amounts of amylose and amylopectin [8]. In general, glass transition temperature (T_g) varies between −50 °C and 110 °C [9] and at temperatures higher than 150 °C and above 250 °C, the glucoside linkages start to break and the granules collapse, respectively [3]. However, at low temperatures, that is, during cooling, a phenomenon known as retrogradation is observed, that consists in a reorganization of hydrogen bonds [5]. As for starch, gelatinization represents a process useful to improve its crystallinity, carried out in presence of plasticizers, such as water and glycerol, which promote a faster water transfer into the molecules, raising the breakage of the amylopectin matrix and the release of the amylose, and to decrease the T_g of the starch.

To solve the problems related to brittleness, water sensibility, and mechanical properties, different modifications have been conducted, including a chemical modification of hydroxyl group by acetylation [10], since its nucleophilic behavior allows substitution reactions to increase the mechanical properties due to a decrease of water retention and to an associated increase of its resistance to thermomechanical shear. Starch acetate, for example, is prepared by acetylation with pyridine/acetic acid and formic acid. The presence of a high content of linear amylose provides an increase of its hydrophobic behavior if compared to starch. Grafting of monomers like styrene and methyl methacrylate to the starch structure could represent another strategy, but it gets worse simultaneously the biodegradability properties [11]. Since these processes imply sometimes a greater product cost compared to the increase of mechanical properties, blending starch with other biodegradable and biocompatible polymers is considered the most effective strategy to overcome these defects [12, 13]. Indeed, these blends present several advantages since the properties of the materials can be adjusted according to the applications. Furthermore, blending is a low-cost process compared to the development of new biodegradable synthetic materials.

It is possible to list different starch-based blends depending on the specific polymer used. Poly(ethylene-co-vinyl alcohol), with different ratios, and native starch are used for blown film production. A higher strain to break and lower tensile strength and modulus were obtained when the blend was processed at 5 °C [5]. Blends with polyvinyl alcohol (PVOH) present particular interest because of its low-cost process and for the possibility to improve mechanical properties of the final material [14], due to the presence of OH groups and the hydrogen bond.

Although PVOH is a synthetic polymer, it is easily degraded by biological organisms and the films obtained from PVOH are fully biodegradable, odorless, transparent, nontoxic, and have useful physical properties, such as, good oxygen aroma barrier properties and transparency [15]. Poly(lactic) acid is another polymer used to manufacture starch-based blends, that has been extensively studied as potential replacer of nondegradable petrochemical polymers, since it is renewable, biocompatible, as well as approved by the Food and Drug Administration for direct contact with biological fluids. It is semicrystalline, such as starch, and the crystallinity degree of films depends on the source and processing conditions. The biodegradation goes from 6 months to 2 years, depending on its stereochemistry and molecular weight [16].

Polycaprolactone (PCL) is one of the earliest biodegradable polymers, synthesized through ring opening polymerization of ε-caprolactone [17]. It is considered to be a major biodegradable polyester, but the limitation for its use is the hydrophobic behavior, which may affects the rate of degradation.

When it is used in blending with hydrophilic polymers, such as starch, an improvement of its hydrophilicity is reported, as well as an adjustment of rheological properties. Blends of starch and PCL have used for many tissue engineering applications, such as bone, cartilage, and spinal cord. However, since the blending efficiency since and starch immiscibility are affected by PCL, several modification methods are required, such as compatibilization and physical modification by its coating or chemical modifications by grafting reactions [12]. Research is also focused on finding solutions to enhance mechanical properties, fracture behavior, biodegradation, compounding starch-PCL blends with other polymers and ceramics.

For example, the use of Montmorillonites Cloisite 30B and Cloisite 10A allows to enhance fracture behavior and toughness, respectively [18]. Biocomposites with bleached sisal fibers have also been investigated to improve mechanical properties because of the good dispersibility of the fibers [19], as well as nano-TiO$_2$ addition, that has a positive influence on the mechanical behavior as an interprenetrating network structure can be formed [20]. Starch blends with others polymers, such as poly(hydroxybutyrate) [21] or poly(butylene succinate) [22], have also been extensively studied.

The evaluation of other properties in addition to mechanical ones is equally interesting. For an example, bacterial adhesion has been reduced adding chitosan [23] while soy protein isolate and sorbitol were added to control starch biodegradation and C/N balance [24]. Anyway, biodegradation of starch is achieved via hydrolysis by enzymes. Amylase and glucosidases attack the α-1,4 and the α-1,6 links, respectively, producing nontoxic compounds. Trade suppliers of starch and the relative blends are reported in Table 5.4.

Table 5.4: Suppliers of starch and blends.

Company	Country
Novamont	Italy
National Starch	USA
Vegemat	France
Limagrain	France
Biotech	Germany
Biotec	England
Plantic Technologies	Australia
Rodenberg	Netherlands
BIOP	Germany
Wuhan Huali Environment Protection Science & Technology	China
Biograde	China
PSM	USA
Livan	Canada

5.2.3 Lignin

Lignin is a complex polymer made up of a network of phenyl propanol units with a highly branched 3-D structure that contains different functional groups: hydroxyl, methoxyl, carbonyl, carboxyl, and three most common phenolic macromolecules, such as cumaryl, sinapyl, and conferyl alcohol connected by different C–C and ether bonds (Figure 5.3).

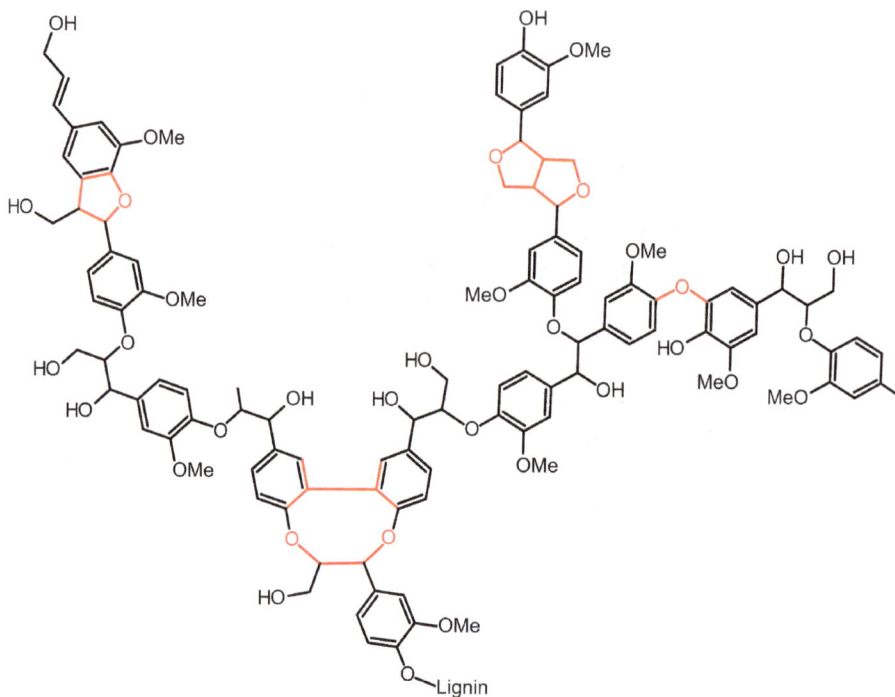

Figure 5.3: Structure of lignin.

Natural lignin is pale yellow, with a molecular mass in the range of 1,000 to 20,000 g/mol and a degree of polymerization difficult to estimate because of fragments as well as several types of sub units. The T_g is strictly related to the amount of water and chemical functionalization. The mobility of lignin molecules will be better at low T_g. It is classified into three major groups depending on its origin, namely softwood, hardwood, and grass lignin [25] and, since it is the most important nonpolysaccharide fraction of lignocellulosic biomass (only about 2% is commercialized, the remaining 98% is burned [26]) and the major aromatic resource of the biobased economy, the interest in its valorization is increasing both in industry and academia [27].

Lignin is extracted from lignocellulosic biomasses following two different approaches: lignin recovery before and after carbohydrate conversion [28]. The first approach consists of extraction or delignification by using mostly chemical methods, where chemicals are used to degrade the highly branched structure of lignin. For example, in soda process, lignocellulosics are pretreated with NaOH, which is able to deprotonate phenolic hydroxyl groups. Kraft pulp is considered an upgrade of soda process, in which the presence of sodium sulfide, beside NaOH, has the aim to accelerate the degradation of lignin. Further processes are represented by sulfite treatment and organosolv. The first one is the main source of commercially available lignin.

The concentration of sulfite, bisulfite, and sulfur dioxide and consequent pH of reaction determine the mechanism of lignin degradation. In the organosolv, the extraction is due to the use of polar-organic solvents like methanol, ethanol, formic acid, and acetic acid [27]. Finally, the biological degradation of lignin is also considered possible by enzymes as well as lignin peroxidase, which degrades nonphenolic lignin units like manganese peroxidase, able to act on phenolic and nonphenolic lignin units through lipid peroxidation reactions and laccases, which catalyze the oxidation of phenolic units to radicals [29]. Micro- and macroorganisms such as *P. chrysosporium*, *Ceriporiala lacerata*, *Cyathus stercoreus*, *C. subvermispora*, *Pycnoporus cinnabarinus*, and *Pleurotus ostreatus* produce enzymes that are involved in lignin degradation [30].

However, the complete exploitation of lignin cannot ignore the needs of new technological developments to break down the chemical structure into commodity chemicals such as benzene, toluene, xylene, phenols, hydroxybenzoic acids, as well as coniferyl, sinapyl, and p-coumaryl compounds. This difficulty could be overcome by employing either heterogeneous or homogeneous catalysts, which can break the complicated network of C–O and C–C bonds in lignin without leaving substantial amounts of tar or gasification [31].

Once lignin is isolated and degraded, the technology to convert these compounds in monomers and polymers is already mature and the combination of new and current approaches would provide novel opportunities for lignin conversion in high-performance advanced materials (Figure 5.4) [26].

Since lignin has several useful properties (biodegradability, antioxidant, antimicrobial, and adhesive), it is used in blend and composite materials, in synthetic polymers functionalized by lignin derivatives, or as reactive component for the preparation of resins and several kinds of polymers.

Lignin blends include polymers with different behaviors according to their polarity (e.g., polyolefins), the presence of aromatic rings and the H-bonds depending interactions. The self-interactions in lignin are very strong because of the large number of polar-functional groups play a decisive role in the determination of the structure and properties of polymer/lignin blends [27].

Polyolefin/lignin blends are affected by apolar properties. Indeed, polyolefin can enter only into weak dispersion and considering the strong polarity and functionality of lignin, immiscibility and poor properties are expected. On the other hand, polymers containing aromatic groups can form π interactions to improve compatibility and partial miscibility. In addition, since the phenolic groups of lignin are able to scavenge radicals, the use as stabilizers and to protect the matrix polymer against the oxidation under UV radiation is strongly suggested, even to allow the biodegradation [27]. Another important issue is represented by the influence of the amount of lignin on the homogenization and compatibility directly related to stabilization of polymer/lignin blends. When lignin is added in small amount, it is not easy to determine the quality of dispersion, but it is even more important when adding large amount to modify mechanical properties of material. Good dispersion compatibility and complete miscibility

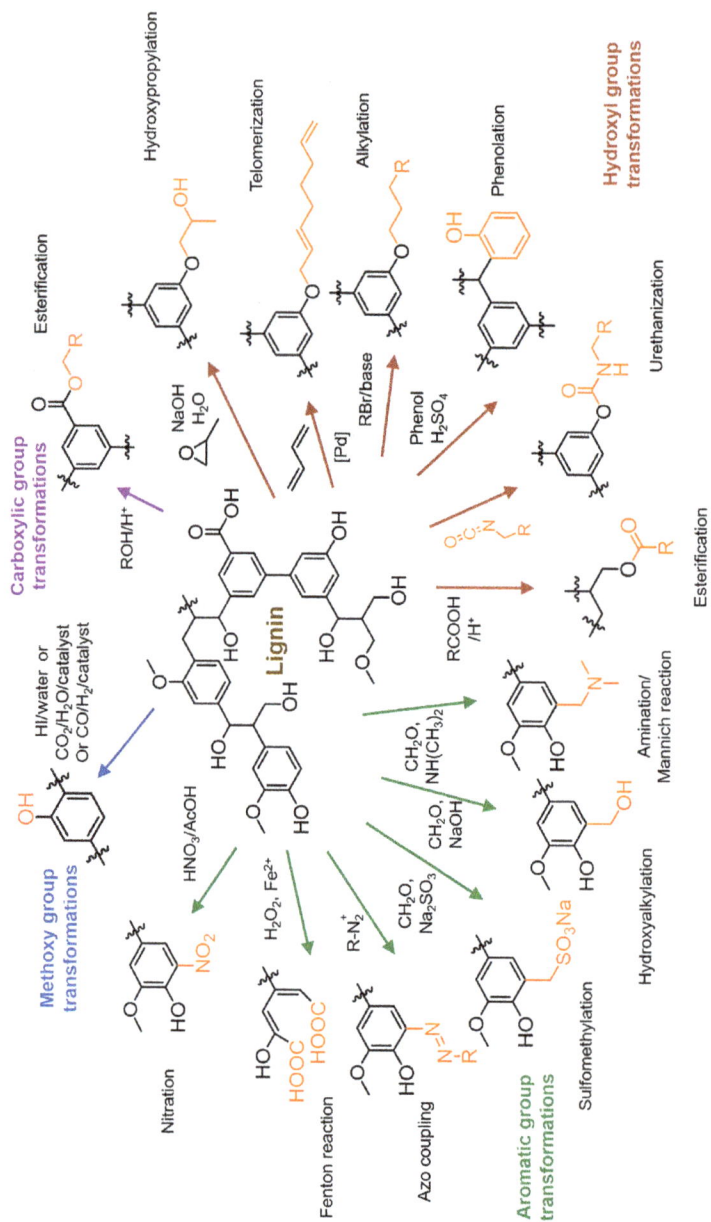

Figure 5.4: A summary of chemical transformations to diversify chemical functionality on lignin. Reproduced from Ganewatta M.S. et al., Lignin Biopolymers in the Age of Controlled Polymerization. Reprinted from: *Polymers* **2019**, *11*, 1176, doi:10.3390/polym11071176.

was claimed by several polymers including low-density polyethylene and polypropylene in presence of soft Kraft lignin and a wide variety of effects on different properties, such as tensile strength (Figure 5.5) [27, 32].

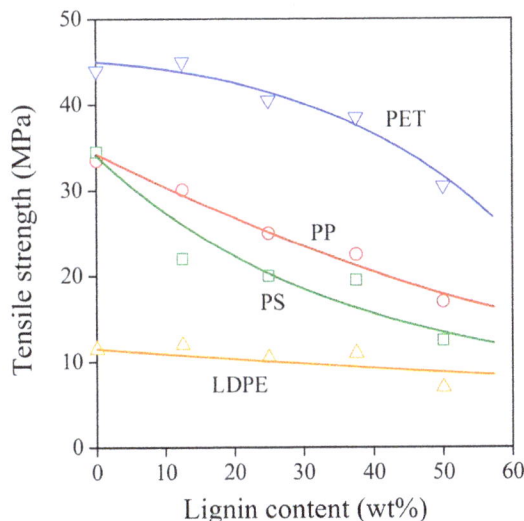

Figure 5.5: Tensile strength of polymer/softwood Kraft lignin blends. Symbols: (△) LDPE (low-density polyethylene), (□) PS (polystyrene), (○) PP (polypropylene), (▽) PET (polyethylene terephthalate). Reproduced from Jeong, H., Park, J., Kim, S., Lee, J., Cho J.W. Use of acetylated softwood kraft lignin as filler in synthetic polymers. *Fiber Polym.* **2012**, *13*, 1310–1318.

However, complete miscibility is rarely achieved, the blends usually have heterogeneous structure and their properties, especially deformability are not exceptionally good [27].

Blends prepared capable of interactions by means of hydrogen bonds are considerably stronger than the interactions discussed above. Polymers such as poly(ethylene oxide) [33], poly(4-vinylpyridine) [34], poly(vinylpyrrolidone) [35], or polyaniline (PANI) [36] are miscible with lignin.

However, in different cases it has been observed that despite the fact that interactions due to the hydrogen bonds are very strong, the structures of the materials clearly show heterogeneity. It is demonstrated quite well by PVOH/lignin blends, which is rather surprising, since the number of active OH groups is considerable in PVOH [27].

For all the reasons discussed above, it seems clear that the properties of polymer/lignin blends still need to be investigated and studied. A different approach to control the structure and the properties of lignin-based blends is achieved by different kinds of modifications (e.g., plasticization and compatibilization).

Another way to use lignin, or rather the products of a derivatization process of lignin is to functionalize different polymers which have applications in different fields such as coating films, adhesive resins, rigid foams, and so on. Lignosulfonate-based polymers, for example, exploit the high content of phenolic groups when compared to carboxylic acid groups in native lignin and the presence of strong sulfonic groups. The applications of these polymers are as corrosion control agents in water delivery systems [25]. In the last years, research activity was focused on finding solutions to reduce the increase of environmental pollution and the depletion of fossil fuel by using energy storage devices such as fuel cells, Li-ion batteries, and supercapacitors [37]. For instance, carbon-based material received attention such as graphene, that when used as electrode materials of supercapacitors, not only inherits the outstanding properties, but improves supercapacitive performances by facilitating the ion diffusion into the inside of electrode materials [38]. Lignosulfonate/graphene oxide sensors were used to detect and remove, by using fast analysis, no sample destruction, and high sensitivity, Pb^{+2} or ferric ion (Fe^{+3}) from industrial wastewater [39] constantly released from industrial processes and could cause several diseases because of its ubiquitous behavior. Lignosulfonate could be used also with PANI [40] or poly (N-methylaniline) [41].

Kraft lignin-based polymers are based on Kraft lignin that is commonly used as fuel for energy recovery after pulping process. It consists of thiol groups, low-molecular weight, and high polydispersity [42]. Lignin/poly(ethylene glycol) (PEG) copolymer is an example of materials able to exploit the characteristics of both Kraft lignin and PEG, in particular low toxic, biocompatible, biodegradable, and amphiphilic, to improve brightness stabilization and strength properties, improving simultaneously bending strength of cement [43]. Kraft lignin was also used as electrically conductive carbon filler with micron size for polymers and its depolymerized version with a low-molecular weight and high-hydroxyl groups could represent a substitute for petroleum-based polyols for rigid polyurethane foam production in engineering applications, that is, automotive parts, structural materials, and insulation materials [44].

Besides it should be considered organosolv lignin, important precursor for carbon fibers spun with high-tensile strength [45]. Electrospinning is a technique that allows the production of nanosized fibers with high modulus and toughness when compared to microscale fibers. Smaller size of diameter affects mechanical properties and the higher specific surface area shows a way to improve poor-interfacial bonding between fibers and matrix [27]. Phenolation is another pathway to improve the reactivity of organosolv lignin. Indeed, phenol formaldehyde, polybenzoxazines, and epoxy resins have huge potential applications due to the excellent performance expressions in dimensional stability, mechanical properties, flame resistance, and chemical durability [46].

The applications of lignin-based materials are different and mostly related to lignosulfonates, that are used as dispersant to improve the workability, but commonly applied as binder [47], as component of adhesive [48], or for the production of chemicals (vanillin, dimethyl sulfide, methyl mercaptane). It could be applied in water-based inks and paints. A Germany compounding company, TECNARO, has already utilized lignin in a product with the trade name ARBOFORM® also named "liquid wood" for its specific properties. Figure 5.6 shows commercial headphones with earcups made of "liquid wood" [27].

Figure 5.6: Headphones with earcups made of ARBOFORM®. Reproduced from Kun, D., Pukánszky, B. Polymer/lignin blends: interactions, properties, applications. European Polymer J. **2017**, 619–639.

References

[1] Stephen, A.M., Phillips, G.O., and Williams, P.A. (2006). Food Polysaccharides and Their Applications. 2nd Edition. Taylor & Francis Group LLC.

[2] Adeodato Vieira, M.G., Altenhofen da Silva, M., Oliveira Dos Santos, L., and Masumi Beppu, M (2011). Natural-based plasticizers and biopolymer films: A review. Eur. Polym. J. 47, 254–263.

[3] Chandra, R., and Rustgi, R. (1998). Biodegradable polymers. Progr. Polym. Sci. 23, 1273–1335.

[4] Kamel, S., Ali, N., Jahangir, K., Shah, S.M., and El-Gendy, A. A. (2008). Pharmaceutical significance of cellulose: a review. Express Polym. Lett. 2, 758–778.

[5] Vroman, I., and Tighzert, L. (2009). Biodegradable Polymers. Materials 2, 307–344.

[6] Wang, S., Lu, A., and Zhang, L. (2016). Recent advances in cellulose materials. Prog. Polym. Sci. 53, 169–206.

[7] Babu, R.P., O'Connor, K., and Seeram, R. (2013). Current progress on bio-based polymers and their future trends. Prog. Biomater. 2, 8–24.

[8] Wang, S., Li, C., Copeland, L., Niu, Q., and Wang, S. (2015). Starch retrogradation: A comprehensive review. Compr. Rev. Food Sci. Food Saf. 14, 568–585.

[9] Jane, J. (1995). Starch properties, modifications and applications. J. Macromolecular Sci. 32, 751–757.

[10] Parandoosh, S.M., and Hudson, S.M. (1993). The acetylation and enzymatic degradation of starch films. J. Appl. Polym. Sci. 48, 787–791.

[11] Beliakova, M.K., Aly, A.A., and Abdel-Mohdy, F.A. (2004). Grafting of poly(methacrylic acid) on starch and poly(vinyl alcohol). Starch – Starke 56, 407–412.

[12] Ghavimi, S.A.A., Ebrahimzadeh, H.M., Solati-Hashjin, M., and Osman, N.A.U.M. (2014). Polycaprolactone/starch composite: Fabrication, structure, properties, and applications. Biomed. Mater. Res. Part A 1–17.

[13] Wang, X., Yang, K., and Wang, Y. (2003). Properties of starch blends with biodegradable polymers. J Macrom. Sci. C-Polym. Rev. 43, 385–409.

[14] Tian, H., Yan, J., Rajulu, A. V., Xiang, A., and Luo, X. (2017). Fabrication and properties of polyvinyl alcohol/starch blend films: Effect of composition and humidity. Int. J. Biol. Macromol. 96, 518–523.

[15] Ramaraj, B. (2007). Crosslinked poly(vinyl alcohol) and starch composite films. II. Physicomechanical, thermal properties and swelling studies. J. Appl. Polym. Sci. 103, 909–916.

[16] Armentano, I., Bitinis, N., Fortunati, E., Mattioli, S., Rescignano, N., Verdejo, R., and Kenny, J.M. (2013). Multifunctional nanostructured PLA materials for packaging and tissue engineering. Prog. Polym. Sci. 38, 1720–1747.

[17] Van Natta, F., Hill, J., and Carothers, W. (1934). Studies of polymerization and ring formation. XXIII. e-caprolactone and its polymers. J. Am. Chem. Soc. 56, 455–457.

[18] Pérez, E., Pérez, C., Alvarez, V., and Bernal, C. (2013). Fracture behavior of a commercial starch/polycaprolactone blend reinforced with different layered silicates. Carbohydr. Polym. 97, 269–276.

[19] Campos, A., Marconcini, J., Martins-Franchetti, S., and Mattoso, L. (2012). The influence of UV-C irradiation on the properties of thermoplastic starch and polycaprolactone biocomposite with sisal bleachd fibers. Polym. Degrad. Stabil. 10, 1948–1955.

[20] Fei, P., Shi, Y., Zhou, M., Cai, J., Tang, S., and Xiong, H. (2013). Effects of nano-TiO_2 on the properties and structures of starch/poly (ε-caprolactone) composites. J. Appl. Polym, Sci. 6, 4129–4136.

[21] Ma, P., Xu, P., Chen, M., Dong, W., Cai, X., Schmit, P., Spoelstra, A.B., and Lemstra, P.J. (2014). Structure–property relationships of reactively compatibilized PHB/EVA/starch blend. Carbohydr. Polym. 108, 299–306.

[22] Suchai-in, K., Koombhongse, P., and Chirachanca, S. (2014). Starch grafted poly(butylene succinate) via conjugating reaction and its role on enhancing the compatibility. Carbohydr. Polym. 102, 95–102.

[23] Alix, S., Mahieu, A., Terrie, C., Soulestin, J., Gerault, E., Feulloley, M.G.J., Gattin, R., Edon, V., Ait-Younes, T., and Leblanc, N. (2013). Active psuedo-multilayered films from polycaprolactone and starch based matrix for food packaging applications. Eur. Polym. J. 49, 1234–1242.

[24] Mariani, P., Allganer, K., Oliveira, F., Cardoso, E., and Innocentini-Mei, L. (2009). Effect of soy protein isolate on the thermal, mechanical and morphological properties of poly (ε-caprolactone) and corn starch blends. Polym. Test 8, 824–829.

[25] Naseem, A., Tabasum, S., Zia, K.M., Zuber, M., Ali, M., and Noreen, A. (2016). Lignin-derivates based polymers, blnds and composites: a review. Intern. J. Biol. Macrom. 93, 296–313.

[26] Ganewatta, M.S., Lokupitya, H.N., Tang, C. (2019) Lignin Biopolymers in the Age of Controlled Polymerization. Polymers 11, 1176–1220.

[27] Kun, D., and Pukánszky, B. (2017). Polymer/lignin blends: interactions, properties, applications. Eur. Polym. J. 619–639.

[28] Ragauskas, A.J., Beckham, G.T., Biddy, M.J., Chandra, R., Chen, F., Davis, M.F., Davison, B.H., Dixon, R.A., Gilna, P., Keller, M. et al. (2014). Lignin valorization: Improving lignin processing in the biorefinery. Science 344, 1246843.

[29] Madadi, M., and Abbas, A. (2017). Lignin degradation by fungus pretreatment: a review. J. Plant Pathol. Microbiol. 8, 1–6.

[30] Kumar, G., Bakonyi, P., Periyasamy, S., Kim, S.H., Nemestóthy, N., et al. (2015). Lignocellulose biohydrogen: Practical challenges and recent progress. Renew. Sustain. Energy Rev. 44, 728–737.

[31] Mathers, R.T. (2012). How well can renewable resources mimic commodity monomers and polymers?. J. Polym. Sci. Pol. Chem. 50, 1–15.

[32] Jeong, H., Park, J., Kim, S., Lee, J., and Cho, J.W. (2012). Use of acetylated softwood kraft lignin as filler in synthetic polymers. Fiber Polym. 13, 1310–1318.

[33] Kadla, J.F., and Kubo, S. (2004). Lignin-based polymer blends: analysis of intermolecular interactions in lignin-synthetic polymer blends. Compos. A 35, 395–400.

[34] Liu, C., Xiao, C., and Liang, H. (2005). Properties and structure of PVP–lignin "Blend Films". J. Appl. Polym. Sci. 95, 1405–1411.

[35] Silva, M.F., da Silva, C.A., Fogo, F.C., Pineda, E.A.G., and Hechenleitner, A.A.W. (2005). Thermal and FTIR study of polyvinylpyrrolidone/lignin blends. J. Therm. Anal. Calorim. 79, 367–370.

[36] Rodrigues, P.C., Cantão, M.P., Janissek, P., Scarpa, P.C.N., Mathias, A.L., Ramos, L.P., and Gomes, M.A.B. (2002). Polyaniline/lignin blends: FTIR, MEV and electrochemical characterization. Eur. Polym. J. 38, 2213–2217.

[37] Simon, P., and Gogotsi, Y. (2008). Materials for electrochemical capacitors. Nat. Mater. 7, 845–854.

[38] Wu, X.L., and Xu, A.W. (2014). Carbonaceous hydrogels and aerogels for supercapacitors. J. Mater. Chem. A 2, 4852–4864.

[39] Zhao, H.B., Wang, W.D., Lü, Q.F., Lin, T.T., Lin, Q., and Yang, H. (2015). Preparation and application of porous nitrogen-doped graphene obtained by co-pyrolysis of lignosulfonate and graphene oxide. Bioresour. Technol. 176, 106–111.

[40] Fu, G.D., Li, G.L., Neoh, K.G., and Kang, E.T. (2011). Hollow polymeric nanostructures-Synthesis, morphology and function. Prog. Polym. Sci. 36, 127–140.

[41] Lü, Q.F., Luo, J.J., Lin, T.T., and Zhang, Y.Z. (2014). Novel Lignin–Poly(N-methylaniline) Composite Sorbent for Silver Ion Removal and Recovery. ACS Sustainable Chem. Eng. 2, 465–471.

[42] Jiang, X., Savithri, D., Du, X., Pawar, S., Jameel, H., Chang, H., and Zhou, X. (2017). Fractionation and Characterization of Kraft Lignin by Sequential Precipitation with Various Organic Solvents. ACS Sustainable Chem. Eng. 5, 835–842.

[43] Lou, H.M., Wang, M.X., Lai, H.R., Lin, X.L., Zhou, M.S., Yang, D.J., and Qiu, X.Q. (2013). Synthesis, structure, and dispersion property of a novel lignin-based polyoxyethylene ether from kraft lignin and poly(ethylene glycol). Bioresour. Technol. 146, 478–484.

[44] Narine, S.S, Kong, X., Bousidi, L., and Sporns, P.J. (2007). Physical properties of polyurethanes produced from polyols from seed oils: II. Foams. Am. Oil Chem. Soc. 84, 65–72.

[45] Baker, D.A., and Rials, T.G. (2013). Recent advances in low-cost carbon fiber manufacture from lignin. J. Appl. Polym. Sci. 130, 713–728.

[46] Lligadas, G., Tüzün, A., Ronda, J.C., Galià, M., and Cádiz, V. (2014). Polybenzoxazines: new players in the bio-based polymer arena. Polym. Chem. 5, 6636–6644, Center for Renewable Carbon, The University of Tennessee, Knoxville, Tennessee.

[47] Knodt, C.B. Feed pelletin process and resulting product, US 3035820 A, 1963.

[48] Particle board, hardboard and plywood produced in combination with lignin sulfonate-phenol formaldehyde glue system, US 3931072 A, 1976.

Angela Marotta, Veronica Ambrogi, Alice Mija

6 Biobased thermosetting materials

Abstract: This chapter highlights the main recent researches on biobased thermosetting resins. In the last two decades, especially the last years, many efforts were dedicated i) to synthesize new biobased building blocks used thereafter to build thermosetting resins and/or ii) to elaborate thermosetting resins from renewable resources. The interest on biobased resins is justified by the latest concerns regarding sustainability and toxicity. These new developed resins should represent a sustainable alternative to the fossil based ones. The chapter presents the main aspects concerning the synthesis, precursors, principal properties or peculiar aspects on each class of thermosets. The chapter dedicated a subchapter to most used commercial resins: unsaturated polyesters, polyurethanes and non-isocyanate polyurethanes, epoxides and phenolics.

Keywords: Thermosets, biobased molecules, unsaturated polyesters, polyurethanes, epoxy, phenolic resins

6.1 Introduction

Polymers are classified as thermoplastic or thermosetting materials depending on their physical behavior in response to temperature variation. Thermoplastics are polymers that exhibit reversible modification of their physical properties when heated (or cooled), for example, they melt (or crystallize). On the other hand, thermosets undergo irreversible modification when heated and can never recover their former properties, even if cooled. This different behavior is due to the chemical structure of these materials: thermoplastics are constituted by long chains interrelated only by physical interactions, whereas in thermosets chemical linkages are present between the main chains, resulting in a continuous three-dimensional structure. As a consequence of their chemical structure, thermosets possess outstanding thermomechanical performances [1, 2].

Even though the majority of polymers produced are thermoplastics, a strong sustainability issue is related to thermosets area. Due to their thermal behavior and chemical structure, and except for vitrimers, it is impossible to dissolve or melt them. For this reason, thermosetting resins are known to be highly stable materials.

Angela Marotta, Veronica Ambrogi, University of Naples Federico II, Department of Chemical, Materials and Production Engineering (DICMAPI), Naples, Italy
Alice Mija, Université Côte d'Azur, Université Nice-Sophia Antipolis, Institut de Chimie de Nice, France

https://doi.org/10.1515/9783110590586-006

This peculiar feature translates into the difficulty to rework thermosets and consequently to recycle them. Beyond the recyclability aspects, an environmental concern appears due to the toxicity of some of their building blocks.

The most used thermosetting classes are epoxy resins, unsaturated polyesters (UPE) and polyurethanes (PU), and phenolic resins. Each of them is produced starting with at least one toxic molecule. For example, phenolic resins are obtained through the polycondensation reactions between phenols and formaldehyde (identified as resoles or novolac depending on phenol/formaldehyde stoichiometric ratio). Both reagents are toxic, and their use is regulated by European Chemical Agency (ECHA) [3] and United States Environmental Protection Agency (EPA) [4].

By reacting isocyanates and polyols, PU are obtained. The isocyanates are harmful volatile molecules that can, in short time, sensitize the operators in polyurethane industry.

As the name suggests, epoxy resins are made by reaction of molecules bearing at least two epoxy rings (namely epoxides), with curing agents of different nature. Thanks to their ability to cure at low temperature and final properties of the derived resins, the most widespread curing agents are amines, generally harmful. In addition, the quite totality of epoxy resins produced are made starting by bisphenol A (BPA), a reprotoxic and carcinogenic molecule.

UPE are generally not harmful but when they are used to produce thermosetting resins by cross-linking of polymers chains on the unsaturated moieties, styrene is used. Styrene is a so-called reactive diluent (RD), with which vinyl groups react to form bridges between the polyester chains.

Currently researches are therefore pushed to the investigation and discovery of novel molecules to substitute building blocks currently under regulation. In addition, not only they are obtained by using generally toxic building blocks but also these are fossil-derived molecules produced by nonrenewable feedstocks. The uncertainty regarding the price and availability of the oil, in addition to the direct political strategies and institutional tendencies toward a sustainable development, have pushed the chemical industry to turn toward sustainable and renewable resources to synthesize biobased products [5]. Biobased products are wholly or partly derived from materials of biological origin, excluding materials embedded in geological formations and/or fossilized. In industrial processes, by using fermentation and biocatalysis instead of traditional chemical synthesis, higher process efficiency can be obtained, resulting in a decrease in energy and water consumption, and a reduction of toxic waste. As they are derived from renewable raw materials such as plants, biobased products can help to reduce the CO_2 emissions and offer other advantages, such as lower toxicity or novel product characteristics (e.g., biodegradable plastic materials) [6]. Bioderived does not directly imply that these materials are biodegradable. A material is biodegradable when it can be degraded by means of microbes or enzymes and this is not related to its origin but to its structure and environmental conditions [7].

In this chapter, a summary of the most recent bioderived molecules envisaged for the production of thermosets is proposed. In particular, attention was paid to the above-mentioned four main thermosetting categories.

6.2 Unsaturated polyester resins

The first example of unsaturated polyester resins (UPRs) dates back to 1900s when the reaction between phthalic anhydride and glycerol resulted in a tough paint. Following this discovery, several unsaturated fatty acids were employed with phthalic anhydride in the production of alkyd resins as surface coatings [8]. Polyesters are commonly thermoplastic materials but when a monomer bearing unsaturation is present in the polymeric chain, it can further react with substances, called reactive diluents (RD), which are able to bind the main chains in a three-dimensional structure, characteristic of thermosetting material. The most used RD is the styrene, which can react in presence of unsaturation as reported in Scheme 6.1 [8].

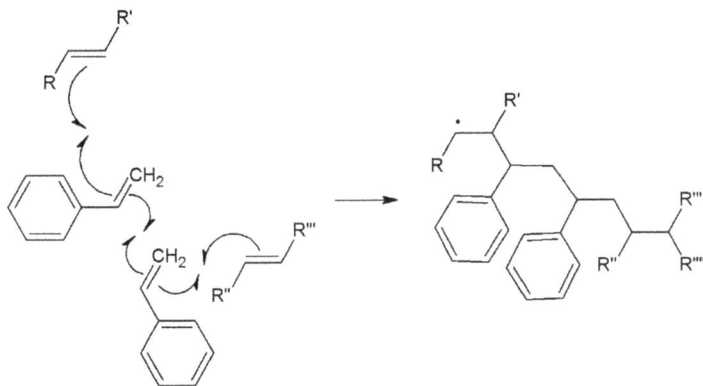

Scheme 6.1: Cross-linking of styrene with unsaturated groups in polyesters during curing of polyester [8].

Due to the use of styrene in UPRs production, the use of those materials are not only restricted in Europe as an effect of Volatile Organic Compounds (VOCs) Solvents Emissions Directive, a regulation aimed to the reduction of industrial emissions of VOCs, but also in the United States according to the Hazardous Air Pollutants regulation [9]. Therefore, less toxic and volatile RDs should be found. The reactants for the UPE production can be also substituted by renewable bioderived molecules. Nowadays, valuable biobased dicarboxylic acids or polyols could be found [10] to produce fully biobased polyesters [11–13]. On the other hand, it is still difficult to identify RD capable of substituting the styrene in the cross-linking process. Therefore, researches are aimed in this way and some satisfactory data reported in literature are herein presented.

Acrylated epoxidized soybean oil (AESO) (Figure 6.1) based thermosets are not so easily to be produced, due to AESO's high viscosity and low cross-linking density; the use of RD is helpful for both decreasing the viscosity and increasing the cross-linking density. To avoid the use of styrene, Fei et al. proposed trimethylolpropane trimethacrylate and 1,4-butanediol dimethacrylate, derived from succinic acid, as RD to prepare fully biobased resins and composites containing bamboo fibers as reinforcement [14]. The presence of double bonds on methacrylate monomers makes the cross-linking reaction more effective than when using the styrene, and the final properties of these fully biobased resins were comparable with the styrene-cured resins. In their study, Wu et al. [15] proposed AESO as a styrene replacement in cross-linking reactions based on the use of different commercial UPEs. First, the polyester chain containing isophthalic acid, maleic anhydride, and then propylene glycol was mixed with AESO, the polyester showing low viscosity and long pot life and used thereafter in the production of kenaf fiber-reinforced composites. The mechanical properties of the resulted composites were competitive with those from styrene-cured analogous. A more recent work analyzed the UPRs obtained reacting AESO with a commercial UPE prepared from propylene glycol and maleic anhydride [16] or ethylene and diethylene glycol with maleic anhydride [17]. Thanks to the good miscibility of AESO with the UPEs, this resin was used as a matrix in glass fiber-reinforced composites. Compared with the styrene, AESO negatively affected the mechanical properties of the material, but the decrease in these properties was mitigated when the material was reinforced. Mechanical properties of UPRs can be improved by inserting rigid moieties such as isosorbide in the UPE chain. It is the case of the study proposed by Liu et al. that synthesized an isosorbide-methacrylate, which was coreacted with methacrylated AESO, giving rise to an UPR with superior mechanical and thermal properties [18].

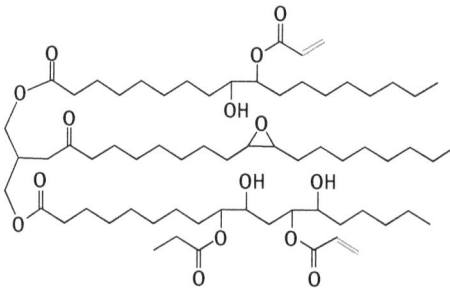

Figure 6.1: Acrylated epoxidized soybean oil (AESO) structure.

Itaconic acid (Figure 6.2) is one of the most promising sugar-derived building block [10] and thanks to the unsaturation present in its structure, it is a worthy molecule for

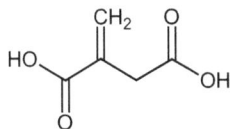

Figure 6.2: Itaconic acid structure.

the biobased UPRs production. It can react, for example, with 1,2-propanediol to produce UPE chains that in turn are cross-linked with its ester's derivatives, namely di-
methyl-, diethyl, di-*n*-butyl-, and diisopropyl itaconate, obtaining UPRs with good
mechanical properties, comparable with the styrene-cured resins [19]. Fully biobased
UPRs can be synthesized, in which the polyester main chain is obtained by reacting at
different molar ratios of the 1,2-propanediol with saturated acids as oxalic, adipic, succinic acids and also with itaconic acid – representing the moiety bearing unsaturation – and the thermosetting resins are produced through the cross-linking reaction
with dimethyl itaconate. Using different percentages of the curing agent, thermosetting resins with tailorable properties have been obtained [20]. Analogous thermosets
were produced when reacting oxalic and itaconic acids with ethylene glycol, and the
same RD of the previous work, and used as matrix for cotton fibers reinforced composites, showing excellent thermomechanical properties [21]. Biobased UPRs from itaconic acid, succinic acid, and propylene glycol, with dimethyl itaconate as RD, can
also be used as matrix for fiber-reinforced composites, where fibers are composed by
itaconyl groups grafted recycled PET [22]. Unfortunately, the mechanical properties
of these composites are not competitive with those of the glass fiber reinforced UPRs
composites.

Four environmentally friendly UPE were synthesized by reacting itaconic acid,
succinic acid, sebacic acid, glycerol, 1,4-butanediol, and 1,6-hexanediol in a work
where the use of 2-hydroxyethyl methacrylate (HEMA), methyl methacrylate, and
isobornyl methacrylate (IBOMA) as substitute of styrene as RD were investigated
[23]. Among them, HEMA was a valuable cross-linker for biobased polyester, resulting in resins with comparable viscosity, reduced shrinkage upon cure, and higher
thermal stability compared with the styrene-based resins. In a recent work, the possibility of enzymatically obtaining UPE by the reaction of dimethyl and dibutyl itaconate, dimethyl and dibutyl fumarate, or dimethyl maleate with polyols with
different chain lengths was proved [24]. The reactivity of these multiple monomers
seemed affected by both solubility in the polymerization system, steric hindrance of
the monomer's pendant group, and reaction's mass transfer.

Xu et al. obtained isosorbide-based UPE by reacting, at different molar ratios,
maleic and succinic anhydride, isosorbide, and 1,4-butanediol, and successively they
cured them with an isosorbide-methacrylate RD, synthesized in their lab [25]. The
synthesized systems showed a better miscibility than common RDs and the thermal
properties of final biobased UPRs material were comparable with those of the petrol-
based analogous. UPRs with improved thermal and mechanical properties were

prepared by using a mixture of itaconic and succinic acid with an aromatic bioderived rigid molecule such as 2,5-furandicarboxylic acid (FDCA). The obtained UPRs
were subsequently cured with a biobased RD based on guaiacol methacrylate [26].

Recently, a study reported the use of veratrole derivatives, namely 1,2-dimethoxy-
4-vinylbenzene, 3-allyl-1,2-dimethoxy-5-vinylbenzene, and 3,5-diallyl-1,2-dimethoxy-
benzene as styrene replacement [27]. These bioderived RD have been tested as
cross-linkers for dimethacrylated-epoxidized-sucrose soyate-based UPRs. The molecular structure of these methacrylate polyesters, with a rigid core and high functionality,
confers good thermomechanical properties compared with the traditional soybean
oil-derived resins. Moreover, the biobased RDs came out to be valuable substitutes of
styrene, imparting to the resin comparable thermomechanical properties. Different
bioderived methacrylate derivatives were also proposed by Cousinet et al. as RD for
commercial UPE, synthesized from a mixture of diols including propylene glycol and
maleic anhydride [28]. These methacrylate derivatives, namely butanediol dimethacrylate, IBOMA, and lauryl methacrylate, were synthesized by reacting different biobased molecules with methacrylic acid, bioderived itself [29], and showed a low
volatility. However, the methacrylate moieties had a tendency to homopolymerize instead of reacting with UPE. For this reason, a phase-separation occurs, generating
heterogeneous morphology with consequent poor mechanical properties.

A summary of bioderived molecules so far proposed for UPRs production is reported in Figure 6.3.

6.3 Thermosetting polyurethanes

PU, derived by polycondensation reactions of polyisocyanates and polyols (Scheme 6.2),
represent a wide category of materials, thanks to the broad range of precursor chemical
structures.

PU are very easily tailorable materials; in fact, in the thermosetting resin structure the urethane moiety confers rigidity, while the polyol represents a soft segment,
so that final properties can be tailored by varying the reactants ratio. Moreover, the
polyol nature (hydrophilicity or hydrophobicity as well as its functionality) can be
suitably chosen to confer desired properties and cross-linking density to the final material. PU were discovered by Otto Bayer in 1930s, who was searching for an alternative to polyamide production, and the invention was already patented by DuPont
[30]. Nowadays, about 70% of total PUs production is for flexible or rigid foams,
mainly used in appliance and buildings construction, thanks to their optimal thermal
and acoustic insulation and good mechanical properties; Moreover, PUs are also employed as adhesives and coatings [30].

Vegetable oils seem to be the ideal biobased molecules in PU production but not
all of them intrinsically possess hydroxyl groups (Figure 6.4). To impart this mandatory

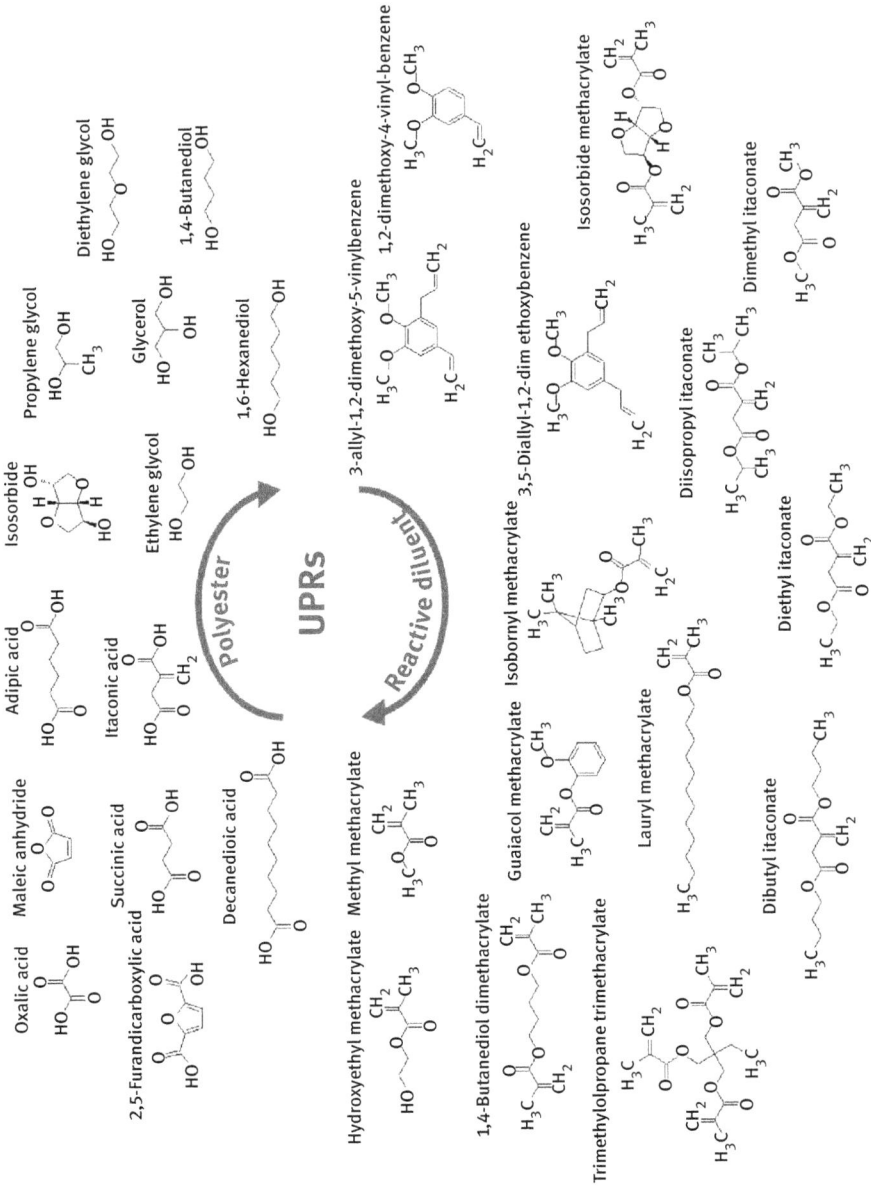

Figure 6.3: Summary of biobased molecules for UPRs production.

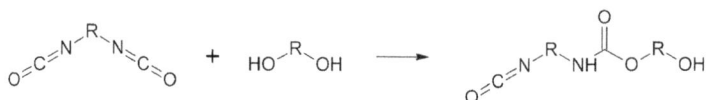

Scheme 6.2: Formation of the urethane group by condensation reaction between isocyanate and polyol.

characteristic, generally the triglyceride structure of vegetable oils is broken up by a transesterification reaction with polyfunctional alcohols, and fatty acid chains or monoglycerides are obtained and reacted with isocyanates (Scheme 6.3). It is the case of linseed oil used in the production of waterborne PU [31] or tung oil, used to produce polyurethane-imides [32]. Analogously, epoxidized rapeseed oil was reacted with diethanolamine, triethanolamine, and diethylene glycol to simultaneously open the epoxy ring and obtain, by transesterification, a polyol; the so obtained polyols were used for the production of rigid PU foams with excellent mechanical properties [33]. In a further work polyols were obtained by thiol-ene photoclick reaction starting from seven different vegetable oils and cross-linked to obtain waterborne PU films with high cross-linking densities and subsequently excellent properties [34].

Castor oil is a quite unique case of vegetable oil naturally bearing hydroxyl groups; therefore, it can directly react with isocyanates. An example is the reaction with isophorone diisocyanate in presence of dimenthylolpropanoic acid and 2-dimethylolbutanoic acid obtaining waterborne polyurethane films [35] or mixed with a sodium alginate water solution to produce composites [36]. Also, cardanol (Figure 6.5), the main constituent of cashew nutshell oil, can be used as polyol in PU synthesis. It is constituted by a metasubstituted phenol with a 15 carbons aliphatic chain that can bear unsaturations to some degrees; on these unsaturations, functionalization can be performed. Cardanol can be epoxidated and the subsequent opening of the epoxy ring by means of amines leads to a polyol available for PU production. Huo et al. obtained cardanol-based PU foams with improved mechanical properties because of the presence of the aromatic ring that imparts rigidity and excellent thermal properties but also of the long aliphatic chain which prevents a fragile behavior [37].

6.3.1 Nonisocyanate polyurethanes

The sole introduction of nontoxic and bioderived polyols is not sufficient to make PU entirely safe. High concerns are present in polyurethane production working environments because of the use of polyisocyanates strictly related with the occurrence of occupational asthma [38, 39]. In addition, the production of isocyanate itself is highly harmful, as it involves the use of phosgene. Broadly, the use of isocyanates has been restricted by ECHA [40] and EPA [41]. Unfortunately, since the toxicity of these molecules

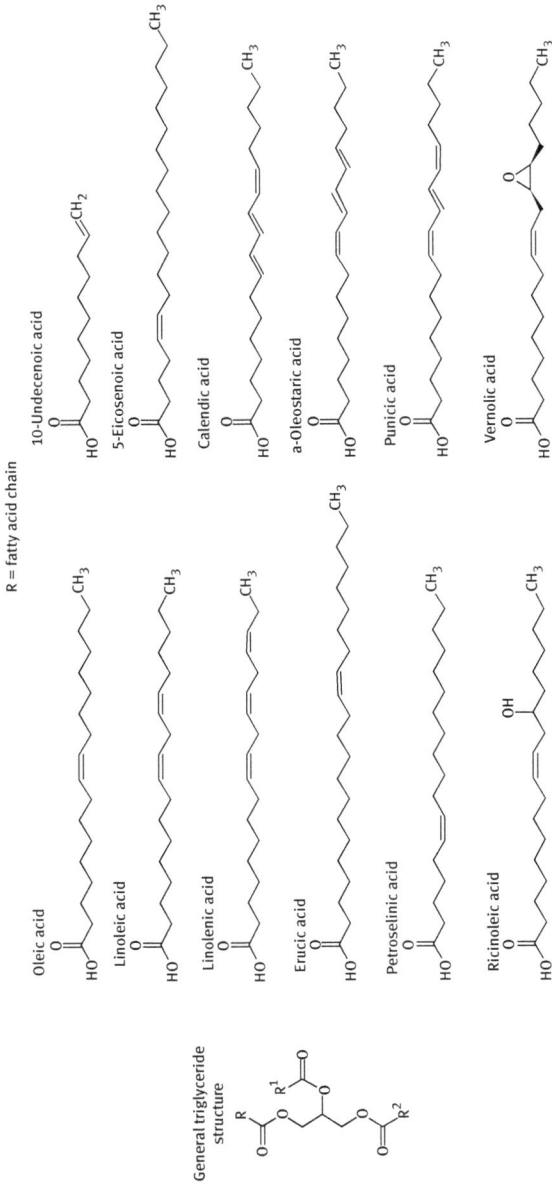

Figure 6.4: General structure of triglycerides and of the most common fatty acids.

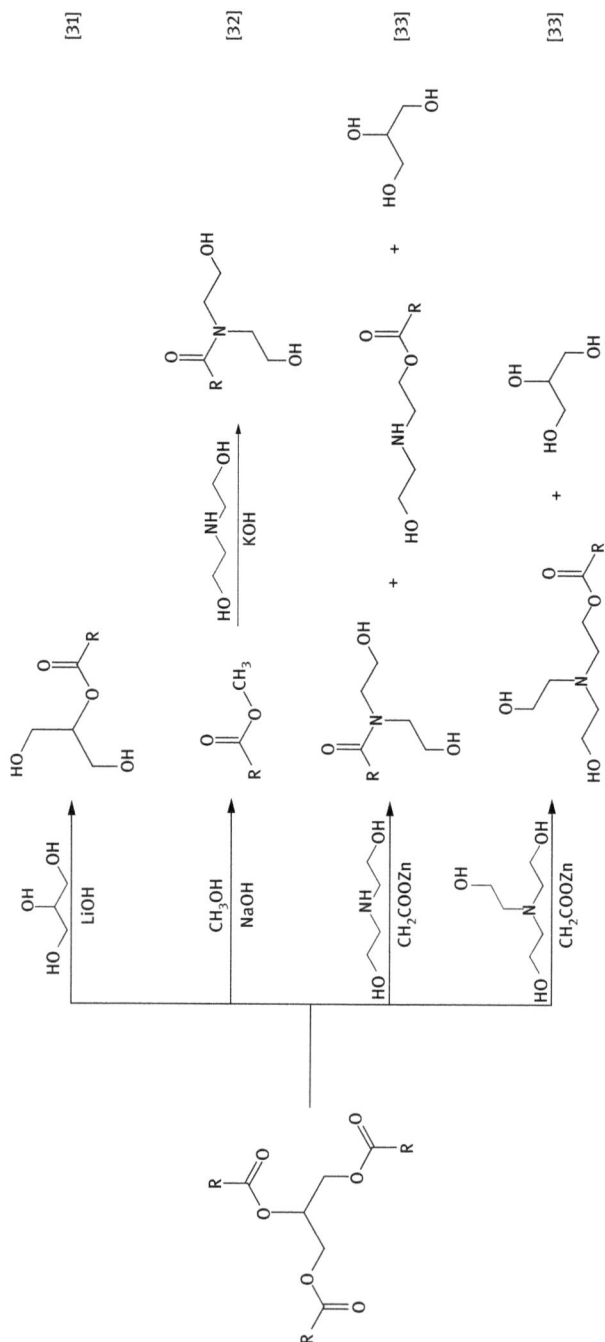

Scheme 6.3: Transesterification reaction of triglycerides with polyfunctional alcohols.

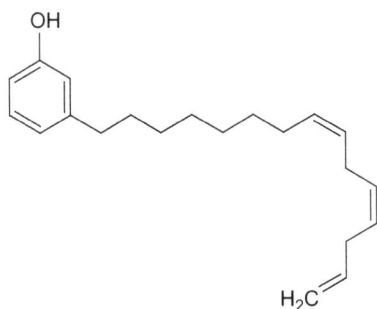

Figure 6.5: Chemical structure of cardanol.

resides in the isocyanate group, the hazard is not removed even if biobased isocyanates are used. A safer option is given by the synthesis of nonisocyanate polyurethanes (NIPU) following different pathways. The first and mainly used are the polyaddition reaction between polycyclic carbonates and polyamines (Scheme 6.4a), and also the polycondensation between polycarbamate and polyol leading to the formation of urethane groups (Scheme 6.4b) [42]. However, the NIPU functionality is not the same as conventional isocyanate-derived PU. Because of the presence of hydroxyl groups deriving from the cyclic carbonates they are generally referred to as polyhydroxyurethanes (PHU) [43]. Several works can be retrieved where not only NIPU thermosets are synthesized, but, as added value, they are also derived from renewable feedstocks.

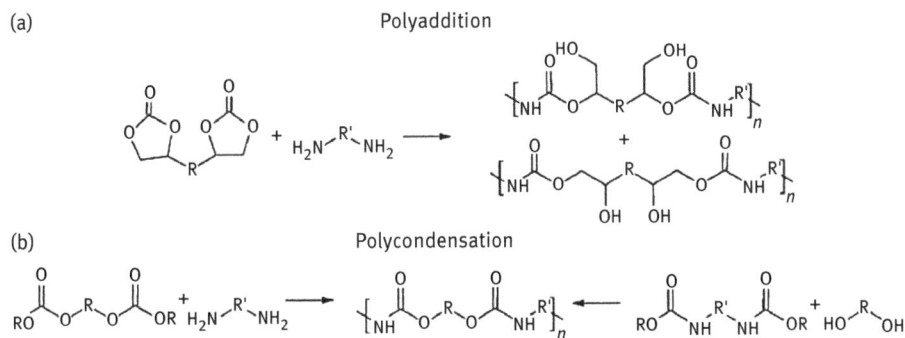

Scheme 6.4: Reaction scheme of (a) polyaddition and (b) polycondensation reactions for NIPU synthesis [43].

Cyclic carbonates generally result from the addition of CO_2 to an epoxy ring, as proposed in a recent work, where five-membered cyclic carbonates are synthesized from biobased limonene oxide [44]. Also epoxidized soybean oil (ESO) and D-sorbitol polyglycidyl ethers were reacted with CO_2, and the respective multifunctional carbonates obtained were further reacted either with diethylene glycol bis(3-aminopropyl) ether (a synthetic diamine) or Priamine™ 1074 (a commercial biobased diamine) obtaining different PHU [45]. Reprocessability of these materials was examined and multiple re-process steps resulted in a full property recovery. Even carbonated soybean oil was cured with short chain amines (as 1,4-butanediamine or 1,5-pentanediamine) to obtain PU thermosets with high cross-linking density, or with oligoamides, resulting in elastomeric thermoplastic PU [46]. However, the so-obtained thermosetting materials showed lower mechanical properties and thermal resistance compared to the thermoplastic analogous. Among vegetable oils, jojoba oil is not a triglyceride but was quite totally composed by monoesters of fatty acids (Figure 6.6) bearing unsaturations. Mokhtari et al. reacted these unsaturations with thioglycolic acid, subsequently esterified with 1,1'-carbonyldiimidazole to obtain a jojoba dicyclic carbonate. Analogously, castor oil was functionalized by means of maleic anhydride, thioglycolic acid, and 1,1'-carbonyldiimidazole to obtain tri- and hexacyclic castor oil carbonates [47]. Different PHU resins were produced by reacting the previously reported carbonates with either 1,4-diaminobutane or m-xylylenediamine, however, due to the relatively low functionality of carbonates, they possess low T_gs and poor mechanical properties.

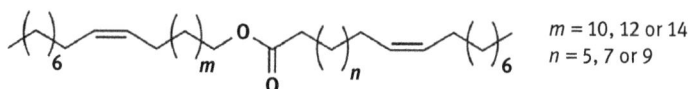

$m = 10, 12$ or 14
$n = 5, 7$ or 9

Figure 6.6: Chemical structure of Jojoba oil [47].

A dimeric fatty acid was instead used by Carré et al. to produce a dimeric biscyclocarbonate after transesterification with glycerol carbonate; analogously, sebacic biscyclocarbonate was also produced and both the carbonates were used in production of NIPUs by reacting with dimeric diamines [48]. The properties of so-obtained NIPUs were compared with that of analogous isocyanate-derived PUs, and it shows that even if the biobased PU show promising properties they must be optimized since their performances are inferior than those of petrol-based materials.

Also epoxidized sucrose soyate, converted into carbonate by supercritical CO_2, was proposed as building block for the production of PHUs for coating applications. The presence of hydroxyl groups in the resin structure was reflected in the good adhesion properties of the coatings. However, low hardness values were determined [49]. With a very easy synthetic procedure, consisting of reacting sucrose first with dimethyl carbonate and then with hexamethylenediamine, sucrose-based PU were synthesized by polycondensation. Sucrose functionalities were also used to introduce further moieties by addition of an epoxy silane coupling agents that, increasing the cross-linking-density, enhanced water and heat resistance of the material, making these resins suitable as wood adhesives [50]. Other sugar derived molecules, as FDCA, were used in production of PHUs with good thermal stability; once the epoxy functionality is introduced in the molecule, CO_2 is added and the dicyclic carbonate obtained is reacted with amines [51].

6.4 Epoxy resins

Epoxy resins are a class of thermosetting polymeric materials obtained from mono-mers containing two or more oxirane groups (epoxides). The high reactivity of the epoxide ring is a consequence of its peculiar triangular planar structure, which makes the ring highly strained and allows the epoxide to react with essentially any strong nucleophile or electrophile. To cross-link epoxies, amines and carboxylic acid anhydrides are the most commonly used hardeners. Epoxy resins are available in a large range of molecular structures and they can react with a large range of curing agents; since there is the possibility of using more than a hundred of curing agents, it is obvious that a lot of different properties can be obtained [52]. The in-vention of epoxy resin is shared by the researcher Castan in Switzerland, which in 1934 produced a low melting amber colored resin, cross-linked with phthalic anhy-dride, and Greenlee in the United States, who in 1939 explored the synthesis of epi-chlorohydrin and BPA to produce resins that could be used as coatings [53].

Diglycidyl ether of bisphenol A (DGEBA) is the monomer mainly used in the epoxy resin industry, filling 75% of the resins used in industrial applications. It is generally prepared by direct glycidylation of BPA with epichlorohydrin. BPA, among other bisphenols, is part of the common class of endocrine disruptors be-cause of the significant hormonal activity, and according to many lines of evi-dence, it acts as an endocrine disruptor even at low doses [54, 55]. The detrimental effects of BPA have led governments to enforce regulations mostly in the European Union, in order to limit the exposition of their citizens to this substance. In 2017, in line with the opinion adopted by the Risk Assessment Committee, BPA has been identified as "a substance of very high concern" [56]. About 20% of the BPA pro-duced accounts for the production of epoxy resins, widely employed as protective linings for a variety of canned foods and beverages, and as a coating on metal lids for glass jars and bottles, including containers used for infant formula. However, part of the unreacted monomer, and some molecules resulting from the breakage of unstable chemical bonds linking BPA to the polymer backbone, can migrate in foods. For this reason, a very strong push to research and development of biobased epoxy resins lies on the requirement to replace, especially in the field of food pack-aging, resins based on BPA.

Initial attempts were made to replace this molecule concerned with the use of epoxidized vegetable oils (EVO), that is, vegetable oils in which the double bonds have been epoxidized. Unfortunately, due to long aliphatic and flexible chains with a relatively low amount of epoxy rings (many of which are internal and less reac-tive) [57], thermal and mechanical properties of EVO-based epoxy resins are not al-ways competitive if compared with DGEBA-based ones.

Effective approaches to go beyond these limits are based on the cross-linking with rigid molecules or on the use of vegetable oils with a higher amount of func-tionality. An example of this strategy was given by Pin et al. who realized highly cross-linked epoxy resins with high T_g improved thermal stability and impact strength by anionic living copolymerization of epoxidized linseed oil (ELO) with methylhexahydrophthalic anhydride and benzophenone-3,3',4,4'-tetracarboxylic dianhydride [58]. Moreover, the same authors proposed a novel synthetic route to obtain fully biobased epoxy resins with improved mechanical properties using ELO copolymerized with polyfurfurylalcohol in presence of boron trifluoride ethylamine complex as catalyst [59]. ELO was also used to synthesize a novel xylitol-functionalized epoxidized linseed oil polyol that could be used in the production of epoxy resins [60]. Furthermore, completely biobased epoxy networks derived from epoxidized linseed oil and castor oil cured with citric acid were developed, but their mechanical properties did not match DGEBA-based resins [61]. Also cottonseed and algae oil can be epoxidized with a very high yield, and then cured with citric or tartaric acid obtaining totally bioepoxy resins. The latter show very good thermal and mechanical properties compared with a petrol-based resin and seed germination and larvicidal activity [62]. Epoxidized cottonseed oil was used also with a mixture of a flexible anhydride, namely dodecenyl succinic anhydride, and a rigid cyclic anhydride (methyl nadic anhydride) [63]. Having different characteristics, the two types of hardeners required different cure temperatures and underwent diverse polymerization processes depending on their length, chain mobility and so on generating different thermosetting structures. Linoleic acid can be esterified with hexamethylol melamine obtaining a six-armed unsaturated molecule, which is epoxidized and cured with 4-methyl hexahydrophthalic anhydride. The properties of the obtained resins, compared to resins obtained by ESO and epoxidized sucrose soyate, resulted in higher T_g and cross-linking density but worse thermal resistance and mechanical properties [64]. Properties of epoxidized linseed oil-based epoxy resins, cured with different anhydride, were evaluated by Boquillon et al. as a function of the cross-linking agent structure, initiator percentage, and epoxy/anhydride ratio [65]. The synthesis of a completely biobased pressure sensitive adhesive (PSA) with tunable viscoelastic properties from mixtures of ESO and sebacic acid was proposed; the rheological profile and gel content of the obtained PSA were greatly dependent on the curing conditions, but the authors found the conditions that guarantee the best balance between tack and cohesion [66].

Epoxy resins obtained by EVOs have, in general, low T_g. Because of it and because the reducing availability of aromatic petrol-based molecules due to the shift from oil to shale gas, the concept of developing biobased aromatic molecules is ever more increasing. The most widespread renewable aromatic source is lignin, a heavy and complex organic polymer consisting mainly of phenolic compounds (Figure 6.7) but that possesses the strong drawback of being difficult to isolate. Its aromatic structure confers relatively good thermostability and good mechanical properties [5]. Thanks to the variety of groups and molecules that can be obtained from lignin, including aldehydes, it is a good candidate to substitute formaldehyde in phenolic resins. In epoxy resins production, lignin can be used by directly mixing it with some epoxy resins, or after some modifications, as direct glycidylation.

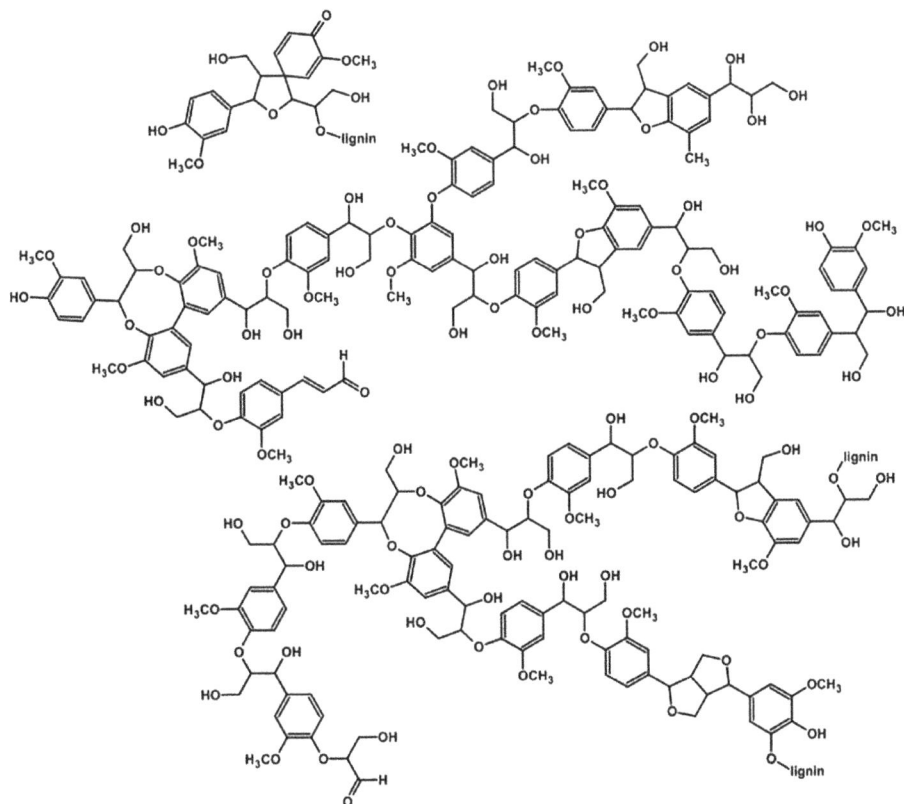

Figure 6.7: Structural motifs of softwood lignin [67].

Fully biobased resins have been obtained starting from partial depolymerized lignins (PDL), subsequently glycidylated and cross-linked using a biobased anhydride as curing agent. The latter was the Diels–Alder adduct produced from condensation of maleic anhydride and methyl ester of oleostearic acid (a major tung oil fatty acid) [68]. Even though PDL-based resins showed a lower reactivity, their mechanical properties and thermal stability were found superior to those of the petrol-derived resins. The kind of feedstock used for the lignin extraction influences its composition and therefore the final properties of the derived epoxy resins. This effect of the lignin origin is clearly showed by Asada et al. who used low molecular weight lignins obtained from different lignocellulosic biomasses (cedar, bamboo, and eucalyptus) further cured with two curing agents: the lignin itself, obtaining high biobased epoxy resins, and a petrol-derived phenol novolac [69]. Lignins from steam-exploded bamboo were also used as curing agents but the resulting material showed poor thermal properties and a decrease in the material transparency [70]. Gioia et al. reported the synthesis of thermosetting resins from low molar mass

Kraft lignins fractions of high functionality, refined by solvent extraction, cross-linked with polyetheramine D2000 [71]. The so-obtained epoxy resins exhibited varied thermal and mechanical properties as a function of the lignin fraction's molecular weight. Lignin-based resins with tunable properties could be obtained by reacting epoxidized lignins with a comacromonomer consisting of polyethylene glycol and two different amines; the PEG chain acts as a soft segment to modulate the polymer characteristics, while the amines are used as chain extender and cross-linkers [72].

Lignins could be depolymerized, and their constitutive elements can be further functionalized, in particular glycidylated. Aouf et al. functionalized gallic acid obtaining a mixture of tri and tetraglycidyl ether derivatives, thereafter cross-linked with isophorone diamine (IPDA) and produced epoxy resins with higher cross-linking density, higher T_g, and a higher char yield than the DGEBA networks cured in the same conditions [73]. Starting from vanillin derivatives, different epoxy monomers, diglycidyl ether of methoxyhydroquinone, diglycidylether of vanillic acid, and diglycidyl ether of vanillyl alcohol can be produced and used in epoxy resins production by reacting with IPDA. Thermomechanical properties of the materials obtained were investigated, in comparison with an industrial DGEBA–IPDA, showing comparable results [74]. An interesting work reports the formation of an epoxy molecule by linking two epoxidated eugenol molecules on p-xylene, subsequently cured with diaminodiphenyl methane [75]. The obtained epoxy resin exhibits a significant increase in Young's module, hardness, and especially char yield, compared with the DGEBA-based resin.

Terpenes are other valuable bioderived platform molecules, with aromatic or unsaturated cyclic structures, that are able (even if hardly) to be epoxidized. For example, p-cymene was used to produce diepoxide and a diamine, which were used in combination with the petrol-derived DGEBA and methylenedianiline to produce epoxy resins with reduced moisture uptake [76].

Further molecules that can lead to epoxy networks with interesting properties are monomers derived from carbohydrates, such as sucrose, sorbitol, maltitol, and isosorbide. The most important polysaccharide, especially in biobased materials production, is cellulose, composed by D-glucose units, that constitutes, together with hemicellulose and lignin, the structural component of the vegetal cell.

Carbohydrate-based polycondensates have the great advantage of a lower toxicity and a higher susceptibility to biodegradation, compared to those coming from petrochemical feedstock; nevertheless, they typically show also increased hydrophilicity. This is the case of 1,4:3,6-dianhydrohexitols based epoxy resins. Epoxidized 1,4:3,6-dianhydrohexitols are the carbohydrate-based monomer most widely employed in the synthesis of epoxy resins thanks to their availability, especially regarding isosorbide. Realization of isosorbide-based epoxy resins was proved by Feng et al. but at the same time the high hydrophilicity of this material was also proved [77]. Typically, this characteristic is not considered as a desirable property, but Park et al. proved that this feature can be used in a proper mode proving the antifogging behavior (depending on thickness of films and lasting only for few seconds) of this water-adsorbing material [78]. A very interesting fully isosorbide-based cross-linked material was obtained by the reaction of diglycidyl ether of isosorbide with an isosorbide diamine [79]. If cured together, these two molecules give a shape-memory fully biobased epoxy resin with a recovery factor (R_f) of about 97% during thermomechanical cycling. Starting from D-glucose and methyl-α-D-glucopyranoside, mixtures of bi-, tri-, and tetraglucopyranoside-based and a trifunctional glucofuranoside-based epoxies have been obtained and then cured with 4,4′-diaminodiphenyl methane resulting in epoxy resins with properties depending on the glucofuranoside functionality and an hardness similar to that of DGEBA-based resins [80]. Darroman et al. used sucrose, isosorbide, and sorbitol epoxy derivatives to improve the properties of materials derived from epoxidized cardanol, which was treated with different amines, and studied to potentially replace DGEBA [81]. Thermogravimetric analyses showed that the thermal stability of DGEBA was higher than that of other reagents; however, epoxidized isosorbide could be a suitable replacement. Improved mechanical properties were obtained in the case of itaconic acid-based epoxy resins, derived from the reaction between trifunctional itaconic-based epoxy and two different curing agents: poly(propylene glycol) bis(2-aminopropyl ether) and methyl hexahydrophthalic anhydride.

Other important categories of sugar-derived molecules are the furans, obtained by dehydration of pentose and hexose sugars, which exhibit great potential as platform chemicals in the biorefinery. The interest on furanic molecules is increasing in the last years leading to ever-improved technology for production of furfural and hydroxymethylfurfural (HMF) from C5 and C6 sugars or polysaccharides, respectively. These platform molecules can undergo further transformation resulting in a widespread variety of molecules with all kinds of functionalities [82]. Nowadays the greatest industrial interest is posed on oxidation of HMF into 2,5-FDCA, considered as one of the most promising platform molecules [10]. FDCA can be epoxidized following different routes: allylation followed by epoxidation, glycidylation, or transesterification [83, 84]. The resulting furan diglycidyl ester, cured with amines or anhydrides, leads to resins with improved T_g and equivalent mechanical properties and thermal stability to those of its petrol-derived counterpart benzene diglycidyl ester [84]. HMF can also be reduced to furan dimethanol, whose glycidylated derivative, furan diglycidyl ether, possess a more flexible structure of the ester. This flexibility is reflected into a T_g decrease when the ether is cured with methyl nadic anhydride, even if the thermal and mechanical properties of the resulting epoxy resins can be tailored by managing the epoxy/anhydride ratio [85]. Feasibility of UV-cured epoxy resins based on the diglycidyl ether of furan as adhesive was tested, and the resulting material exhibited good tensile strengths, but its properties were affected by its hydrophobic nature [86]. Multiring bisphenol–furan molecules can be synthesized from lignin-derived phenols, such as 4-methylcatechol, and carbohydrate-derived furanic molecules, such as 5-HMF and furfural, used as polymerization precursors [87]. After curing with diethylenetriamine, the furan-based epoxy resins show good mechanical and thermal properties.

6.5 Phenolic resins

The phenolic resins are not only the earliest thermosetting resins but also the first synthetic polymers ever produced. In 1907, Baekland obtained Bakelite by the poly-condensation reaction between phenol and formaldehyde [88]. The most common phenolic resin is the phenol-formaldehyde (PF) made by condensation of the simplest phenolic and aldehydic molecules (i.e., phenol and formaldehyde) but in general they can be obtained also from phenol derivatives and any aldehyde. They are used as coatings, adhesives, parts of appliances, and thermal and acoustic insulators, thanks to their good stiffness, shape stability, electrical characteristics, and excellent heat and solvent resistance. Depending on the reaction conditions, PF resins are conventionally classified in resols and novolacs. The resols are synthesized in alkaline conditions with a formaldehyde/phenol molar ratio > 1 (usually 1.5) yielding insoluble thermosets, whereas the novolacs are obtained in acidic medium using a molar ratio < 1 (generally 0.5–0.8) and are thermoplastic materials [1]. Unfortunately, despite their good properties such resins are obtained from toxic and/or carcinogenic molecules: phenol and cresols are toxic by inhalation and skin contact while formaldehyde is classified as a carcinogenic–mutagenic–reprotoxic) substance. In particular, formaldehyde is very toxic because of its high volatility, and reduction of exposure to this substance has been recommended by ECHA [89]. Over the years, greatest efforts have been made to replace the phenolic molecules in this type of resins; the easiest solution is to use the most widespread aromatic structure in nature, that is lignin. Fewer solutions were found instead to substitute the formaldehyde. Lignins reactivity is lower compared to the phenol, but if chemically modified, its reactivity can be increased resulting in lignin-PF resol with higher thermal stability [90]. Li et al. proposed a modification of lignins by means of NaOH/urea solution which induced both depolymerization and functionalization, with final effects of increasing the curing rate and bonding strength together with a reduction in formaldehyde emission compared to classical alkali lignin-PF [91]. The effect of both acidic and alkaline organosolv treatments of pulping of bamboo were compared in a preliminary study where the extracted lignins were used for the synthesis of lignin-PF resins [92]. A solid acid catalytic phenolation treatment for enzymatic hydrolysis of lignins is a valuable process for the lignin depolymerisation, which also increases its reactivity with formaldehyde in foams production [93].

The excellent property of heat resistance typical of phenolic resins is fully exploited in the production of foams for thermal insulation application. As an example, lignin-based birch bark-derived oil was used to substitute up to 50 wt% of lignin in the production of PF foams resulting in good mechanical properties, cell structure, and thermal conductivity [94]. Analogously, raw kraft lignins were used, in the percentages as above, in pentane/hexane blown foams yielding comparable fire resistance as PF foams but reduced emission of carbon monoxide on combustion [95]. In a very recent paper, improved cell morphology of larch tannin-based rigid composite foams were achieved by adding cork powder, the obtained material having improved not only the morphology but also the thermal insulation and thermal stability [96].

Lignins properties strongly depend on their feedstocks, so various papers are presented in literature in which petrol-derived phenols are partially substituted by different kinds of lignin-derived phenols. Kraft and soda lignins from palm fruit bunch (a waste of palm oil industry) have been proposed as partial substitute of phenols in PF resins; kraft lignins, bearing a high quantity of hydroxyl groups, have high reactivity with formaldehyde resulting in resins with a high thermal stability [97]. Tannin-based phenolic resins have been obtained by partially substituting phenol with depolymerized *Acacia mangium* tannin to obtain a good adhesive mimicking the chemical structure of mussel adhesive proteins, with fast curing rate and high thermal resistance [98].

Even more difficult is to find a valuable substitute to the highly reactive formaldehyde. Glyoxal, a dialdehyde obtainable from different natural resources, can be used in PF adhesive production where also phenol is substituted by organosolv lignins from oil palm fronds, which improves the adhesion properties of the material [99], or by soda and kraft lignins from pulping of Kenaf, improving mechanical properties [100]. Biobased composites with sisal fibers have been also produced, where the matrix is given by a glyoxal-phenol resin [101]. Foyer et al. set up functionalization reactions for the preparation of aromatic dialdehydes starting by 4-hydroxybenzaldehyde and vanillin, but this system showed low reactivity, due to the presence of hydroxyl groups in paraposition to the aldehyde function, both in basic and acid conditions for the synthesis of resol and novolac resins, respectively. The resulting aldehydes, reacted with phenol with an appropriate curing program, produce black-colored rigid polymers almost totally insoluble with enhanced thermal stability and high char yield [102, 103]. Furfural was also evaluated as partial substitute of formaldehyde in order to increase the cross-linking density of biooil-PF adhesives [104].

6.6 Conclusions

Thermosetting polymers are a class of materials characterized by high versatility, thanks to the variety of suitable chemical structures that make possible tuning of their properties based on the application field. Nowadays, their use and production are strongly affected by sustainability and toxicity concerns. Therefore, in the last two decades there has been a significant growth in using bioderived materials both in academic and industrial environments, driven by the necessity of replacing toxic precursors and reducing fossil fuel consumption. Significant advancements in this field are possible thanks to the availability of abundant resources from biomasses and the ever-increasing activity and specificity of cleaner synthetic routes, which are expected to provide better results in this field in the years to come.

Abbreviations

UPE	unsaturated polyester
UPR	unsaturated polyester resin
RD	reactive diluents
VOC	Volatile Organic Compounds
ECHA	European Chemicals Agency
EPA	United States Environmental Protection Agency
HAS	Hazardous Air Pollutants
PU	polyurethane
NIPU	non-isocyanate polyurethane
PHU	polyhydroxyurethanes
PF	phenol-formaldehyde

References

[1]   Guo, Q. (2019). Thermosets. Elsevier.
[2]   Pascalut, J.-P., Verdu, J., and Williams, R. J. J. (2002). Thermosetting Polymers. MarcelDekker, Inc.
[3]   European Chemicals Agency. https://echa.europa.eu/legislation.
[4]   United States Environmental Protection Agency. https://www.epa.gov/laws-regulations.
[5]   Auvergne, R., Caillol, S., David, G., Boutevin, B., and Pascault, J. P. (2014). Biobased thermosetting epoxy: present and future. Chem. Rev. 114(2), 1082–1115.
[6]   Pan, X., Sengupta, P., and Webster, D. C. (2011). High biobased content epoxy -anhydride thermosets from epoxidized sucrose esters of fatty acids. Biomacromolecules 12(6), 2416–2428.
[7]   Witt, U., Yamamoto, M., Seeliger, U., Müller, R.-J., and Warzelhan, V. (1999). Biodegradable polymeric materials-not the origin but the chemical structure determines biodegradability. Angew. Chem. Int. Ed. 30(10), 1438–1442.
[8]   Jones, F. R. (2017). Unsaturated polyester resins. In: Brydson's Plastics Materials: Eighth Edition. Elsevier, 743–772.
[9]   EPA – Hazardous Air Pollutants. https://www.epa.gov/haps.
[10]   Werpy, T., and Petersen, G. (2004). Top Value Added Chemicals from Biomass Volume I – Results of Screening for Potential Candidates from Sugars and Synthesis Gas. U.S. Department of Energy.
[11]   Sadler, J. M., Toulan, F. R., Nguyen, A.-P. T., Kayea, R. V. I., Ziaee, S., Palmese, G. R., and La Scala, J. J. (2014). Isosorbide as the structural component of bio-based unsaturated isosorbide as the component of bio-based unsaturated. Carbohydr. Polym. 100, 97–106.
[12]   Can, E., Kinaci, E., and Palmese, G. R. (2015). Preparation and characterization of novel vinyl ester formulations derived from cardanol. Eur. Polym. J. 72, 129–147.
[13]   Campanella, A., Zhan, M., Watt, P., Grous, A. T., Shen, C., and Wool, R. P. (2015). Triglyceride-based thermosetting resins with different reactive diluents and fiber reinforced composite applications. Composites: Part A 72, 192–199.
[14]   Fei, M.-e., Liu, W., Jia, A., Ban, Y., and Qiu, R. (2018). Bamboo fibers composites based on styrene -free soybean-oil thermosets using methacrylates as reactive diluents. Composites Part A 114, 40–48.

[15] Wu, Y., and Li, K. (2016). Replacement of styrene with acrylated epoxidized soybean oil in an unsaturated polyester resin from propylene glycol, isophthalic acid, and maleic anhydride. J. Appl. Polym. Sci. 133, 43052.

[16] Wu, Y., and Li, K. (2017). Replacement of styrene with acrylated epoxidized soybean oil in an unsaturated polyester resin from propylene glycol and maleic anhydride. J. Appl. Polym. Sci. 134, 45056.

[17] Wu, Y., and Li, K. (2018). Acrylated epoxidized soybean oil as a styrene replacement in a dicyclopentadiene-modified unsaturated polyester resin. J. Appl. Polym. Sci. 135, 46212.

[18] Liu, W., Xie, T., and Qiu, R. (2017). Biobased thermosets prepared from rigid isosorbide and flexible soybean oil derivatives. ACS Sustainable Chem. Eng. 5, 774–783.

[19] Panic, V. V., Seslija, S. I., Popovic, I. G., Spasojevic, V. D., Popovic, A. R., Nikolic, V. B., and Spasojevic, P. M. (2017). Simple one-pot synthesis of fully biobased unsaturated polyester resins based on itaconic acid. Biomacromolecules 18, 3881–3891.

[20] Fidanovski, B. Z., Spasojevic, P. M., Panic, V. V., Seslija, S. I., Spasojevic, J. P., and Popovic, I. G. (2018). Synthesis and characterization of fully bio-based unsaturated polyester resins. J. Mater. Sci. 53, 4635–4644.

[21] Dai, Z., Yang, Z., Chen, Z., Zhao, Z., Lou, Y., Zhang, Y., Liu, T., Fu, F., Fu, Y., and Liu, X. (2018). Fully biobased composites of an itaconic acid derived unsaturated polyester reinforced with cotton fabrics. ACS Sustainable Chem. Eng. 6, 15056–15063.

[22] Fidanovski, B. Z., Popovic, I. G., Radojevic, V. J., Radisavljevic, I. Z., Perisic, S. D., and Spasojevic, P. M. (2018). Composite materials from fully bio-based thermosetting resins and recycled waste poly(ethylene terephthalate). Composites Part B 153, 117–123.

[23] Mehta, L. B., Wadgaonkar, K. K., and Jagtap, R. N. (2019). Synthesis and characterization of high bio-based content unsaturated polyester resin for wood coating from itaconic acid: effect of various reactive diluents as an alternative to styrene. J. Dispersion Sci. Technol. 40 (5), 756–765.

[24] Pellis, A., Hanson, P. A., Comeford, J. W., Clark, J. H., and Farmer, T. J. (2019). Enzymatic synthesis of unsaturated polyesters: functionalization and reversibility of the aza-Michael addition of pendants. Polym. Chem. 10, 843.

[25] Xu, Y., Hua, G., Hakkarainen, M., and Odelius, K. (2018). Isosorbide as core component for tailoring biobased unsaturated polyester thermosets for a wide structure–property window. Biomacromolecules 19, 3077–3085.

[26] Dai, J., Ma, S., Teng, N., Dai, X., Shen, X., Wang, S., Liu, X., and Zhu, J. (2017). 2,5-furandicarboxylic acid- and itaconic acid-derived fully biobased unsaturated polyesters and their cross-linked networks. Ind. Eng. Chem. Res. 56, 2650–2657.

[27] Yu, A. Z., Serum, E. M., Renner, A. C., Sahouani, J. M., Sibi, M. P., and Webster, D. C. (2018). Renewable reactive diluents as practical styrene replacements in biobased vinyl ester thermosets. ACS Sustainable Chem. Eng. 6, 12586–12592.

[28] Cousinet, S., Ghadban, A., Fleury, E., Lortie, F., Pascault, J.-P., and Portinha, D. (2015). Toward replacement of styrene by bio-based methacrylates in unsaturated polyester resins. Eur. Polym. J. 67, 539–550.

[29] Le Nôtre, J., Witte-van Dijk, S. C. M., van Haveren, J., Scott, E. L., and Sanders, J. P. M. (2014). Synthesis of bio-based methacrylic acid by decarboxylation of itaconic acid and citric acid catalyzed by solid transition-metal catalysts. Chem. Sus. Chem. 7, 2712–2720.

[30] Heath, R. (2016). Isocyanate-based polymers: polyurethanes, polyureas, polyisocyanurates, and their copolymers. In: Brydson's Plastics Materials. Elsevier Ltd, 799–835.

[31] Cheng, Z., Li, Q., Yan, Z., Liao, G., Zhang, B., Yu, Y., Yi, C., and Xu, Z. (2019). Design and synthesis of novel aminosiloxane crosslinked linseed oil -based waterborne polyurethane composites and its physicochemical properties. Prog. Org. Coat. 127, 194–201.

[32] Feng, Y., Man, L., Hu, Y., Chen, L., Xie, B., Zhang, C., Yuan, T., and Yang, Z. (2019). One-pot synthesis of polyurethane-imides with tailored performance from castor and tung oil. Prog. Org. Coat. 132, 62–69.

[33] Kirpluks, M., Kalnbunde, D., Benes, H., and Cabulis, U. (2018). Natural oil based highly functional polyols as feedstock for rigid polyurethane foam thermal insulation. Ind. Crops Prod. 122, 627–636.

[34] Liang, H., Yechang, F., Lu, J., Liu, L., Yang, Z., Luo, Y., Zhang, Y., and Zhang, C. (2018). Bio-based cationic waterborne polyurethanes dispersions prepared from different vegetable oils. Ind. Crops Prod. 122, 448–455.

[35] Liang, H., Wang, S., He, H., Wang, M., Liu, L., Lu, J., Zhang, Y., and Zhang, C. (2018). Aqueous anionic polyurethane dispersions from castor oil. Ind. Crops Prod. 122, 182–189.

[36] Wang, X., Zhang, Y., Liang, H., Zhou, X., Fang, C., Zhang, C., and Luo, Y. (2019). Synthesis and properties of castor oil-based waterborne polyurethane/sodium alginate composites with tunable properties. Carbohydr. Polym. 208, 391–397.

[37] Huo, S., Wu, G., Chen, J., Liu, G., and Kong, Z. (2016). Constructing polyurethane foams of strong mechanical property and thermostability by two novel environment friendly bio-based polyols. Korean J. Chem. Eng. 33(3), 1088–1094.

[38] Vandenplas, O., Malo, J.-L., Saetta, M., Mapp, C. E., and Fabbri, L. M. (1993). Occupational asthma and extrinsic alveolitis due to isocyanates: current status and perspectives. Br. J. Ind. Med. 50, 213–228.

[39] Jarvis, J., Sed, M. J., Elton, R. A., Sawyer, L., and Agius, R. M. (2005). Relationship between chemical structure and the occupational asthma hazard of low molecular weight organic compounds. Occup. Environ. Med. 62, 243–250.

[40] ECHA – Diisocyanate Restrictions. https://echa.europa.eu/registry-of-restriction-intentions /-/dislist/details/0b0236e180876053.

[41] https://www.epa.gov/sites/production/files/2019-03/documents/ry_2018_tri_chemical_ list.pdf.

[42] Suryawanshi, Y., Sanap, P., and Wani, V. (2019). Advances in the synthesis of non-isocyanate polyurethanes. Polym. Bull. 76, 3233.

[43] Błażek, K., and Datta, J. (2019). Renewable natural resources as green. Crit. Rev. Environ. Sci. Technol 49(3), 173–211.

[44] Rehman, A., López Fernández, A. M., Gunam Resul, M., and Harvey, A. (2019). Highly selective, sustainable synthesis of limonene cyclic carbonate from bio-based limonene oxide and CO2: a kinetic study. J. CO2 Util. 29, 126–133.

[45] Hu, S., Chen, X., and Torkelson, J. M. (2019). Bio-based reprocessable polyhydroxyurethane networks: full recovery of cross-link density with three concurrent dynamic chemistries. ACS Sustainable Chem. Eng.

[46] Poussard, L., Mariage, J., Grignard, B., Detrembleur, C., Jérôme, C., Calberg, C., Heinrichs, B., De Winter, J., Gerbaux, P., Raquez, J.-M., Bonnaud, L., and Dubois, P. (2016). Non-isocyanate polyurethanes from carbonated soybean oil using monomeric or oligomeric diamines to achieve thermosets or thermoplastics. Macromolecules 49(6), 2162–2171.

[47] Mokhtari, C., Malek, F., Manseri, A., Caillol, S., and Negrell, C. (2019). Reactive jojoba and castor oils-based cyclic carbonates for biobased polyhydroxyurethanes. Eur. Polym. J. 113, 18–28.

[48] Carré, C., Bonnet, L., and Avérous, L. (2015). Solvent- and catalyst-free synthesis of fully biobased nonisocyanate polyurethanes with different macromolecular architectures. RSC Adv 5, 100390.

[49] Yu, A. Z., Setie, R. A., Sahounani, J. M., Docken, J. J., and Webster, D. C. (2019). Catalyzed non-isocyanate polyurethane (NIPU) coatings from bio-based poly(cyclic carbonates). J. Coat. Technol. Res. 16(1), 41–57.

[50] Xi, X., Wu, Z., Pizzi, A., Gerardin, C., Lei, H., Zhang, B., and Du, G. (2019). Non-isocyanate polyurethane adhesive from sucrose used for particleboard. Wood Sci. Technol 53(2), 393–405.

[51] Zhang, L., Luo, X., Qin, Y., and Li, Y. (2017). A novel 2,5-furandicarboxylic acid -based bis (cyclic carbonate) for the synthesis of biobased non-isocyanate polyurethanes. RSC Adv 7, 37–46.

[52] Lee, H., and Neville, K. (1967). Handbook of Epoxy Resins. McGraw-Hill.

[53] Henry, L., and Kris, N. (1967). Handbook of Epoxy Resins. McGraw-Hill.

[54] Vandenberg, L. N., Ehrlich, S., Belcher, S. M., Ben-Jonathan, N., Dolinoy, D. C., Hugo, E. R., Hunt, P. A., Newbold, R. R., Rubin, B. S., Saili, K. S., Soto, A. M., Wang, H.-S., and Vom Saal, F. S. (2013). Low dose effects of bisphenol A. Endocr. Disruptors 1(1).

[55] Vandenberg, L. N., Hauser, R., Marcus, M., Olea, N., and Welshons, W. V. (2007). Human exposure to bisphenol A (BPA). Reprod. Toxicol. 24(2), 139–177.

[56] ECHA – Agreement of the Member State Committee on the Identification of 4,4'-Isopropylidenediphenol (Bisphenol A) as a Substance of Very High Concern. https://echa.eu ropa.eu/documents/10162/13638/svhc_msc_agreement_bpa_ii_en.pdf/ac9efb97-c06b-d1a7-2823-5dc69208a238.

[57] Radojčić, D., Hong, J., Ionescu, M., Wan, X., Javni, I., and Petrović, Z. S. (2016). Study on the reaction of amines with internal epoxides. Eur. J. Lipid Sci. Technol 118.

[58] Pin, J.-M., Sbirrazuoli, N., and Mija, A. (2015). From epoxidized linseed oil to bioresin: an overall approach of epoxy /anhydride cross-linking. Chem. Sus. Chem. 8(7), 1232–1243.

[59] Pin, J.-M., Guigo, N., Vincent, L., Sbirrazzuoli, N., and Mija, A. (2015). Copolymerization as a strategy to combine epoxidized linseed oil and furfuryl alcohol: the design of a fully bio-based thermoset. Chem. Sus. Chem. 8, 4149–4161.

[60] Albarrán-Preza, E., Corona-Becerril, D., Vigueras-Santiago, E., and Hernández-López, S. (2016). Sweet polymers: synthesis and characterization of xylitol-based epoxidized linseed oil resins. Eur. Polym. J. 75, 539–551.

[61] Sahoo, S. K., Khandelwal, V., and Manik, G. (2018). Development of completely bio-based epoxy networks derived from epoxidized linseed and castor oil cured with citric acid. Polym. Adv. Technol. 29(7), 2080–2090.

[62] Pawar, M., Kadam, A., Yemul, O., Thamke, V., and Kodam, K. (2016). Biodegradable bioepoxy resins based on epoxidized natural oil (cottonseed & algae) cured with citric and tartaric acids through solution polymerization: a renewable approach. Ind. Crops Prod. 89, 434–447.

[63] Carbonell-Verdu, A., Bernardi, L., Garcia-Garcia, D., Sanchez-Nacher, L., and Balart, R. (2015). Development of environmentally friendly composite matrices from epoxidized cottonseed oil. Eur. Polym. J. 63, 1–10.

[64] Liu, R., Zhang, X., Gao, S., Liu, X., Wang, Z., and Yan, J. (2016). Bio-based epoxy -anhydride thermosets from six- armed linoleic acid-derived epoxy resin. RSC Adv 6, 52549–52555.

[65] Boquillon, N., and Fringant, C. (2000). Polymer networks derived from curing of epoxidised linseed oil: influence of different catalysts and anhydride hardeners. Polymer 41, 8603–8613.

[66] Ciannamea, E. M., and Ruseckaite, R. A. (2018). Pressure sensitive adhesives based on epoxidized soybean oil: correlation between curing conditions and rheological properties. J. Am. Oil Chem. Soc. 95(4), 525–532.

[67] Holladay, J. E., White, J. F., Bozell, J. J., and Johnson, D. (2007). Top Value-Added Chemicals from Biomass Volume II – Results of Screening for Potential Candidates from Biorefinery Lignin. U.S. Department of Energy.

[68] Xin, J., Li, M., Li, R., Wolcott, M. P., and Zhang, J. (2016). A green epoxy resin system based on lignin and tung oil and its application in epoxy asphalt. ACS Sustainable Chem. Eng. 4, 2754–2761.

[69] Asada, C., Basnet, S., Otsuka, M., Sasaki, C., and Nakamura, Y. (2015). Epoxy resin synthesis using low molecular weight lignin separated from various lignocellulosic materials. Int. J. Biol. Macromol. 74, 413–419.

[70] Sasaki, C., Wanaka, M., Takagi, H., Tamura, S., Asada, C., and Nakamura, Y. (2013). Evaluation of epoxy resins synthesized from steam-exploded bamboo lignin. Ind. Crops Prod. 43(1), 757–761.

[71] Gioia, C., Lo Re, G., Lawoko, M., and Berglund, L. (2018). Tunable thermosetting epoxies based on fractionated and well-characterized lignins. J. Am. Chem. Soc. 140, 4054–4061.

[72] Salanti, A., Zoia, L., Simonutti, R., and Orlandi, M. (2018). Epoxidized lignin derivatives as bio-based cross-linkers used in the preparation of epoxy Resins. BioResources 13(2), 2374–2396.

[73] Aouf, C., Nouailhas, H., Fache, M., Caillol, S., Boutevin, B., and Fulcrand, H. (2013). Multi-functionalization of gallic acid. Synthesis of a novel bio-based epoxy resin. Eur. Polym. J. 49 (6), 1185–1195.

[74] Fache, M., Auvergne, R., Boutevin, B., and Caillol, S. (2015). New vanillin -derived diepoxy monomers for the synthesis of biobased thermosets. Eur. Polym. J. 67, 527–538.

[75] Wan, J., Gan, B., Li, C., Molina-Aldareguia, J., Kalali, E. N., Wang, X., and Wang, D.-Y. (2016). A sustainable, eugenol -derived epoxy resin with high biobased content, modulus, hardness and low flammability: synthesis, curing kinetics and structure-property relationship. Chem. Eng. J. 284, 1080–1093.

[76] Garrison, M. D., and Harvey, B. G. (2016). Bio-based hydrophobic epoxy -amine networks derived from renewable terpenoids. J. Appl. Polym. Sci. 43621, 1–12.

[77] Feng, X., East, A. J., Hammond, W. B., Zhang, Y., and Jaffe, M. (2011). Overview of advances in sugar-based polymers. Polym. Adv. Technol. 22(1), 139–150.

[78] Park, S., Park, S., Jang, D. H., Lee, H. S., and Park, C. H. (2016). Anti-fogging behavior of water-absorbing polymer films derived from isosorbide -based epoxy resin. Mater. Lett. 180, 81–84.

[79] Li, C., Dai, J., Liu, X., Jiang, Y., Ma, S., and Zhu, J. (2016). Green synthesis of a bio-based epoxy curing agent from isosorbide in aqueous condition and shape memory properties investigation of the cured resin. Macromol. Chem. Phys. 217(13), 1439–1447.

[80] Rapi, Z., Szolnoki, B., Bakó, P., Niedermann, P., Toldy, A., Bodzay, B., Keglevich, G., and Marosi, G. (2015). Synthesis and characterization of biobased epoxy monomers derived from d-glucose. Eur. Polym. J. 67, 375–382.

[81] Darroman, E., Durand, N., Boutevin, B., and Caillol, S. (2015). New cardanol/sucrose epoxy blends for biobased coatings. Prog. Org. Coat. 83, 47–54.

[82] Lewkowski, J. (2005). Synthesis, chemistry and applications of 5-hydroxymethyl-furfural and its derivatives. Arkivoc 2001(1), 17–54.

[83] Marotta, A., Ambrogi, V., Cerruti, P., and Mija, A. (2018). Green approaches in the synthesis of furan-based diepoxy monomers. RSC Adv 8, 16330–16335.

[84] Deng, J., Liu, X., Li, C., Jiang, Y., and Zhu, J. (2015). Synthesis and properties of a bio-based epoxy resin from 2,5-furandicarboxylic acid (FDCA). RSC Adv 5, 15930–15939.

[85] Marotta, A., Faggio, N., Ambrogi, V., Cerruti, P., Gentile, G., and Mija, A. (2019). Curing behavior and properties of sustainable furan-based epoxy /anhydride resins. Biomacromolecules 20, 3831–3841.

[86] Cho, J. K., Lee, J.-S., Jeong, J., and Kim, B. (2013). Synthesis of carbohydrate biomass-based furanic compounds bearing epoxide end group (s) and evaluation of their feasibility as adhesives. J. Adhes. Sci. Technol. 27(18–19), 2127–2138.

[87] Jiang, Y., Ding, D., Zhao, S., Zhu, H., Kenttämaa, H. I., and Abu-Omar, M. M. (2018). Renewable thermoset polymers based on lignin and carbohydrate derived monomers. Green Chem. 20(5), 1131–1138.

[88] Hirano, K., and Asami, M. (2013). Phenolic resins-100 years of progress and their future. React. Funct. Polym. 73, 256–269.

[89] ECHA-Formaldehyde Restriction Intentions. https://echa.europa.eu/it/registry-of-restriction-intentions/-/dislist/details/0b0236e182439477.

[90] Domínguez, J. C., Oliet, M., Alonso, M. V., Rojo, E., and Rodríguez, F. (2013). Structural, thermal and rheological behavior of a bio-based phenolic resin in relation to a commercial resol resin. Ind. Crops Prod. 42, 308–314.

[91] Li, J., Zhang, J., Zhang, S., Gao, Q., Li, J., and Zheng, W. (2018). Alkali lignin depolymerization under eco-friendly and cost-effective NaOH/urea aqueous solution for fast curing bio-based phenolic resin. Ind. Crops Prod. 120, 25–33.

[92] Pang, B., Yang, S., Fang, W., Yuan, T.-Q., Argyropoulos, D. S., and Sun, R.-C. (2017). Structure-property relationships for technical lignins for the production of lignin -phenol-formaldehyde resins. Ind. Crops Prod. 108, 316–326.

[93] Wang, G., Liu, X., Zhang, J., Sui, W., Jang, J., and Si, C. (2018). One-pot lignin depolymerization and activation by solid acid catalytic phenolation for lightweight phenolic foam preparation. Ind. Crops Prod. 124, 216–225.

[94] Li, B., FEng, H., Niasar, H. S., Zhang, Y. S., Yuan, Z. S., Schmidt, J., and Xu, C. (2016). Preparation and characterization of bark-derived phenol formaldehyde foams. RSC Adv. 6 (47), 40975–40981.

[95] Li, B., Yuan, Z., Schmidt, J., and Xu, C. C. (2019). New foaming formulations for production of bio-phenol formaldehyde foams using raw kraft lignin. Eur. Polym. J. 111, 1–10.

[96] Li, J., Zhang, A., Zhanhg, S., Gao, Q., Zhang, W., and Li, J. (2019). Larch tannin-based rigid phenolic foam with high compressive strength, low friability, and low thermal conductivity reinforced by cork powder. Composites Part B 156, 368–377.

[97] Ibrahim, M. N. M., Zakaria, N., Sipaut, C. S., Sulaiman, O., and Hashim, R. (2011). Chemical and thermal properties of lignins from oil palm biomass as a substitute for phenol in a phenol formaldehyde resin production. Carbohydr. Polym. 86, 112–119.

[98] Li, J., Zhu, W., Zhang, S., Gao, Q., Xia, C., Zhang, W., and Li, J. (2019). Depolymerization and characterization of Acacia mangium tannin for the preparation of mussel-inspired fast-curing tannin-based phenolic resins. Chem. Eng. J. 370, 420–431.

[99] Hussin, M. H., Samad, N. A., Latif, N. H. A., Rozuli, N. A., Yusoff, S. B., Gambier, F., and Brosse, N. (2018). Production of oil palm (Elaeis guineensis) fronds lignin -derived non-toxic aldehyde for eco-friendly wood adhesive. Int. J. Biol. Macromol. 113, 1266–1272.

[100] Hussin, M. H., Aziz, A. A., Iqbal, A., Ibrahim, M. N. M., and Latif, N. H. A. (2019). Development and characterization novel bio-adhesive for wood using kenaf core (Hibiscus cannabinus) lignin and glyoxal. Int. J. Biol. Macromol. 122, 713–722.

[101] Ramires, E. C., Megiatto, J. D. J., Gardrat, C., Castellan, A., and Frollini, E. (2010). Biobased composites from glyoxal–phenolic resins and sisal fibers. Biores. Technol. 101, 1998–2006.

[102] Foyer, G., Chanfi, B.-H., Boutevin, B., Caillol, S., and David, G. (2016). New method for the synthesis of formaldehyde -free phenolic resins from lignin -based aldehyde precursors. Eur. Polym. J. 74, 296–309.

[103] Foyer, G., Chanfi, B.-H., Virieux, D., David, G., and Caillol, S. (2016). Aromatic dialdehyde precursors from lignin derivatives for the synthesis of formaldehyde -free and high char yield phenolic resins. Eur. Polym. J. 77, 65–74.

[104] Cheng, Y., Sui, G., Liu, H., Wang, X., Yang, X., and Wang, Z. (2019). Preparation of highly phenol substituted bio-oil–phenol–formaldehyde adhesives with enhanced bonding performance using furfural as crosslinking agent. J. Appl. Polym. Sci. 136, 46995.

Leire Ruiz-Rubio, Julia Sánchez-Bodón, Isabel Moreno,
Leyre Pérez-Álvarez, José Luis Vilas-Vilela

7 Polyester-based biodegradable polymers for commodities

Abstract: The ever-increasing use of polymers in commodities due to their ease of processing and low price, has arisen as one of the current concerns for society. The non-biodegradability of many of the conventional polymers, like polyethylene and polypropylene, have led to serious ecological problems. In contrast, biodegradable polymers (BPs) are naturally recycled, because they can be degraded by the enzymatic action of microorganisms and by abiotic processes. Degradable polymers are sustainable, recyclable, and environmentally friendly plastic materials, which are considered an alternative to traditional polymers. Among them, biodegradable polyesters such as poly(lactic acid) (PLA), poly(glycolic acid) (PGA), polycaprolactone (PCL), poly(hydroxyalkanoates) (PHA) have emerged as examples of technical and economical alternatives that are able to substitute traditional polymers in order to improve the biodegradability of these kind of materials. In this context, this work aims to emphasize the importance of BPs as promoters of environmental responsibility and searching for new and improved disposable materials and biomaterials. The main synthetic routes used for the development of these polymers and the different degradation mechanism described for them are summarized, in addition to their main applications.

Keywords: Biopolyesters, PLA, PCL, biodegradation, biodegradable polymers

7.1 Introduction

Nowadays, polymers play a basic role in everyday life due to their unique properties such as durability, ease of processing, and low price, which confer them a great importance for current society. Conventional polymers, like polyethylene and polypropylene (PP), have led to serious ecological problems due to their total nonbiodegradability,

---

Julia Sánchez-Bodón, Grupo de Química Macromolecular (LABQUIMAC), Departamento de Química Física, Facultad de Ciencia y Tecnología, Universidad del País Vasco UPV/EHU, Leioa, Spain

Leire Ruiz-Rubio, Leyre Pérez-Álvarez, José Luis Vilas-Vilela, Grupo de Química Macromolecular (LABQUIMAC), Departamento de Química Física, Facultad de Ciencia y Tecnología, Universidad del País Vasco UPV/EHU, Leioa, Spain; BCMaterials, Basque Centre for Materials, Applications and Nanostructures, UPV/EHU Science Park, Leioa, Spain

Isabel Moreno, Grupo de Química Macromolecular (LABQUIMAC), Departamento de Química Orgánica II, Facultad de Ciencia y Tecnología, Universidad del País Vasco UPV/EHU, Leioa, Spain

https://doi.org/10.1515/9783110590586-007

which makes their recycling profitless and impractical. Thus, these petroleum-based polymers are not appropriate for short time usages and then being disposed. In contrast, biodegradable polymers (BPs) are naturally recycled, because they can be degraded by the enzymatic action of microorganisms and by nonenzymatic processes like chemical hydrolysis that convert them to CO_2, CH_4, water, and biomass.

Consequently, degradable polymers have gained increasing interest as a part of sustainable, recyclable, and environmentally friendly plastic materials, representing an alternative to traditional polymers. Biodegradable polymers can be classified according to their origin as polymers from renewable sources and as synthetic polymers of nonrenewable/fossil resources. These last ones are mainly obtained from mineral oil derived polymers that are biodegradable. Among biodegradable synthetic polymers, poly(lactic acid) (PLA), poly(glycolic acid) (PGA), polycaprolactone (PCL), poly(hydroxyalkanoates) (PHA), and starch-based blends have attracted much attention in the last decades.

Aliphatic polyesters are one of the characteristic examples of BPs, owing to their hydrolyzable ester linkage. These polymers have excellent biodegradability, biocompatibility, are water resistant, and can be easily extruded into a wide range of shapes. This has enabled the presence of synthetic polyesters in the current markets. In addition, several aliphatic polyesters get similar materials properties to conventional polymers despite their poor mechanical and physical properties that can be overcome by blending or copolymerization.

The biodegradation of aliphatic polyesters was discovered in the earliest 1960s, when the poor thermal and hydrolytic stability of PGA was evidenced [1]. Later, this disadvantage was exploited to become an advantageous characteristic, and PGA was employed to develop sutures that degraded in the human body [2]. At the same time, it was shown that, PCL was degraded in soil, whereas PHA commercial production did not take place until 1980s, when PHA produced by microbial fermentation was used to fabricate shampoo bottles. The high price of PHA has been diminished along the 1990s since the genetic alteration of plants whose metabolic pathway lead to PHA accumulation has been succeeded [2].

Degradation time and mechanical behavior are some of the most important properties of BPs. Degradation time is expected to be as short as possible after disposal, typically, it can range from months to years, and can be strongly affected by properties of the polymer, such as crystallinity [3]. Nevertheless, it must be optimized because high-degradation capacities can be detrimental to corrosion resistance. Regarding tensile properties there are great differences between polyesters, for instance, PGA and PLA are especially strong while PCL is a soft polymer.

In this review, the most important synthetic methods and degradation mechanisms of aliphatic polyesters (in particular polylactides) as well as current and promising applications in various fields are described. This work aims to emphasize the importance of BPs as promoters of environmental responsibility and searching for new and improved disposable materials and biomaterials.

7.2 Synthesis

There are two typical approaches for the synthesis of polyesters. The first one, the traditional method, consists of the condensation of a compound with an acid group, or an ester derivative, on each end with a diol, or the polymerization of a hydroxyl acid derivative. This method suffers from different shortcomings, which frequently can be successfully overcome by the second methodology, that is, the polymerization through the ring opening of a lactide or a lactone.

7.2.1 Polycondensation of carboxylic acids or derivatives with alcohols

Direct polycondensation of combinations of diols and carboxylic acids, with or without catalyst, can be employed to obtain biodegradable polyesters, although the principal disadvantages of this methodology are the low-molecular weight of the polymer, due to the difficulty of removing water from the highly viscous reaction mixture, the high temperatures required and the broad molecular weight distribution. Consequently, the polymer obtained by this methodology possesses inferior-mechanical properties.

Different strategies have been reported in the literature to improve the low-molecular weight of the PLA obtained by polycondensation. As an example, Chen et al. reported the employment of titanium(IV) butoxide as catalyst while the decompression of the reaction pressure went on, and so, the esterification time was extended to 7 h [4]. Kimura and coworkers, described a new synthetic route for high molecular weight poly(L-lactic acid) by melt/solid polycondensation of L-lactic acid. First, a typical polycondensation process using tin(II) chloride hydrate/p-toluenesulfonic acid system was carried out to obtain a polymer with a molecular weight of 2×10^4 g · mol^{-1}. The so-obtained polycondensate was crystallized and heated at 150 °C for further polycondensation since solid state the polymerization reaction is favored over side reactions such as depolymerization [5].

Kobayashi et al. reported the first enzyme-catalyzed polymerization using a sugar alcohol as monomer. In this work, sorbitol and divinyl sebacate were employed to obtain regioselectively a sugar-based polyester. The process was successfully catalyzed by lipase derived from *Candida antarctica* (Scheme 7.1) [6].

In this context, stereoisomeric 1,4:3,6-dianhydrohexitols DAG, DAM, and DAI (Figure 7.1), obtained by double dehydration of the associated sugar, have been used as diol components in the synthesis of a variety of biodegradable polyesters [7].

The incorporation of aromatic moieties in the polyester chain is one of the effective approaches to improve the thermal and mechanical properties of aliphatic polyesters [8]. In this context, direct polycondensation of succinic acid, 1,4-butanediol, and dimethyl terephthalate was employed to prepare a series of copolyesters [9].

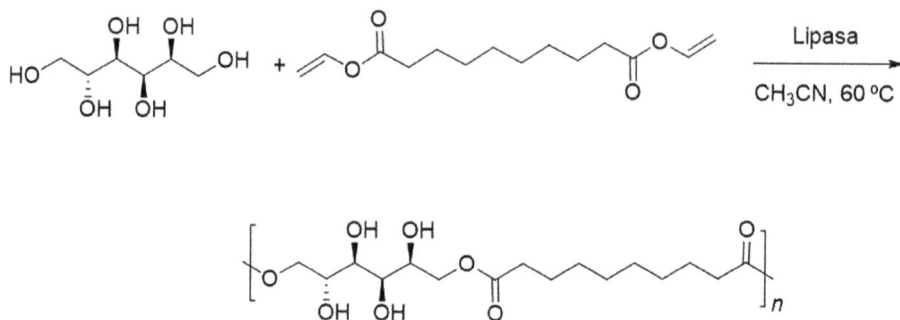

Scheme 7.1: Enzyme-catalyzed polymerization using sorbitol as a monomer.

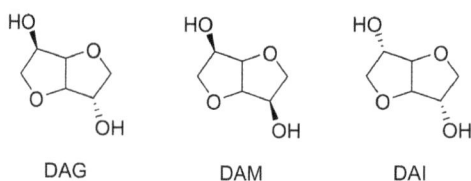

DAG   DAM   DAI

Figure 7.1: Structures of DAG, DAM, and DAI.

The difuranic diesters bis(5-(ethoxycarbonyl)-2-furyl)methane (BFM) and 1,1-bis (5-(ethoxycarbonyl)-2-furyl)ethane (BFE) have been used in the synthesis of diverse polycondensates. Therefore, a two-step transesterification procedure was applied to combinations of BFM and BFE and both aliphatic and furanic diols (Scheme 7.2), DAG and oligo(ethylene glycol)s [10, 11].

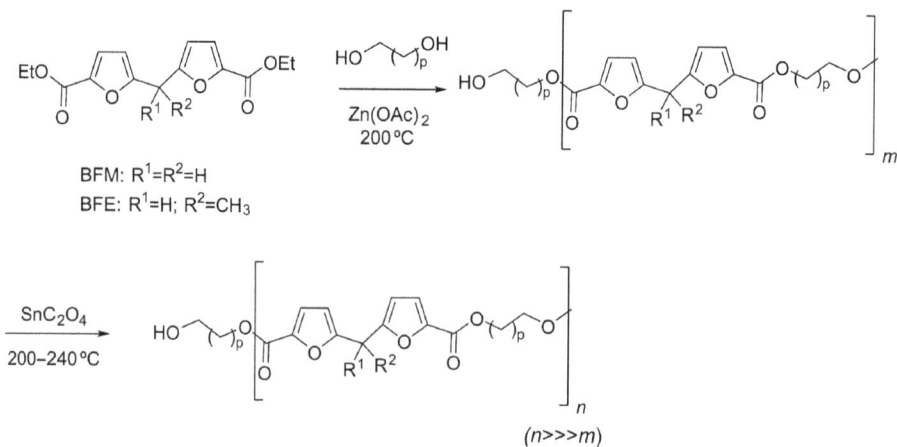

BFM: R^1=R^2=H
BFE: R^1=H; R^2=CH$_3$

(n>>>m)

Scheme 7.2: Polycondensate obtained by a two-step transesterification procedure.

On the other hand, 2,5-furanedicarboxylic acid has gained much interest in poly-condensates. In particular, it was found to be a possible substitute of terephthalic acid in aromatic polyesters such as poly(ethyleneglycol terephthalate) (PET), poly (butyleneglycol terephthalate) (PBT), or PTT.

Furthermore, poly(alkenedicarboxylate) polyesters are obtained by polyconden-sation reactions of glycols. Aliphatic dicarboxylic acid such as succinic acid and adipic acid reacted with ethylene glycol or 1,4-butanediol [12]. Yang and coworkers described the synthesis of poly(butylene succinate-co-adipate) copolymer using 1,4-butanediol [13]. As shown in Scheme 7.3, the synthesis consists of two main steps/reactions, the first one is an esterification reaction and the second is deglycolization.

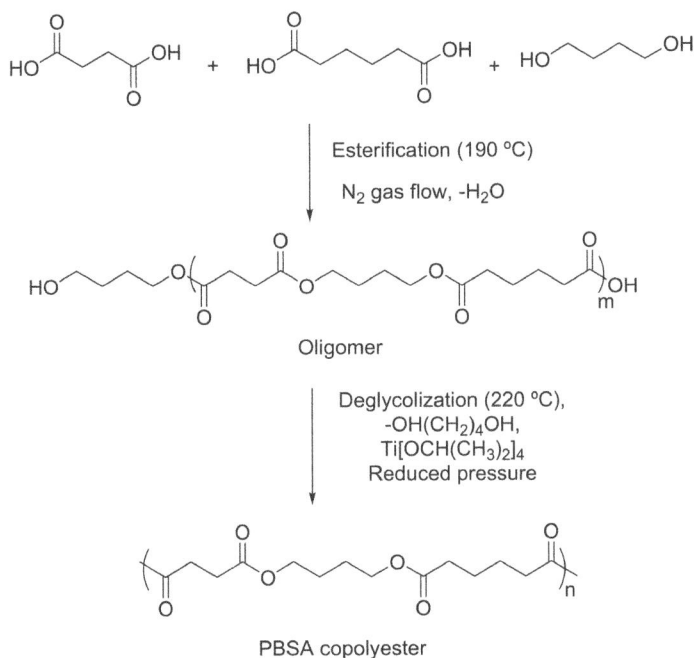

Scheme 7.3: Synthesis of PBSA copolyesters from succinic acid and adipic acid with 1,4-butanediol.

7.2.2 Polyesters by ring-opening polymerization of lactones and lactides

Ring-opening polymerization (ROP) of lactones and lactides is an effective method for synthesizing biodegradable polyesters. The first work concerning the use of this type of polymerization was reported by Carothers and coworkers in the third decade of the past century [14]. The principal advantage of this methodology comparing to polycondensation of acids or derivatives is the no requirement of equimolecular balance of functional groups to obtain a high–molecular-weight polymer. In addition, ROP takes place without side reactions and under milder reaction conditions, for example, the polymerization can be conducted at room temperature because the activation energy for chain growth is generally rather low, and, therefore, the variation of the rate of propagation with temperature is not very large.

The general reaction of the polymerization by ring opening of lactones with a catalyst or initiator is depicted in Scheme 7.4. The so-obtained polymer contains a functional group derived from the initiator and, in the other end of the chain, another functional group derived from the termination step of the polymerization process. These ending groups determine the thermal and hydrolytic stabilities of the resultant polyester. So, by modifying the catalyst and the conditions of the termination step, a series of different polymers with different structures and so many applications can be synthesized [15, 16].

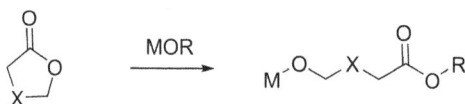

Scheme 7.4: General reaction of ROP.

The polymerization reaction can be explained by three principal mechanisms depending on the catalyst: anionic, cationic, and coordination insertion ROP. Additional mechanisms have been described such as the radical pathway or the zwitterionic mechanism, however, the low molecular weight of the obtained polyester explains the small number of applications of these methods.

In the anionic pathway, the carbonyl carbon of the monomer is attacked by an anionic species, inducing the acyl-oxygen bond cleavage, and therefore, the opening of the cycle and the formation of the alkoxide intermediate as growing species. Regarding the molecular weight of the polymer, good results are obtained when the anionic ROP is carried out in polar solvents [17]. However, the "back-biting," that is, the intramolecular transesterification in the later stages of the polymerization, constitutes the main disadvantage of this methodology. In fact, this side reaction results in the broadening of the polydispersity of the polymer and the loss of control of the process.

This mechanism takes place for example when alkali metal-based catalysts are employed. However, principally, due to the above-mentioned transesterification side reaction, the polymerization is not well controlled when these types of catalysts are used. Moreover, a diminished solubility of these compounds in the reaction medium due to their tendency to form aggregates has been described. Bhaw-Luximon and coworkers reported the polymerization or ε-caprolactone in dioxane employing lithium diisopropylamide. A medium-molecular-weight polymer was obtained in a few minutes and at room temperature. Higher molecular weights were obtained when phenyl lithium was employed providing longer reaction times, few hours, and high temperatures (170 °C) were used [18].

With respect to the cationic mechanism, a cationic species is attacked by the monomer through a S_N2 pathway. Different cationic complexes have been employed to induce this type of polymerization [19], although relatively few studies have been carried out mainly due to the difficulty of controlling the process and the low molecular weight of the so-obtained polymers.

The coordination–insertion ROP is the most common form of ROP. This mechanism takes place when metal alkoxides having a covalent metal-oxygen bond and weak-Lewis acid character are used as catalyst. In Scheme 7.5, the initiation step of coordination–insertion polymerization using aluminum triisopropoxide as catalyst is depicted. As shown in Scheme 7.5, considering the initiation step, this method can be actually considered a *pseudo*-anionic ROP.

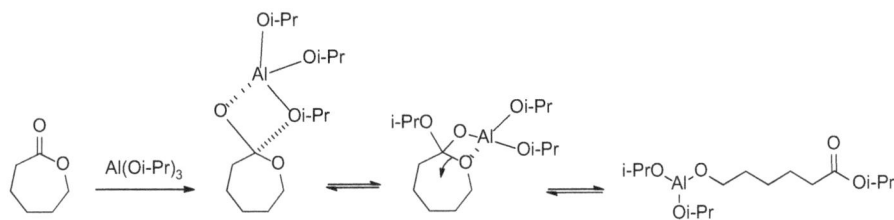

Scheme 7.5: Coordination–insertion polymerization using aluminum triisopropoxide.

Nomura et al. reported the polymerization of ε-caprolactone catalyzed by Sc(OTf)$_3$ [20]. The plausible mechanism of this polymerization is shown in Scheme 7.6.

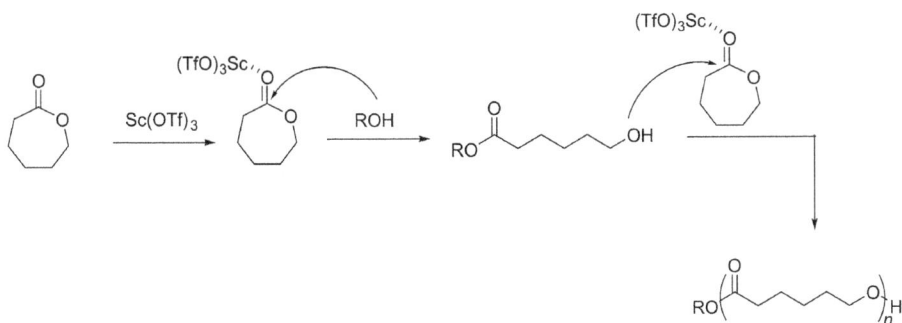

Scheme 7.6: Coordination–insertion polymerization using Sc(OTf)$_3$.

Poly(ε-caprolactone) was obtained in 99% yield by the polymerization of toluene at 25 °C for 33 h employing 0.16 mol% of the catalyst. Similar results were obtained with ε-valerolactone as monomer. Chujo et al. [21] described poly(glycolide) synthesis via ROP in the presence of various catalysts. Employing antimony trifluoride as a catalyst, they observed that polymerization yield increased linearly with the polymerization time. Milione et al. [22] have reported the polymerization of poly(hydroxybutyrate) (PHB) from *rac-β*-butyrolactone employing ROP methodology catalyzed by pentacoordinated aluminum complex.

Another polyester synthesized by ROP is the poly(*p*-dioxanone) (PPDO). This aliphatic polyester is marketed by the name Monoplus (B. Braun Surgical) and PDS and PDS II (Ethicon, INC.) [23]. Yang and coworkers described the synthesis of PPDO via ROP [24]. Similarly, Kimura et al. [25] reported the poly(α-malic acid) synthesis via ROP of a carboxylic group-protected six-membered cyclic malide and a successive hydrogenation to eliminate the corresponding protection groups (Scheme 7.7).

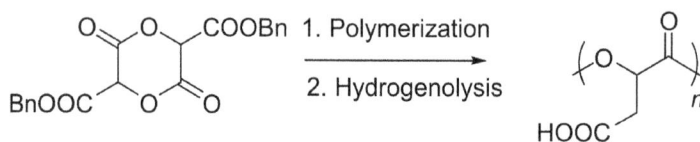

Scheme 7.7: Ring opening polymerization of poly(α-malic acid).

7.2.3 ROP of lactides

Four different mechanisms have been described to explain the ROP of lactides depending on the catalyst employed to induce the polymerization.

In the cationic mechanism, the lactide is protonated by a strong acid or alkylated by a carbocation species [26]. As a consequence of the electrophilic activation, the carbonyl group is attacked by another nucleophilic monomer inducing the opening of the cycle. This process is repeated in the propagation step until a monofunctional nucleophile, like water, causes the termination step. Two factors make this methodology unappealing from a preparative point of view. First, the cationic polymerization is quite slow when carried out below 50 °C, and second, at higher temperatures, racemization occurs which dramatically affects the physical and mechanical properties of the polymer.

Bourissou and coworkers described an efficient cationic polymerization of lactide employing trifluoromethanesulfonic acid (HOTf) and a protic reagent at room temperature (Scheme 7.8).

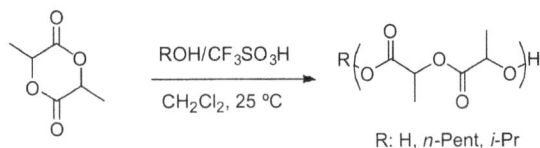

Scheme 7.8: Cationic ring opening polymerization of PLA using trifluoromethanesulfonic acid at room temperature.

In this case, the synthesized PLA was obtained in 70% yield and with >96% conversion by the polymerization in dichloromethane at 25 °C for 3 h employing 2-propanol as initiator [27]. In Scheme 7.9, the mechanism of the first stages of this transformation is depicted.

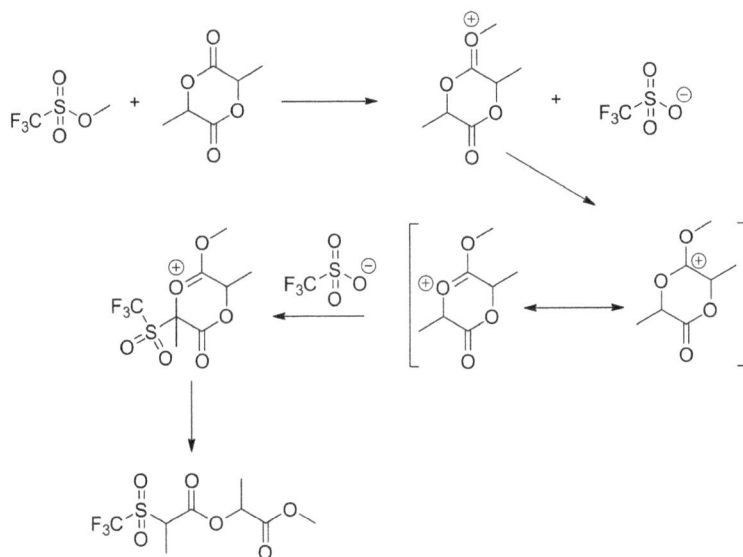

Scheme 7.9: Cationic ring opening of L-lactide.

The anionic polymerization of lactides is initiated by strong nucleophiles such as metal alkoxides, although at higher temperatures phenoxides and carboxylates can be employed as initiators. As shown in Scheme 7.10, the initiation and the subsequent propagation step, results in the nucleophilic attack on the carbonyl group, followed by the ring opening due to the C–O cleavage. However, the alkoxide used as initiator is basic enough to affect the α-deprotonation of the monomer and, therefore, the racemization is, again, an unavoidable side reaction.

Kricheldorf et al. reported the anionic polymerization of L-lactide in solution using strong bases [28]. McGuiness and coworkers also analyzed anionic ROP using iron(II) alkoxides as initiator (Scheme 7.11) [29].

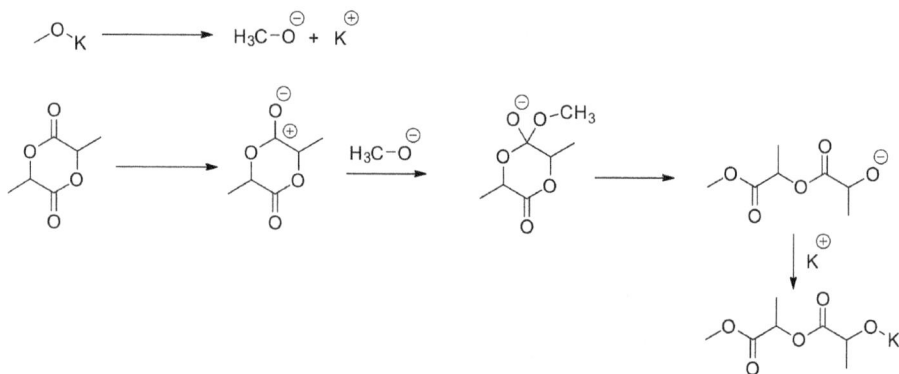

Scheme 7.10: Anionic ring opening polymerization of lactide.

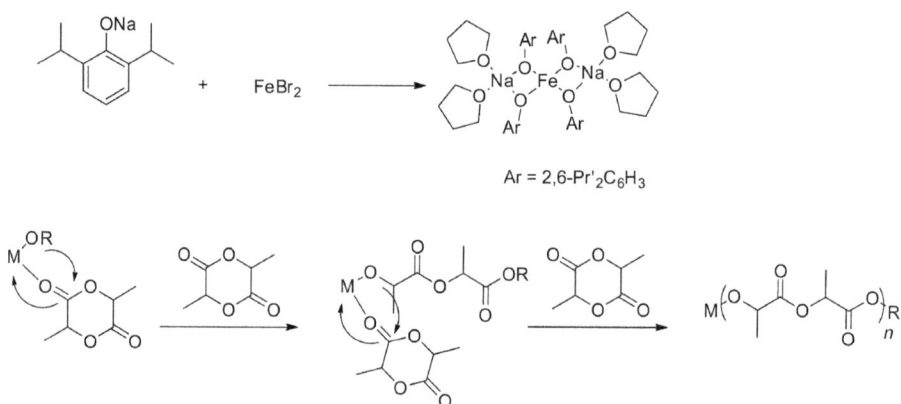

Ar = 2,6-Pr'$_2$C$_6$H$_3$

Scheme 7.11: Anionic ring-opening polymerization of L-lactide using a metal alkoxide.

The polymerization of lactides can be induced by alkoxides having a covalent metal–oxygen bond and weak Lewis acid character. In this case, the polymerization takes place through a coordination–insertion mechanism. The coordination of the metal atom with the oxygen enhances the electrophilicity of the carbonyl group and the nucleophilicity of the alkoxy groups, consequently, the insertion of the lactone into the metal–O bond occurs.

Enzymatic polymerization of lactides is a "greener" alternative to metal-based catalysts; it is based on using renewable resources as starting materials such as lipases. Yoshizawa-Fujita et al. reported the synthesis of L-lactide employing lipase-catalyzed polymerization in ionic liquids (Scheme 7.12) [30].

The synthesized PLLA was obtained in 35.2% yield with the total conversion when they used [C$_4$mim] [BF$_4$] as ionic liquid at 120 °C.

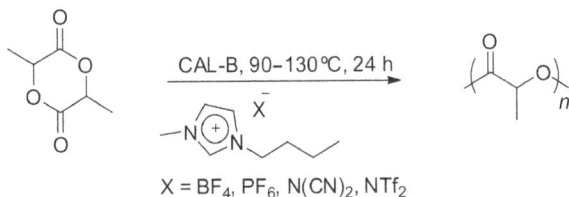

X = BF$_4$, PF$_6$, N(CN)$_2$, NTf$_2$

Scheme 7.12: Lipase-catalyzed poly(ʟ-lactide) synthesis in ionic liquids.

7.3 Degradation of biopolyesters

There is an increasing concern regarding the accumulation of plastic waste in the oceans and landfills. In this context, the biodegradable polyesters have been arising as suitable materials to substitute synthetic polyesters in order to minimize this problem. When a biodegradable polyester is exposed to the outdoor environment (weather, aging, and burying), photodegradation as well as biodegradation, thermal and hydrolytic degradation take place. In addition, the chemical structure of these BPs contain oxygen linkages which make them susceptible to suffer enzymatic hydrolysis, and prone to natural recycling by biological processes [31, 32]. In addition, the degradation could determine the cost, recyclability, and their environmental impact. On the other hand, for commodities and industrial applications, where biopolyesters are used as alternatives to petroleum-based polymeric materials, degradation processes such as hydrolytic degradation has an unfavorable effect due to the decrease of mechanical properties.

7.3.1 Photodegradation

Several materials are photosensitive, in particular polymers exposed to light (or radiation) suffer from degradation, and as a result a discoloration and/or brittle fracture could appear. This degradation process is maximum outdoors where the interaction of sunlight, which includes visible and UV radiation, accelerates the photodegradation of the polymer materials. The usual structure of biopolyesters presents C–O and C=O bonds and the energy of the photons can create an excited state of the C=O by a n–π* transition. Once the carbonyl photoexcitation occurs, several chemical reactions could be induced: α and β cleavage, atom abstraction, radical addition, and electron abstraction or transfer. Some examples are shown in Scheme 7.13 [33].

Photodegradation of biopolyesters can occur via Norrish reactions (Type I and II) [34–36]. Usually, the degradation process for polyesters takes place by photoionization (Norrish I) and chain scission (Norrish II), and/or crosslinking reactions, or

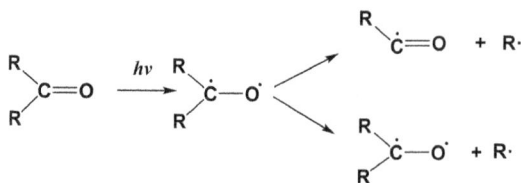

Scheme 7.13: Example of photochemical reaction of carbonyl group.

oxidative processes. Sadi et al. studied the photodegradation process of PHB and concluded that it decreased the melting temperature and mechanical properties of the material [37]. In this case, the photodegradation is initiated mainly by a free radical. Even if Norrish I and crosslinking reactions occur, they are not predominant. In this case, however, their results indicated that Norrish II reaction probably did not occur. Similar behavior was also described for poly(3-hydroxybutyrate-co-3-hydroxyhexanoate) [38]. Scheme 7.14 shows the possible reactions present in the PHB photodegradation.

Contrarily, some authors reported that the main chain scission occurs via Norrish II reaction during the photodegradation of PLLA and PCL, being this reaction for both polymers depicted in Scheme 7.15 [39, 40]. In addition, Tsuji et al. reported that the photodegradability of PCL is higher than PLLA, which suggests that the chemical structure of the ester oxygen adjacent groups has more influence on this degradation than the density of ester group itself [39].

For PLLA, Copinet and coworkers [41, 42] reported that UV radiation could induce an enhancing effect on the autocatalysis of the hydrolysis process at different relative humidity (RH) and temperature conditions. An increase in the RH and temperature increases the degradation rate due to the reduction on the molecular weight, crystallinity, and glass transition temperature.

7.3.2 Hydrolytic degradation

The hydrolytic degradation of these polymers could be undesired during processing or storage periods, but highly interesting for composting one-use commodities or packages. The general structure of polyesters contains hydrolyzable covalent bonds present in the characteristic esters group. The hydrolytic degradation, rate, and mechanism depend on many factors such as the material nature (molecular weight and structure) and medium (temperature, pH, and catalytic species). Scheme 7.16 shows a general hydrolysis reaction of ester group in acid and basic environment.

Several authors studied the basic hydrolysis of PLLA reporting that a random scission occurs, but under acidic conditions the degradation takes place mainly via chain-end scissions (Scheme 7.17) [43, 44]. On the one hand, in the end-chain

(1) Initiation by alkyl radical

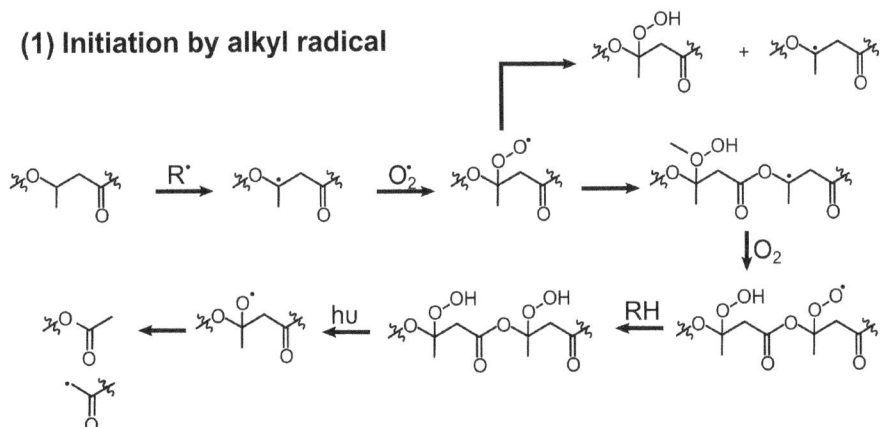

(2) Cross-linking

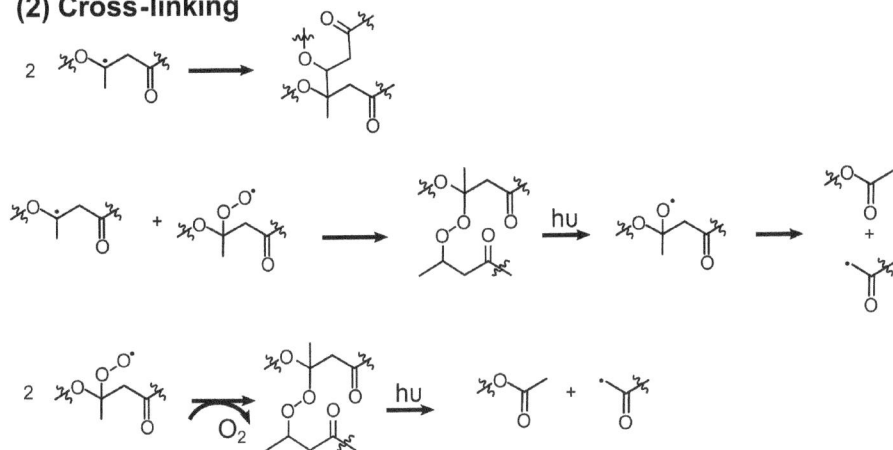

(3) Norrish I

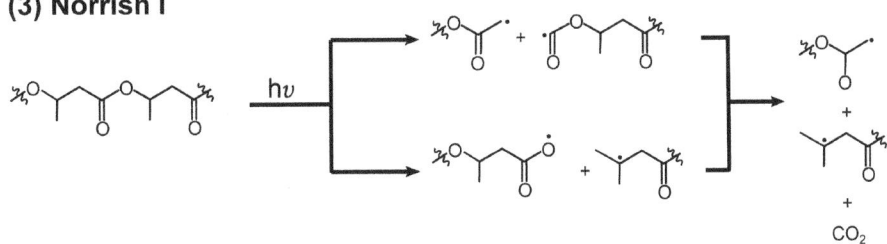

Norrish II

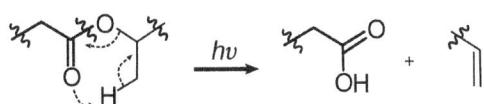

Scheme 7.14: Possible reactions of the photodegradation of PHB according to Sadi et al. [37].

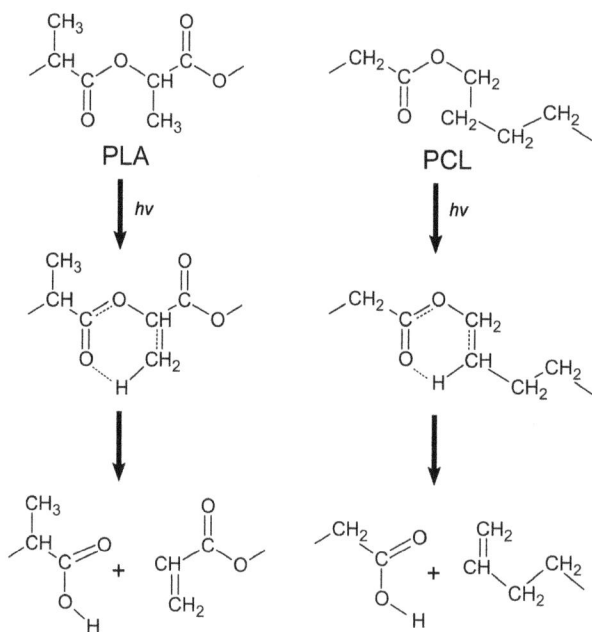

Scheme 7.15: Norrish II reaction for PLLA and PCL.

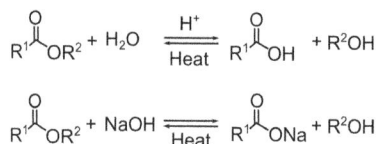

Scheme 7.16: General hydrolysis reaction for ester group in (top) acid and (bottom) basic medium.

degradation due to an intramolecular transesterification, an electrophilic attack of the terminal OH group on the secondary carbonyl group takes place, forming a ring which is subsequently hydrolyzed to lactide. Afterwards, lactide molecules are hydrolyzed forming two lactic acid molecules from each lactide. On the other hand, the intramolecular degradation could occur by a random alkaline attack and the hydrolysis of the ester link, reducing the molecular weight of the polymer [44].

Moreover, in the acid hydrolysis of PLLA (Scheme 7.18) terminal OH groups form an intramolecular hydrogen bond. Lactic acid molecules are released, due to the hydrolysis of the ester group, decreasing the molecular weight of the polymer chains. In addition, an intramolecular random protonation of the ester group could lead to the hydrolysis of ester linkages [44].

Renard and coworkers reported the study of PHAs on a basic hydrolytic degradation (pH 10). Poly(3-hydroxybutyrate-co-3-hydroxyvalerate) (PHBHV) and poly

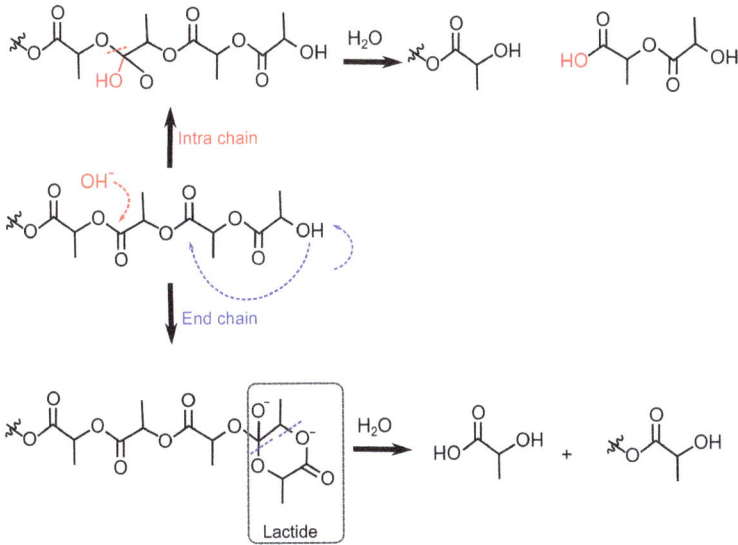

Scheme 7.17: Basic hydrolysis of PLLA.

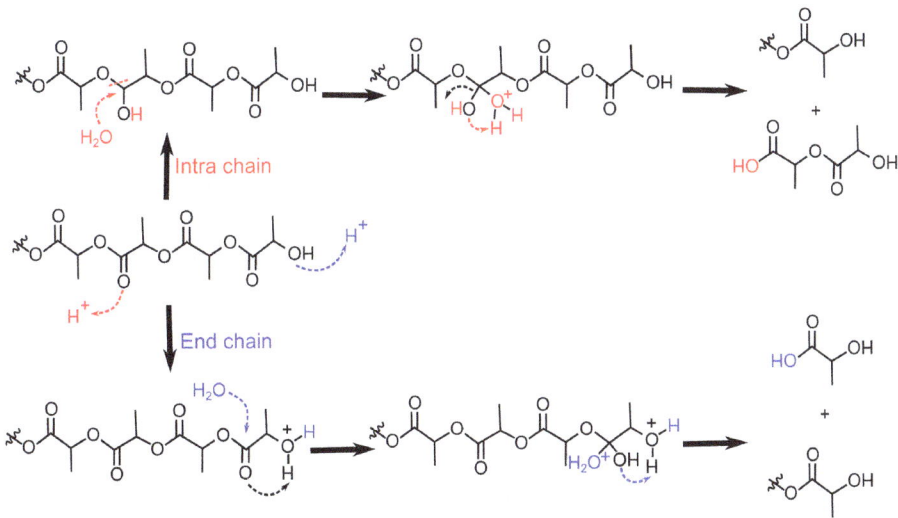

Scheme 7.18: Acid hydrolysis of PLLA.

(3-hydroxyoctanoate) (PHO) blends with PLLA were studied. The presence of PLLA accelerated the hydrolytic process compared to PHBHV. Besides, PHO/PLLA blends did not vary, and this behavior was attributed to the incompatibility of PLLA with PHO. In addition, a study of hydrolytic degradation was carried out by Doi et al. under neutral conditions (PBS buffer, pH 7.4) with P3HB, P(3HB-co-4HB), and P(3HBco-4HV). They reported a two-stage degradation for this process; in the first stage, a random hydrolytic chain scission of the ester groups takes please. In the second stage, the onset of weight loss occurs. Also, they reported that 4HB units present in these biopolyesters accelerate the chain scission [45].

A significant pH dependence has been observed in the hydrolytic degradation of this kind of materials. Sailema-Palate and coworkers reported a slow degradation in a neutral pH for PCL, whereas a higher degradation rate was observed in basic media (pH 13) than in acidic pH (pH 1) [46]. Similarly, Jung et al. compared the acid and basic hydrolysis of PLLA and PCL reporting that, in acidic pH, the degradation increasing order is PLLA < PCL, while, in basic pH, the order changed, PCL ≪ PLLA. Also, an important dependence on the temperature has been observed for the hydrolysis, being the degradation rate increased at higher temperatures due to the increase of the chain mobility above the glass transition temperature [47]. In addition, the crystallinity is another factor which varies the hydrolysis of the biopolyesters, since this process first occurs in the amorphous regions [48].

7.3.3 Thermal degradation

The thermal degradation of the biopolyesters has a common mechanism due to the similar structures of PLLA, PCL, PHA, or PBS. The thermal degradation mechanism of this kind of polymers is complex, and consequently, many studies have been devoted to research on the thermal degradation of PLLA [49–53]. McNeill et al. [54] described hydroxyl end-initiated ester interchange degradation process yielding cyclic oligomers, including lactide, and other by-products such as carbon dioxide, ketene, acetaldehyde, and carbon monoxide. At high temperatures (over 270 °C), the homolysis of PLLA increased [55]. Studies reported by Kopinke and coworkers reported a multistep degradation for this polymer, describing the intramolecular transesterification as the main degradation step [49]. The main degradation pathways described for PLLA are shown in Scheme 7.19.

Several researchers have studied the thermal degradation of PCL, however, the mechanism is debatable [56–61]. A two-stage random cleavage degradation mechanism was proposed by Persenaire and coworkers. In the first stage, a *cis*-elimination occurs, and in the second stage an unzipping depolymerization from the hydroxyl end of the polymer chain takes place [57]. However, a different degradation pathway was suggested by Aoyagi et al. [58]. They proposed a single stage mechanism in which it degrades by specific monomer scission from the end groups.

Thermal degradation of PHB has been described as more simple mechanism than PLLA being by a random chain scission reaction (*cis*-elimination) with a six-membered ring ester intermediate stage (Scheme 7.20), this mechanism is also called McLafferty rearrangement [62–66].

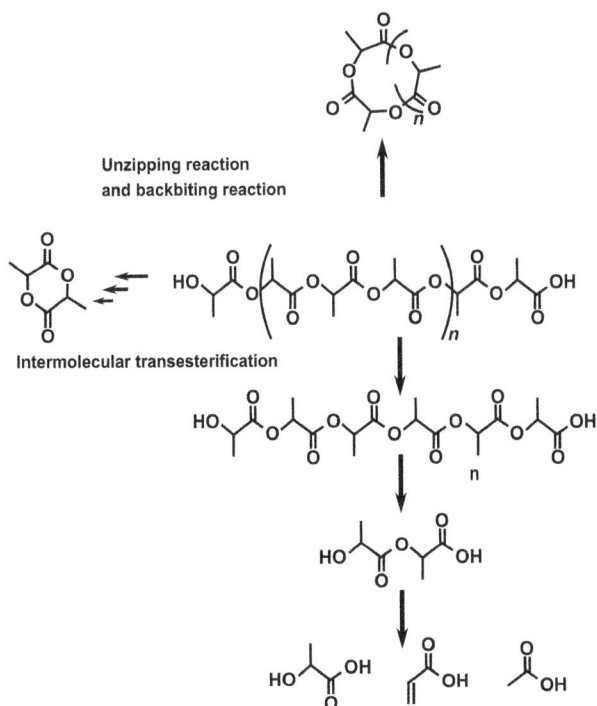

Scheme 7.19: Mechanisms for the thermal degradation of PLLA.

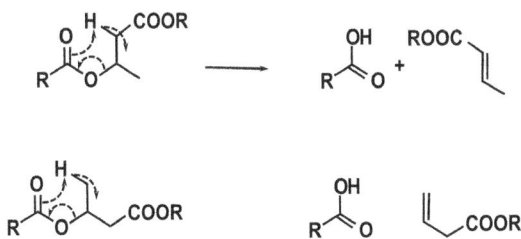

Scheme 7.20: *Cis*-elimination mechanism of PHB degradation.

Similarly, the thermal degradation of poly(butylene succinate) occurs by a cyclization degradation mechanism to obtain succinic acid [67, 68].

In addition, hydrolyzed monomers (or oligomers) and residual catalytic metals can affect thermal stability of biopolyesters .The degradation of PLLA can be affected by different chain-end groups such as a metal ion linked to end groups or capped to them. As an example, Nishida and coworkers described that the presence of tin catalyst lowered the degradation temperature and increased unzipping reactions, obtaining more lactide residues [69, 70]. Moreover, PLLA nanocomposites formed by metal oxides (MgO, CaO, and SnO), nanoparticles, or clays have also been studied in the last years [71–76]. In a recent study, combined unzipping depolymerization and transesterification processes (Scheme 7.21) involved in the thermal degradation of PLLA/ZnO nanocomposites have been described.

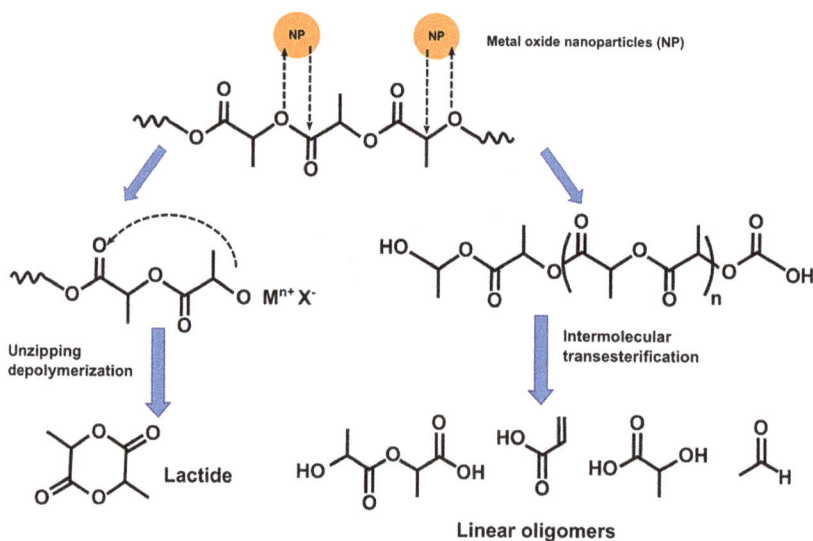

Scheme 7.21: Expected thermal degradation for PLLA/metallic nanoparticles nanocomposites.

Finally, in the case of PCL nanocomposites, Chrissafis et al. reported the thermal degradation mechanism, concluding that nanoparticles affect the degradation rate and the thermal stability of the materials but did not affect the degradation mechanism. Modified montmorillonite and silica nanoparticles used in their study accelerated the thermal degradation of PCL [60].

7.3.4 Environmental biodegradation

Conventional synthetic polymers [i.e., polystyrene (PS), polyethylene] usually persist for many years after being used and their recycling is not so massive when compared to their accumulation. The amount of packaging waste disposed in landfills remains enormous all over the world followed by recycling, incineration, and composting. Even if several members of the European Union have banned landfilling applications, the half of this type of plastic waste are still disposed in landfills. Nonetheless, landfilling could lead to generation of greenhouse gases and leachate. On the contrary, BPs disposed in bioactive environments could be degraded by either enzymatic or nonenzymatic action [77]. The biodegradation of polyesters has been studied in different environments, such as seawater and other aquatic environments (lake water or lake water).

Biodegradation is controlled by several important factors: polymer characteristics (chemical structure, thermal properties (T_g, T_m, crystallinity, elasticity modulus, etc.), surface conditions of the material, type of microorganisms, and the nature of the pretreatment [78, 79]. Several types of microorganisms could degrade bioplastics, such as aerobes, anaerobes, photosynthetic bacteria archaebacterial, and lower eukaryotic. Several studies have been devoted to analyze the biodegradation processes of biopolyesters during composting.

7.3.5 Composting

Composting is a natural process in which organic material is decomposed into humus, a soil-like substance and CO_2 due to the presence of microorganisms such as bacteria, fungi, and actinomycetes. Composting process presents three stages: (1) an active composting stage (mesophilic phase), (2) a curing period (thermophilic phase), and (3) cooling and maturation (second mesophilic phase). In the first stage, with the presence of oxygen, the temperature rises and remains elevated inducing the microbial activity. In the second stage, the temperature decreases and compositing process continues at a lower rate. This process does not stop, it continues until the nutrient present in the materials are consumed by the microorganisms and almost all the carbon is transformed into carbon dioxide [80].

The American Society for Testing and Materials (ASTM) defined compostable plastic as: "a plastic that undergoes degradation by biological processes during composting to yield carbon dioxide, water, inorganic compounds, and biomass at a rate consistent with other known compostable materials and leaves no visually distinguishable or toxic residues" (ASTM D6400-04, 2004). Hence, compostable plastics are biodegradable whereas biodegradable plastic are not always compostable [81]. The main degradation processes present in a compost are mechanical, thermal, and chemical degradation. However, photodegradation only occurs on the surface of the compost pile since the materials is only exposed to the radiation in this area [3]. Many biopolyesters could be susceptible to biodegradation by compost under specific environmental conditions (such as temperature, pH, and moisture content). Several studies evaluated the composting process on biopolyesters, some of them are summarized in Table 7.1.

Rudinik and Briassoulis have reported a significant difference between home (low temperature) and industrial (high temperature) composting processes for PLLA

Table 7.1: Composting process in different biopolyesters.

Source of the biopolymer	Biopolymer	References
Biobased	PLLA	[3, 81–86]
	PHA	[78, 83, 87–90]
Petroleum based	PBS	[91–95]
	PCL	[92, 96–99]

Figure 7.2: PLLA bottle exposed at 30 days of composting environment (reproduced from [3] with permission of Springer. Copyright 2006).

samples. This difference was observed in the second stage (curing period) where the compost temperature oscillates between 35 and 65 °C, and the industrial composting usually reaches 60 °C for PLLA [82]. Biodegradation rate in a composting environment could be affected both by the exposure conditions and the polymer itself. Yang and coworkers have studied the effect of the specific surface area of PLLA, PCL, and PBS on the degradation rate. They described a higher degradation rate, at early stage, for powder shaped samples in these polymers. However, when films are analyzed, at the late stage, PCL is almost independent of the shape, owing to fast disintegration and the fragmentation in smaller pieces increasing the surface area. On the contrary, PLLA and PBS presented a shape dependent degradation, faster for powder than film-shaped samples [95, 100]. Similarly, Kale et al. [3] evaluated the composting process for three PLLA packages (a bottle, a tray, and a meal container). PLLA trays and meal containers degraded before 30 days under composting conditions. Analyzing the degradation kinetics, the bottle and tray undergo a first order degradation kinetics. In addition, they reported that the lactide percentage affects the degradation rate, as the samples with lower lactide degrade faster than the higher ones. Figure 7.2 shows the composting process for a PLLA bottle at 30 days.

7.4 Applications

Nowadays, disposable products are used in many quick service applications due to convenience and low cost. However, disposable materials create a tremendous amount of harmful waste. Every year tons of paper, plastic, and styrofoam products are discarded in landfills. Recent research has described that the weight of all plastic pieces floating in the oceans could be over 250,000 metric tons, being the main source of microplastics due to their fragmentation [101–103].

Unfortunately, many disposable products fabricated with petroleum-based polymers do not decompose until hundreds of years later, and in some cases they can even deposit toxins into the earth. In this context, the biodegradability of biopolyesters added to their adequate properties makes them highly suitable alternatives for conventional plastics in many applications such as packaging and biomedical disposable devices [104, 105].

7.4.1 Food packaging

In today's global economy, packaging has become crucial to allow effective distribution and preservation of food and other consumer products, and to facilitate their end-use convenience and communication at the consumer levels. Therefore, packaging has turned into the third largest industry in the world and it represents about 2% of gross national product in developed countries [106, 107].

Preserving, containing, and protecting food during the whole shelf life is one of the most important characteristics that a packaging material must possess. The choice of packaging material depends on several factors. The possible shape could be between a rigid (bottle, jar, can, cap, tray, and tank), a flexible (bag, foamy trays, shrink, bubble, cling wrap, squeezable tube, stand-up packet, and vacuum bag), and a semi-flexible packaging (caps and closure, box, tetrapack, and multimaterial). Some of the main functions the material should show are the protection of the food against oxygen, temperature fluctuation, moisture, light, and the capability to preserve foods from biological microorganism's attack, the physical protection from damage while reporting information about the food product and its identification. Thus, for choosing the most suitable material there are different factors to consider: chemical–physical properties, mechanical performances, gas barrier behavior, and optical characteristics [108].

Aliphatic polyesters have properties similar to PP, PET, and PS, they are biodegradable but they present poorer thermal and mechanical properties. These types of materials are obtained from renewable resources, by polycondensation reaction of glycol and aliphatic dicarboxylic acid, as has been described earlier [109, 110]. They are odorless and can be used for beverage bottles and they biodegrade in soil and in water giving carbon dioxide and water, in a period of 2 months [3].

One of the most used polyester is PLA, which is commercialized by different companies with different commercial names such as Natureworks (Ingeo), Corbion (EcoPLam), Futerro (Purac), or Symbra (Styrex BioFoam). Even if PLA could be used in a wide range of applications, the major PLA application today is in packaging (nearly 70%); the estimation for 2020 shows the increase PLA food packaging applications [108, 111, 112]. This polymer is ideal for fresh products and those whose quality is not damaged by PLA oxygen permeability. PLA is a growing alternative as "green" food packaging polymer. New applications have been claimed in the field of fresh products, where thermoformed PLA containers are used in retail markets for fruits, vegetables, and salads. The market capacity of products packaged in PLA is unlimited.

In spite of being PLLA, a sustainable substitute of petrochemical-derived synthetic polymers to be used as packaging materials, it presents some limitations in food packaging materials. Among them, the difficulty to be used lies in its flexible film, sheets or injected parts, its limited thermomechanical resistance due to poor crystallization, its restricted use for hot filling liquid food packaging because of its high hydrolysis rate and low thermal resistance, and its low gas barrier properties to O_2, CO_2, H_2O, which are susceptible to react or damage the food [113–115].

The traditional approaches to overcome these limitations have been based on the addition of plasticizers (monomeric, oligomeric, and polymeric), blending, and copolymerization with other polymers [116]. For example, PLA is copolymerized with renewable resources as lactide and aliphatic polyesters like dicarboxylic acid or glycol. A characteristic example is poly(lactide-*co*-glycolide) (PLGA), an aliphatic linear copolymer of PLA and PGA, which possesses hard properties like PS and is soft and flexible like PP. It is easy to process, and its incineration does not produce toxic substances [117]. However, although PLGA studies have been focused mainly on biomedical applications, attractive results have been obtained applied in the food industry [118, 119].

Nowadays, new approaches are being carried out to improve these limitations. Some of these approaches are the hybrid organic–inorganic biobased nanocomposites. Nanocomposites promise to expand the use of edible and biodegradable polyesters.

These nanocomposites can be developed by the combination of layered silicates in a matrix of biodegradable polyesters. These nanohybrid composites present enhanced properties very different from the original polymers such as improved mechanical and oxidation stability, decreased solvent uptake, and even tunable biodegradability [120]. In the case of PLA bionanocomposites, recent investigations try to combine the effect of toughening additives with nanoparticles in order to prepare toughened PLA with improved mechanical and optical properties [121].

PLLA has also been combined with polysaccharides, such as starch [122, 123] or chitosan [124], leading to multilayer built up of bioactive food packaging materials. Sometimes blending and copolymerization are not able to meet specific requirements for food packaging and multilayered materials are being developed for their successful commercialization [125].

Concerning the toxicology that can cause PLA for human, Conn and coworkers reported the migration of small molecules coming from the hydrolytic degradation phenomena of PLA-polymer films of food contact articles. Lactic acid (a safe food substance), lactide monomer, linear dimer of lactic acid can be found. They studied that dimers and oligomers can hydrolyze in aqueous systems yielding lactic acid, the last component has demonstrated to be safe in food at levels far in excess of any small amount that might result from intended uses of PLA. They also found that in any case, migrants from PLA other than lactic acid were represented in a very small and safe amounts [126].

Apart from PLLA, poly(ε-caprolactone) is a biodegradable polyester widely studied for food packaging applications. In fact, an example of improvement of gas barriers properties and brittleness of PLA is the blending with PCL. Poly(ε-caprolactone) is a semicrystalline polymer and possess good chemical resistance to water, oil, solvents, and chlorine. It shows high elongation at break, low-tensile strength, and it is easy to process, showing processing temperature lower than PLA [127]. Due to its low-melting point, PCL is easily blended with other polymers [128]. Also, blending with other polymers have been used in the manufacture of shopping bags [149]. Therefore, it has been exploited to act as plasticizing agent when blending it with PLA polymer [129]. In addition, some PCL-based nanocomposites have been developed to improve its properties [130, 131], and in some cases to add new properties such as antibacterial capacity [132].

Other BPs produced from renewable resources are PHA, these polymers are produced in nature by bacterial fermentation of sugar and lipids [133, 134]. These polymers are also suitable for packaging films both alone or in combination with synthetic plastic or starch [135, 136]. The most common type is the PHB, it is produced by the polymerization of 3-hydroxybutyrate monomer and it shows properties similar to PP but it is more stiffer and brittle [137]. PHB is an active member of PHAs family known for its good biodegradability, biocompatibility, and being bio-derived from renewable resources by bacterial synthesis [138]. However, the polyhydroxybutyrate-valerate copolymer is less stiff and tough, but it is used as packaging material too, the price of this copolymer is very high but it degrades in 5–6 week in microbiology active environments [139]. PHB can be used as microcapsules in therapy or as materials for cell and tablet packaging. It can also be used for packaging applications for deep drawing articles in the food industry, for example, bottles, laminated foils, fishnets, flowerpots, sanitary goods, fast foods, disposable cups, agricultural foils, and fibers in textiles [140, 141].

Current technologies have focused on antimicrobial packages typically prepared by mixtures of BPs with antibacterial agents. As example it is worth to mention a commercial antimicrobial packaging product called AntipackTM manufactured by Handary in Belgium, that consists of PLA and starch films that incorporate antibacterial polymers and active agents whose controlled release delay bacterial proliferation [123]. Considering the lack of commercially available materials, several studies have been focused on the development of biopolyesters-based antibacterial materials as potential packages [132, 142–145].

7.4.2 Fiber and nonwoven

Biodegradable textiles are producing enormous attention in various domains of technological fields. For this, a single polymer may be used in many applications by simple modifications of its physicochemical structure. In some cases, the polymer can be blended with other polymeric or nonpolymeric components to obtain the desired behavior for textile applications [146]. Due to the unique properties of PLA, such as biodegradability, thermoplastic processability and ecofriendly features, PLA is a suitable polymer for a wide variety of technical textile fiber applications, especially for apparel and performance apparel applications [147]. However, the transformation of PLA into textile structures is complicated and depends on structural changes in the polymer during processing. The thermoplastic character of PLA is very useful in allowing transformation of the polymer materials into various shapes [113].

Composite materials combining PLA and natural fiber that could be easily degraded have arisen as highly interesting materials for many industrial applications [148]. Huda et al. studied the physicomechanical properties and microstructure of wood-fiber-reinforced PLA composites, being their mechanical properties higher than those of pure PLA. In addition, the mechanical properties were comparable to similar PP-based composites [149]. Similarly, composites of PLA with Cordenka rayon fibers and flax fibers were studied by Bax and coworkers describing good mechanical properties. Especially, the good-impact properties make this material a potential alternative to petroleum-based composites [150]. However, both composites present poor adhesion when they are analyzed by SEM.

The biodegradation of biopolyesters in biological media is a highly attractive property to design resorbable devices [151, 152]. Among the resorbable polymers can stand out PGA and PLA, these polymers can be formed to nonwovens with high-internal-surface tension and broad variations in their mechanical and degradation behavior. In addition, it has been demonstrated that these polymer nonwovens are a promising tool to guide tissue formation in vivo and in vitro [152]. On the other hand, biopolymers such as PLA and blend of PLA and PHAs have been employed for novel agricultural mulches by using textile technology. As a consequence, many investigations have focused in a completely biobased agricultural mulch making this sector grow [153].

7.4.3 Biomedical applications

In general, biodegradable polyesters are used in two main areas, packaging and biomedicine. Their properties, such as their controlled biodegradability, suitable mechanical properties and air permeability made them a very versatile materials for developing wide amount of biomedical devices [154–156].

In medical field, a large number of biodegradable polyesters can be found [157–160]. Among these, it is worth to mention PHB, which is compatible with the blood and tissues of mammals, and its monomer is a normal metabolic in the human blood. Thus, PHB can be reabsorbed by the body and it could be used as a surgical implant, in surgery, as seam threads for the healing of wounds and blood vessels [161, 162]. On the other hand, polylactide-based polymers are also available for controlled drug release applications, as implantable composites and bone fixation parts, among others [159, 163–166]. They have been also used as internal body components, mainly in the restricted load applications, for example, interference screws in ankle, knee and hand; tacks and pins for ligament attachment; rods and pins in bone; and plates and screws for craniomaxillofacial bone fixation [167, 168]. Otherwise, PCL fibers have demonstrated to be useful for various applications in biomedical and tissue engineering including blood vessels, skin grafts, and tendons, as they possess suitable mechanical properties [169–171]. PLA can be blended with other monomers, such as glycolic acid, and such copolymer is used in drug release [117]. PLA is often mixed with starch to increase biodegradability and to reduce its price and to increase the water absorption [172]. Furthermore, other biopolyesters such as polyanhydrides and poly(orthoesters) have been used for orthopedic (load-bearing) applications [173, 174].

Moreover, in recent years BPs have gained attention as drug delivery systems [175]. Nanocarriers based on biocompatible poly(D,L-lactide-*co*-glycolide) and PLA have shown good potential for intracellular delivery of different drugs and therapeutic agents, such as doxorubicin, insulin or estrogen, among others [176–181]. Lima and Rodrigues Junior developed a biocompatible delivery system using poly (D,L-lactide-*co*-glycolide) microspheres as a controlled release antigen for parenteral administration, which offers several advantages in terms of immune adjuvanticity over other compounds. They found that microspheres were more stable in contrast to other carriers, and thus they permit the oral or parental route administration [182]. Similarly, Eerink et al. [183] reported the release of levonorgestrenol that was filled with PLA fibers. This group produced biodegradable fiber from PLA by dry–wet phase-inversion spinning from PLLA/PVP-dioxane spinning dope. On the other hand, polyanhydrides have demonstrated to be useful in the delivery of chemotherapeutics or antibiotics [184]. As an example, Laurencin et al. reported drug delivery system for the treatment of clinical osteomyelitis in rat model by releasing gentamicin loaded in a polyanhydride matrix. The bioerosion of polyanhydride makes this drug delivery system a promising device for the treatment for these kinds of infections in bone [185].

In general, polymeric fibers, either natural or synthetic, are used for sutures. Natural fibers such as collagen, cotton, silk or linen cause a more intense inflammatory reaction than synthetic materials [186]. Instead synthetic fibers such as PP, PET, PBT, or polyamide, some BPs have been used. PLA-based polymers are commonly used in wound closure. The biodegradable sutures are mainly used for healing internal wounds to avoid secondary removal due to their capacity of being reabsorbed by the body [187]. PGA, polyglycolate, and poly(L-lactide-*co*-ε-caprolactone are the most commonly used polymers for biodegradable sutures [188, 189]. It is important to notice that some examples of shape-memory sutures have been reported based on biodegradable polyesters capable to self-knot in addition to their biodegradation [190, 191].

In addition, PLLA has also been extensively utilized in tissue engineering applications ranging from scaffolds for bone, cartilage, tendon, neural, and vascular regeneration. Sanders and coworkers reported a rat subcutaneous implantation using PLA microfibers for tissue response. On the other hand, Kellomäki et al. [192] designed and manufactured different bioabsorbable scaffolds for guide bone regeneration. The copolymer of PLA (both L- and D,L-lactide forms) and PGA, known as PLGA, is the most investigated degradable polymer for biomedical applications and has been used in sutures, drug delivery devices, and tissue engineering scaffolds [193].

References

[1] Chujo, K., Kobayashi, H., Suzuki, J., and Tokuhara, S. (n.d.). Physical and chemical characteristics polyglycolide. Die Makromol. Chemie. 100, 267–270. doi:10.1002/macp.1967.021000129.

[2] Schmitt, E. E., and Polistina, R. A. Surgical dressings of absorbable polymers. US3875937A, 1975

[3] Kale, G., Auras, R., and Singh, S. P. (2006). Degradation of commercial biodegradable packages under real composting and ambient exposure conditions. J. Polym. Environ. 14, 317–334. doi:10.1007/s10924-006-0015-6.

[4] Chen, G., Kim, H.-S., Kim, E., and Yoon, Y. (2006). Synthesis of high-molecular-weight poly(l-lactic acid) through the direct condensation polymerization of l-lactic acid in bulk state. Eur. Polym. J. 42, 468–472. doi:10.1016/j.eurpolymj.2005.07.022.

[5]   Moon, S.-I., Lee, C.-W., Taniguchi, I., Miyamoto, M., and Kimura, Y. (2001). Melt/solid
        polycondensation of l -lactic acid: an alternative route to poly(l-lactic acid) with high
        molecular weight. Polymer (Guildf) 42, 5059–5062. doi:10.1016/S0032-3861(00)00889-2.
[6]   Uyama, H., Klegraf, E., Wada, S., and Kobayashi, S. (2000). Regioselective polymerization of
        sorbitol and divinyl sebacate using lipase catalyst. Chem. Lett. 29, 800–801. doi:10.1246/
        cl.2000.800.
[7]   Jacquel, N., Saint-Loup, R., Pascault, J.-P., Rousseau, A., and Fenouillot, F. (2015).
        Bio-based alternatives in the synthesis of aliphatic–aromatic polyesters dedicated to
        biodegradable film applications. Polymer (Guildf) 59, 234–242. doi:10.1016/j.
        polymer.2014.12.021.
[8]   Kint, D., and Muñoz-Guerra, S. (1999). A review on the potential biodegradability of poly
        (ethylene terephthalate). Polym. Int. 48, 346–352. doi:10.1002/(SICI)1097-0126(199905)
        48:5<346::AID-PI156>3.0.CO;2-N.
[9]   Lee, S. H., Lim, S. W., and Lee, K. H. (1999). Properties of potentially biodegradable
        copolyesters of (succinic acid-1,4-butanediol)/(dimethyl terephthalate-1,4-butanediol).
        Polym. Int. 48, 861–867. doi:10.1002/(SICI)1097-0126(199909)48:9<861::AID-PI233>3.0.
        CO;2-9.
[10]   Okada, M., Okada, Y., and Aoi, K. (1995). Synthesis and degradabilities of polyesters from 1,
        4:3,6-dianhydrohexitolsand aliphatic dicarboxylic acids. J. Polym. Sci. Part A Polym. Chem.
        33, 2813–2820. doi:10.1002/pola.1995.080331615.
[11]   Okada, M., Tachikawa, K., and Aoi, K. (1999). Biodegradable polymers based on renewable
        resources. III. Copolyesters composed of 1, 4:3,6-dianhydro-D-glucitol,1,1-bis(5-carboxy-2-
        furyl)ethane and aliphatic dicarboxylic acid units. J. Appl. Polym. Sci. 74, 3342–3350.
        doi:10.1002/(SICI)1097-4628(19991227)74:14<3342::AID-APP7>3.0.CO;2-U.
[12]   Fujimaki, T. (1998). Processability and properties of aliphatic polyesters, 'BIONOLLE',
        synthesized by polycondensation reaction. Polym. Degrad. Stab. 59, 209–214. doi:10.1016/
        S0141-3910(97)00220-6.
[13]   Ahn, B. D., Kim, S. H., Kim, Y. H., and Yang, J. S. (2001). Synthesis and characterization of the
        biodegradable copolymers from succinic acid and adipic acid with 1,4-butanediol. J. Appl.
        Polym. Sci. 82, 2808–2826. doi:10.1002/app.2135.
[14]   Carothers, W. H., Dorough, G. L., and van Natta, F. J. (1932). Studies of polymerization and
        ring formation. X. The reversible polymerization of six-membered cyclic esters. J. Am. Chem.
        Soc. 54, 761–772. doi:10.1021/ja01341a046.
[15]   Labet, M., and Thielemans, W. (2009). Synthesis of polycaprolactone: a review. Chem. Soc.
        Rev. 38, 3484. doi:10.1039/b820162p.
[16]   Sisson, A. L., Ekinci, D., and Lendlein, A. (2013). The contemporary role of ε-caprolactone
        chemistry to create advanced polymer architectures. Polymer (Guildf) 54, 4333–4350.
        doi:10.1016/j.polymer.2013.04.045.
[17]   Jedlinski, Z., Kurcok, P., and Kowalczuk, M. (1985). Polymerization of beta-lactones initiated
        by potassium solutions. Macromolecules 18, 2679–2683. doi:10.1021/ma00154a052.
[18]   Bhaw-Luximon, A., Jhurry, D., Motala-Timol, S., and Lochee, Y. (2005). Polymerization of ε-
        caprolactone and its copolymerization with γ-butyrolactone using metal complexes.
        Macromol. Symp. 231, 60–68. doi:10.1002/masy.200590025.
[19]   Jitonnom, J., and Meelua, W. (2017). Cationic ring-opening polymerization of cyclic
        carbonates and lactones by group 4 metallocenes: a theoretical study on mechanism and
        ring-strain effects. J. Theor. Comput. Chem. 16, 1750003. doi:10.1142/S0219633617500031.
[20]   Nomura, N., Taira, A., Nakase, A., Tomioka, T., and Okada, M. (2007). Ring-opening
        polymerization of lactones by rare-earth metal triflates and by their reusable system in ionic
        liquids. Tetrahedron 63, 8478–8484. doi:10.1016/j.tet.2007.05.073.

[21] Chujo, K., Kobayashi, H., Suzuki, J., Tokuhara, S., and Tanabe, M. (n.d.). Ring-opening polymerization of glycolide. Die Makromol. Chemie. 100, 262–266. doi:10.1002/macp.1967.021000128.

[22] García-Valle, F. M., Tabernero, V., Cuenca, T., Mosquera, M. E. G., Cano, J., and Milione, S. (2018). Biodegradable PHB from rac- β-butyrolactone: highly controlled ROP mediated by a pentacoordinated aluminum complex. Organometallics 37, 837–840. doi:10.1021/acs.organomet.7b00843.

[23] Yang, H., Bai, T., Xue, X., Huang, W., Chen, J., and Jiang, B. (2016). Synthesis of metal-free poly(p-dioxanone) by phosphazene base catalyzed ring-opening polymerization. J. Appl. Polym. Sci. 133. doi:10.1002/app.43030.

[24] Yang, K.-K., Wang, X.-L., and Wang, Y.-Z. (2002). Poly(p-dioxanone) and its copolymers. J. Macromol. Sci. Part C Polym. Rev. 42, 373–398. doi:10.1081/MC-120006453.

[25] Kimura, Y., Shirotani, K., Yamane, H., and Kitao, T. (1988). Ring-opening polymerization of 3 (S)-[(benzyloxycarbonyl)methyl]-1,4-dioxane-2,5-dione: a new route to a poly(.alpha.-hydroxy acid) with pendant carboxyl groups. Macromolecules 21, 3338–3340. doi:10.1021/ma00189a037.

[26] Kricheldorf, H. R., and Dunsing, R. (n.d.). Polylactones, 8. Mechanism of the cationic polymerization of L,L-dilactide. Die Makromol. Chemie. 187, 1611–1625. doi:10.1002/macp.1986.021870706.

[27] Bourissou, D., Martin-Vaca, B., Dumitrescu, A., Graullier, M., and Lacombe, F. (2005). Controlled cationic polymerization of lactide. Macromolecules 38, 9993–9998. doi:10.1021/ma051646k.

[28] Kricheldorf, H. R., and Kreiser-Saunders, I. (n.d.). Polylactones, 19. Anionic polymerization of L-lactide in solution. Die Makromol. Chemie. 191, 1057–1066. doi:10.1002/macp.1990.021910508.

[29] McGuinness, D. S., Marshall, E. L., Gibson, V. C., and Steed, J. W. (2003). Anionic iron(II) alkoxides as initiators for the controlled ring-opening polymerization of lactide. J. Polym. Sci. Part A Polym. Chem. 41, 3798–3803. doi:10.1002/pola.10946.

[30] Yoshizawa-Fujita, M., Saito, C., Takeoka, Y., and Rikukawa, M. (2008). Lipase-catalyzed polymerization of L-lactide in ionic liquids. Polym. Adv. Technol. 19, 1396–1400. doi:10.1002/pat.1217.

[31] Vroman, I., and Tighzert, L. (2009). Biodegradable polymers. Materials (Basel) 2, 307–344. doi:10.3390/ma2020307.

[32] Pillai, C. K. S. (2014). Recent advances in biodegradable polymeric materials. Mater. Sci. Technol. 30, 558–566. doi:10.1179/1743284713Y.0000000472.

[33] Wataru, S., and Naoto, T. (2010). Photodegradation and radiation degradation. In: Poly(Lactic Acid) (pp. 413–421). Wiley-Blackwell. doi:10.1002/9780470649848.ch24.

[34] Norrish, R. G. W., and Bamford, C. H. (1937). Photo-decomposition of aldehydes and ketones. Nature 140, 195–196. doi: 10.1038/140195b0.

[35] Norrish, R. G. W., and Bamford, C. H. (1936). Photodecomposition of aldehydes and ketones. Nature 138, 1016. doi:10.1038/1381016a0.

[36] Plooard, P. I., Guillet, E., Guillet, J. E., Guillet, E., and Guillet, J. E. (1972). Photochemistry of ketone polymers. VI. Photolysis of keto diesters and their copolyesters. Macromolecules 5, 405–410. doi:10.1021/ma60028a015.

[37] Sadi, R. K., Fechine, G. J. M., and Demarquette, N. R. (2010). Photodegradation of poly(3-hydroxybutyrate). Polym. Degrad. Stab. 95, 2318–2327. doi:10.1016/j.polymdegradstab.2010.09.003.

[38] Lim, J., and Kim, J. (2016). UV-photodegradation of poly(3-hydroxybutyrate-co-3-hydroxyhexanoate) (PHB-HHx). Macromol. Res. 24, 9–13. doi:10.1007/s13233-016-4004-x.

[39] Tsuji, H., Echizen, Y., and Nishimura, Y. (2006). Photodegradation of biodegradable polyesters: a comprehensive study on poly(l-lactide) and poly(ε-caprolactone). Polym. Degrad. Stab. 91, 1128–1137. doi:10.1016/j.polymdegradstab.2005.07.007.

[40] Ikada, E. (1997). Photo-and bio-degradable polyesters. Photodegradation behaviors of aliphatic polyesters. J. Photopolym. Sci. Technol. 2, 2–265. doi:10.2494/photopolymer.10.265.

[41] Copinet, A., Bertrand, C., Longieras, A., Coma, V., and Couturier, Y. (2003). Photodegradation and biodegradation study of a starch and poly (lactic acid) coextruded material 11.

[42] Copinet, A., Bertrand, C., Govindin, S., Coma, V., and Couturier, Y. (2004). Effects of ultraviolet light (315 nm), temperature and relative humidity on the degradation of polylactic acid plastic films. Chemosphere 55, 763–773. doi:10.1016/j.chemosphere.2003.11.038.

[43] Shih, C. (1995). Chain-end scission in acid catalyzed hydrolysis of poly (d,l-lactide) in solution. J. Control. Release. 34, 9–15. doi:10.1016/0168-3659(94)00100-9.

[44] De Jong, S. J., Arias, E. R., Rijkers, D. T. S., Van Nostrum, C. F., Kettenes-Van den Bosch, J. J., and Hennink, W. E. (2001). New insights into the hydrolytic degradation of poly(lactic acid): participation of the alcohol terminus. Polymer (Guildf) 42, 2795–2802. doi:10.1016/S0032-3861(00)00646-7.

[45] Doi, Y., Kanesawa, Y., Kawaguchi, Y., and Kunioka, M. (1989). Hydrolytic degradation of microbial poly(hydroxyalkanoates). Macromol. Chem. Rapid Commun. 10, 227–230. doi:10.1002/marc.1989.030100506.

[46] Sailema-Palate, G. P., Vidaurre, A., Campillo, A. F., and Castilla-Cortázar, I. (2016). A comparative study on poly(ε-caprolactone) film degradation at extreme pH values. Polym. Degrad. Stab. 130, 118–125. doi:10.1016/j.polymdegradstab.2016.06.005.

[47] Gilding, D. K., and Reed, A. M. (1979). Biodegradable polymers for use in surgery – polyglycolic/poly(actic acid) homo- and copolymers: 1. Polymer (Guildf) 20, 1459–1464. doi:10.1016/0032-3861(79)90009-0.

[48] Fischer, E. W., Sterzel, H. J., and Wegner, G. (1973). Investigation of the Structure of Solution Grown Crystals of Lactide Copolymers by Means of Chemical Reactions (pp. 980–990). doi:10.1007/978-3-642-47050-9_24.

[49] Kopinke, F., Remmler, M., Mackenzie, K., and Milder, M. (1996). Thermal decomposition of biodegradable polyesters-11. Poly(lactic acid). Polym. Degrad. Stab. 43, 329–342. doi:10.1016/0141-3910(96)00102-4.

[50] Arrieta, M. P., Parres, F., López, J., and Jiménez, A. (2013). Development of a novel pyrolysis-gas chromatography/mass spectrometry method for the analysis of poly(lactic acid) thermal degradation products. J. Anal. Appl. Pyrolysis. 101, 150–155. doi:10.1016/j.jaap.2013.01.017.

[51] McNeill, I. C., and Leiper, H. A. (1985). Degradation studies of some polyesters and polycarbonates – 1. Polylactide: general features of the degradation under programmed heating conditions. Polym. Degrad. Stab. 11, 267–285. doi:10.1016/0141-3910(85)90050-3.

[52] Undri, A., Rosi, L., Frediani, M., and Frediani, P. (2014). Conversion of poly(lactic acid) to lactide via microwave assisted pyrolysis. J. Anal. Appl. Pyrolysis. 110, 55–65. doi:10.1016/j.jaap.2014.08.003.

[53] Carrasco, F., Pags, P., Gámez-Pérez, J., Santana, O. O., and Maspoch, M. L. (2010). Kinetics of the thermal decomposition of processed poly(lactic acid). Polym. Degrad. Stab. 95, 2508–2514. doi:10.1016/j.polymdegradstab.2010.07.039.

[54] McNeill, I. C., and Leiper, H. A. (1985). Degradation studies of some polyesters and polycarbonates – 1. Polactide: general features of the degradation under programmed heating conditions. Polym. Degrad. Stab. 11, 267–285.

[55] Tsuji, H., Fukui, I., Daimon, H., and Fujie, K. (2003). Poly(L-lactide) XI. Lactide formation by thermal depolymerisation of poly(L-lactide) in a closed system. Polym. Degrad. Stab. 81, 501–509. doi:10.1016/S0141-3910(03)00150-2.

[56] Sivalingam, G., Karthik, R., and Madras, G. (2003). Kinetics of thermal degradation of poly(ε-caprolactone). J. Anal. Appl. Pyrolysis. 70, 631–647. doi:10.1016/S0165-2370(03)00045-7.

[57] Persenaire, O., Alexandre, M., Degée, P., and Dubois, P. (2001). Mechanisms and kinetics of thermal degradation of poly(ε-caprolactone). Biomacromolecules 2, 288–294. doi:10.1021/bm0056310.

[58] Aoyagi, Y., Yamashita, K., and Doi, Y. (2002). Thermal degradation of poly [(R)-3-hydroxybutyrate], poly [e-caprolactone], and poly [(S)-lactide]. Polym. Degrad. Stab. 76, 53–59. doi:10.1016/S0141-3910(01)00265-8.

[59] Su, T.-T., Jiang, H., and Gong, H. (2008). Thermal stabilities and the thermal degradation kinetics of poly(ε-caprolactone). Polym. Plast. Technol. Eng. 47, 398–403. doi:10.1080/03602550801897695.

[60] Chrissafis, K., Antoniadis, G., Paraskevopoulos, K. M., Vassiliou, A., and Bikiaris, D. N. (2007). Comparative study of the effect of different nanoparticles on the mechanical properties and thermal degradation mechanism of in situ prepared poly(ε-caprolactone) nanocomposites. Compos. Sci. Technol. 67, 2165–2174. doi:10.1016/j.compscitech.2006.10.027.

[61] Sivalingam, G., and Madras, G. (2003). Thermal degradation of poly (ε-caprolactone). Polym. Degrad. Stab. 80, 11–16. doi:10.1016/S0141-3910(02)00376-2.

[62] Nguyen, S., Yu, G. E., and Marchessault, R. H. (2002). Thermal degradation of poly(3-hydroxyalkanoates): preparation of well-defined oligomers. Biomacromolecules 3, 219–224. doi:10.1021/bm0156274.

[63] Kopinke, F. D., and Mackenzie, K. (1997). Mechanistic aspects of the thermal degradation of poly(lactic acid) and poly(β-hydroxybutyric acid). J. Anal. Appl. Pyrolysis 40–41, 43–53. doi:10.1016/S0165-2370(97)00022-3.

[64] Morikawa, H., and Marchessault, R. H. (1981). Pyrolysis of bacterial polyalkanoates. Can. J. Chem. 59, 2306–2313. doi:10.1139/v81-334.

[65] Carrasco, F., Dionisi, D., Martinelli, A., and Majone, M. (2006). Thermal stability of polyhydroxyalkanoates. J. Appl. Polym. Sci. 100, 2111–2121. doi:10.1002/app.23586.

[66] Grassie, N., Murray, E. J., and Holmes, P. A. (1984). The thermal degradation of poly(-(d)-β-hydroxybutyric acid): part 1 – identification and quantitative analysis of products. Polym. Degrad. Stab. 6, 47–61. doi:10.1016/0141-3910(84)90016-8.

[67] Bai, Z., Liu, Y., Su, T., and Wang, Z. (2018). Effect of hydroxyl monomers on the enzymatic degradation of poly(ethylene succinate), poly(butylene succinate), and poly(hexylene succinate). Polymers (Basel) 10. doi:10.3390/polym10010090.

[68] Chrissafis, K., Paraskevopoulos, K. M., and Bikiaris, D. N. (2005). Thermal degradation mechanism of poly(ethylene succinate) and poly(butylene succinate): comparative study. Thermochim. Acta. 435, 142–150. doi:10.1016/j.tca.2005.05.011.

[69] Nishida, H., Mori, T., Hoshihara, S., Fan, Y., Shirai, Y., and Endo, T. (2003). Effect of tin on poly(l-lactic acid) pyrolysis. Polym. Degrad. Stab. 81, 515–523. doi:10.1016/S0141-3910(03)00152-6.

[70] Mori, T. (2004). Effects of chain end structures on pyrolysis of poly(lactic acid) containing tin atoms. Polym. Degrad. Stab. 84, 243–251. doi:10.1016/j.polymdegradstab.2003.11.008.

[71] Ogata, N., Jimenez, G., Kawai, H., and Ogihara, T. (1997). Structure and thermal/mechanical properties of poly (l-lactide) -clay blend. J. Polym. Sci. Part B Polym. Phys. 35, 389–396. doi:10.1002/(SICI)1099-0488(19970130)35:2<389::AID-POLB14>3.0.CO;2-E.

[72] Fan, Y. (2003). Racemization on thermal degradation of poly(L-lactide) with calcium salt end structure. Polym. Degrad. Stab. 80, 503–511. doi:10.1016/S0141-3910(03)00033-8.

[73] Fan, Y., Nishida, H., Mori, T., Shirai, Y., and Endo, T. (2004). Thermal degradation of poly(l-lactide): effect of alkali earth metal oxides for selective l,l-lactide formation. Polymer (Guildf) 45, 1197–1205. doi:10.1016/j.polymer.2003.12.058.

[74] Motoyama, T., Tsukegi, T., Shirai, Y., Nishida, H., and Endo, T. (2007). Effects of MgO catalyst on depolymerization of poly-L-lactic acid to L,L-lactide. Polym. Degrad. Stab. 92, 1350–1358. doi:10.1016/j.polymdegradstab.2007.03.014.

[75] Meng, Q., Heuzey, M. C., and Carreau, P. J. (2012). Control of thermal degradation of polylactide/clay nanocomposites during melt processing by chain extension reaction. Polym. Degrad. Stab. 97, 2010–2020. doi:10.1016/j.polymdegradstab.2012.01.030.

[76] Zhou, Q., and Xanthos, M. (2009). Nanosize and microsize clay effects on the kinetics of the thermal degradation of polylactides. Polym. Degrad. Stab. 94, 327–338. doi:10.1016/j.polymdegradstab.2008.12.009.

[77] Gross, R. A., and Kalra, B. (2002). Biodegradable polymers for the environment. Science 297, 803–807. doi:10.1126/science.297.5582.803.

[78] Weng, Y. X., Wang, X. L., and Wang, Y. Z. (2011). Biodegradation behavior of PHAs with different chemical structures under controlled composting conditions. Polym. Test. 30, 372–380. doi:10.1016/j.polymertesting.2011.02.001.

[79] Tokiwa, Y., and Calabia, B. P. (2006). Biodegradability and biodegradation of poly (lactide). 244–251. doi:10.1007/s00253-006-0488-1.

[80] Kale, G., Kijchavengkul, T., Auras, R., Rubino, M., Selke, S. E., and Singh, S. P. (2007). Compostability of bioplastic packaging materials: an overview. Macromol. Biosci. 7, 255–277. doi:10.1002/mabi.200600168.

[81] Kale, G., Auras, R., Singh, S. P., and Narayan, R. (2007). Biodegradability of polylactide bottles in real and simulated composting conditions. Polym. Test. 26, 1049–1061. doi:10.1016/j.polymertesting.2007.07.006.

[82] Rudnik, E., and Briassoulis, D. (2011). Degradation behaviour of poly(lactic acid) films and fibres in soil under Mediterranean field conditions and laboratory simulations testing. Ind. Crops Prod. 33, 648–658. doi:10.1016/j.indcrop.2010.12.031.

[83] Tabasi, R. Y., and Ajji, A. (2015). Selective degradation of biodegradable blends in simulated laboratory composting. Polym. Degrad. Stab. 120, 435–442. doi:10.1016/j.polymdegradstab.2015.07.020.

[84] Ahn, H. K., Huda, M. S., Smith, M. C., Mulbry, W., Schmidt, W. F., and Reeves, J. B. (2011). Biodegradability of injection molded bioplastic pots containing polylactic acid and poultry feather fiber. Bioresour. Technol. 102, 4930–4933. doi:10.1016/j.biortech.2011.01.042.

[85] Mihai, M., Legros, N., and Alemdar, A. (2014). Formulation-properties versatility of wood fiber biocomposites based on polylactide and polylactide/thermoplastic starch blends. Polym. Eng. Sci. 54, 1325–1340. doi:10.1002/pen.23681.

[86] Sedničková, M., Pekařová, S., Kucharczyk, P., Bočkaj, J., Janigová, I., Kleinová, A., Jochec-Mošková, D., Omaníková, L., Perďochová, D., Koutný, M., Sedlařík, V., Alexy, P., and Chodák, I. (2018). Changes of physical properties of PLA-based blends during early stage of biodegradation in compost. Int. J. Biol. Macromol. 113, 434–442. doi:10.1016/j.ijbiomac.2018.02.078.

[87] Weng, Y. X., Wang, Y., Wang, X. L., and Wang, Y. Z. (2010). Biodegradation behavior of PHBV films in a pilot-scale composting condition. Polym. Test. 29, 579–587. doi:10.1016/j.polymertesting.2010.04.002.

[88] Gilmore, D. F., Antoun, S., Lenz, R. W., Goodwin, S., Austin, R., and Fuller, R. C. (1992). The fate of "biodegradable" plastics in municipal leaf compost. J. Ind. Microbiol. 10, 199–206. doi:10.1007/BF01569767.

[89] Mergaert, J., Anderson, C., Wouters, A., and Swings, J. (1994). Microbial degradation of poly (3-hydroxybutyrate) and poly(3-hydroxybutyrate-co-3-hydroxyvalerate) in compost. J. Environ. Polym. Degrad. 2, 177–183. doi:10.1007/BF02067443.

[90] Rutkowska, M., Krasowska, K., Heimowska, A., Adamus, G., Sobota, M., Musioł, M., Janeczek, H., Sikorska, W., Krzan, A., Žagar, E., and Kowalczuk, M. (2008). Environmental degradation of blends of atactic poly[(R,S)-3-hydroxybutyrate] with natural PHBV in Baltic sea water and compost with activated sludge. J. Polym. Environ. 16, 183–191. doi:10.1007/s10924-008-0100-0.

[91] Anstey, A., Muniyasamy, S., Reddy, M. M., Misra, M., and Mohanty, A. (2014). Processability and biodegradability evaluation of composites from poly(butylene succinate) (PBS) bioplastic and biofuel co-products from Ontario. J. Polym. Environ. 22, 209–218. doi:10.1007/s10924-013-0633-8.

[92] Yang, H. S., Yoon, J. S., and Kim, M. N. (2004). Effects of storage of a mature compost on its potential for biodegradation of plastics. Polym. Degrad. Stab. 84, 411–417. doi:10.1016/j.polymdegradstab.2004.01.014.

[93] Kim, H. S., Kim, H. J., Lee, J. W., and Choi, I. G. (2006). Biodegradability of bio-flour filled biodegradable poly(butylene succinate) bio-composites in natural and compost soil. Polym. Degrad. Stab. 91, 1117–1127. doi:10.1016/j.polymdegradstab.2005.07.002.

[94] Liu, L., Yu, J., Cheng, L., and Yang, X. (2009). Biodegradability of poly(butylene succinate) (PBS) composite reinforced with jute fibre. Polym. Degrad. Stab. 94, 90–94. doi:10.1016/j.polymdegradstab.2008.10.013.

[95] Zhao, J. H., Wang, X. Q., Zeng, J., Yang, G., Shi, F. H., and Yan, Q. (2005). Biodegradation of poly(butylene succinate) in compost. J. Appl. Polym. Sci. 97, 2273–2278. doi:10.1002/app.22009.

[96] Nakasaki, K., Matsuura, H., Tanaka, H., and Sakai, T. (2006). Synergy of two thermophiles enables decomposition of poly-ε- caprolactone under composting conditions. FEMS Microbiol. Ecol. 58, 373–383. doi:10.1111/j.1574-6941.2006.00189.x.

[97] Nakasaki, K., Ohtaki, A., and Takano, H. (2000). Biodegradable plastic reduces ammonia emission during composting. Polym. Degrad. Stab. 70, 185–188. doi:10.1016/S0141-3910(00)00104-X.

[98] Ohtaki, A., Sato, N., and Nakasaki, K. (1998). Biodegradation of poly-ε-caprolactone under controlled composting conditions. Polym. Degrad. Stab. 61, 499–505. doi:10.1016/S0141-3910(97)00238-3.

[99] De Kesel, C., Vander Wauven, C., and David, C. (1997). Biodegradation of polycaprolactone and its blends with poly(vinylalcohol) by micro-organisms from a compost of house-hold refuse. Polym. Degrad. Stab. 55, 107–113. doi:10.1016/0141-3910(95)00138-7.

[100] Yang, H. S., Yoon, J. S., and Kim, M. N. (2005). Dependence of biodegradability of plastics in compost on the shape of specimens. Polym. Degrad. Stab. 87, 131–135. doi:10.1016/j.polymdegradstab.2004.07.016.

[101] Harrison, J. P., Boardman, C., O'Callaghan, K., Delort, A.-M., and Song, J. (2018). Biodegradability standards for carrier bags and plastic films in aquatic environments: a critical review. R. Soc. Open Sci. 5, 171792. doi:10.1098/rsos.171792.

[102] Eriksen, M., Lebreton, L. C. M., Carson, H. S., Thiel, M., Moore, C. J., Borerro, J. C., Galgani, F., Ryan, P. G., and Reisser, J. (2014). Plastic pollution in the world's oceans: more than 5 trillion plastic pieces weighing over 250,000 tons afloat at sea. PLoS One 9, e111913. doi:10.1371/journal.pone.0111913.

[103] O'Brine, T., and Thompson, R. C. (2010). Degradation of plastic carrier bags in the marine environment. Mar. Pollut. Bull. 60, 2279–2283. doi:10.1016/j.marpolbul.2010.08.005.

[104] Gross, R. A. (2002). Biodegradable polymers for the environment. Science 297, 803–807. doi:10.1126/science.297.5582.803.

[105] Vink, E. T. H., Rábago, K. R., Glassner, D. A., and Gruber, P. R. (2003). Applications of life cycle assessment to NatureWorks™ polylactide (PLA) production. Polym. Degrad. Stab. 80, 403–419. doi:10.1016/S0141-3910(02)00372-5.

[106] Mihindukulasuriya, S. D. F., and Lim, L.-T. (2014). Nanotechnology development in food packaging: a review. Trends Food Sci. Technol. 40, 149–167. doi:10.1016/j.tifs.2014.09.009.

[107] Robertson, G. L. (2005). Food Packaging: Principles and Practice. CRC Press.

[108] Jamshidian, M., Tehrany, E. A., Imran, M., Jacquot, M., and Desobry, S. (2010). Poly-lactic acid: production, applications, nanocomposites, and release studies. Compr. Rev. Food Sci. Food Saf. 9, 552–571. doi:10.1111/j.1541-4337.2010.00126.x.

[109] Vilela, C., Sousa, A. F., Fonseca, A. C., Serra, A. C., Coelho, J. F. J., Freire, C. S. R., and Silvestre, A. J. D. (2014). The quest for sustainable polyesters – insights into the future. Polym. Chem. 5, 3119–3141. doi:10.1039/C3PY01213A.

[110] Robertson, G. (2008). State-of-the-art biobased food packaging materials. In: Environmentally Compatible Food Packaging (pp. 3–28). Elsevier. doi:10.1533/9781845694784.1.3.

[111] Farah, S., Anderson, D. G., and Langer, R. (2016). Physical and mechanical properties of PLA, and their functions in widespread applications – a comprehensive review. Adv. Drug Deliv. Rev. 107, 367–392. doi:10.1016/j.addr.2016.06.012.

[112] Castro-Aguirre, E., Iñiguez-Franco, F., Samsudin, H., Fang, X., and Auras, R. (2016). Poly(lactic acid) – mass production, processing, industrial applications, and end of life. Adv. Drug Deliv. Rev. 107, 333–366. doi:10.1016/j.addr.2016.03.010.

[113] Auras, R. A., Lim, L.-T., Selke, S. E. M., and Tsuji, H. (2011). Poly (Lactic Acid): Synthesis, Structures, Properties, Processing, and Applications. John Wiley & Sons.

[114] Sinclair, R. G. (1996). The case for polylactic acid as a commodity packaging plastic. J. Macromol. Sci. Part A. 33, 585–597. doi:10.1080/10601329608010880.

[115] Urayama, H., Moon, S.-I., and Kimura, Y. (2003). Microstructure and thermal properties of polylactides with different L- and D-unit sequences: importance of the helical nature of the L-sequenced segments. Macromol. Mater. Eng. 288, 137–143. doi:10.1002/mame.200390006.

[116] Liu, H., and Zhang, J. (2011). Research Progress in Toughening Modification of Poly (Lactic Acid) (pp. 1051–1083). doi:10.1002/polb.22283.

[117] Makadia, H. K., and Siegel, S. J. (2011). Poly lactic-co-glycolic acid (PLGA) as biodegradable controlled drug delivery carrier. Polymers (Basel) 3, 1377–1397. doi:10.3390/polym3031377.

[118] Gomes, C., Moreira, R. G., and Castell-Perez, E. (2011). Poly (DL-lactide-co-glycolide) (PLGA) nanoparticles with entrapped trans-cinnamaldehyde and eugenol for antimicrobial delivery applications. J. Food Sci. 76, N16–N24. doi:10.1111/j.1750-3841.2010.01985.x.

[119] Prabhakaran, M. P., Zamani, M., Felice, B., and Ramakrishna, S. (2015). Electrospraying technique for the fabrication of metronidazole contained PLGA particles and their release profile. Mater. Sci. Eng. C. 56, 66–73. doi:10.1016/j.msec.2015.06.018.

[120] Sorrentino, A., and Vittoria, V. (2007). Potential perspectives of for food packaging applications. 18, 84–95. doi:10.1016/j.tifs.2006.09.004.

[121] Armentano, I., Bitinis, N., Fortunati, E., Mattioli, S., Rescignano, N., Verdejo, E., Lopez-manchado, M. A., and Kenny, J. M. (2013). Progress in Polymer Science Multifunctional nanostructured PLA materials for packaging and tissue engineering. Prog. Polym. Sci. 38, 1720–1747. doi:10.1016/j.progpolymsci.2013.05.010.

[122] Mallardo, S., De Vito, V., Malinconico, M., Volpe, M. G., Santagata, G., and Di Lorenzo, M. L. (2018). Biodegradable Poly(butylene succinate)-based composites for food packaging. In: Proceedings of the International Conference on Microplastic Pollution in the Mediterranean Sea (Cocca, M., Di Pace, E., Errico, M.E., Gentile, G., Montarsolo, A., and Mossotti, R., Eds., pp. 199–204). Springer International Publishing, Cham.

[123] Muller, J. (2017). Biodegradable Food Packaging (pp. 1–22). doi:10.3390/ma10080952.

[124] Gartner, H., Li, Y., and Almenar, E. (2015). Applied surface science improved wettability and adhesion of polylactic acid/chitosan coating for bio-based multilayer film development. Appl. Surf. Sci. 332, 488–493. doi:10.1016/j.apsusc.2015.01.157.

[125] Andrade-Del Olmo, J., Pérez-Álvarez, L., Hernáez, E., Ruiz-Rubio, L., and Vilas-Vilela, J. L. (2019). Antibacterial multilayer of chitosan and (2-carboxyethyl)- β-cyclodextrin onto polylactic acid (PLLA). Food Hydrocoll. 88, 228–236. doi:10.1016/j. foodhyd.2018.10.014.

[126] Conn, R. E., Kolstad, J. J., Borzelleca, J. F., Dixler, D. S., Filer, L. J., Ladu, B. N., and Pariza, M. W. (1995). Safety assessment of polylactide (PLA) for use as a food-contact polymer. Food Chem. Toxicol. 33, 273–283. doi:10.1016/0278-6915(94)00145-E.

[127] Woodruff, M. A., and Hutmacher, D. W. (2010). The return of a forgotten polymer – polycaprolactone in the twenty-first century. Prog. Polym. Sci. 35, 1217–1256. doi:10.1016/j. progpolymsci.2010.04.002.

[128] Rešček, A., Ščetar, M., Hrnjak-Murgić, Z., Dimitrov, N., and Galić, K. (2016). Polyethylene/ polycaprolactone nanocomposite films for food packaging modified with magnetite and casein: oxygen barrier, mechanical, and thermal properties. Polym. Plast. Technol. Eng. 55, 1450–1459. doi:10.1080/03602559.2016.1163606.

[129] Cabedo, L., Luis Feijoo, J., Pilar Villanueva, M., Lagarón, J. M., and Giménez, E. (2006). Optimization of biodegradable nanocomposites based on aPLA/PCL blends for food packaging applications. Macromol. Symp. 233, 191–197. doi:10.1002/masy.200690017.

[130] Benhacine, F., Ouargli, A., and Hadj-Hamou, A. S. (2018). Preparation and characterization of novel food packaging materials based on biodegradable PCL/Ag-kaolinite nanocomposites with controlled release properties. Polym. Plast. Technol. Eng. 1–13. doi:10.1080/ 03602559.2018.1471714.

[131] Sorrentino, A., Gorrasi, G., and Vittoria, V. (2007). Potential perspectives of bio-nanocomposites for food packaging applications. Trends Food Sci. Technol. 18, 84–95. doi:10.1016/j.tifs.2006.09.004.

[132] Cesur, S., Köroğlu, C., and Yalçın, H. T. (2017). Antimicrobial and biodegradable food packaging applications of polycaprolactone/organo nanoclay/chitosan polymeric composite films. J. Vinyl Addit. Technol. doi:10.1002/vnl.21607.

[133] Li, Z., Yang, J., and Loh, X. J. (2016). Polyhydroxyalkanoates: opening doors for a sustainable future. NPG Asia Mater. 8, e265–e265. doi:10.1038/am.2016.48.

[134] Zhang, M., and Thomas, N. L. (2011). Blending polylactic acid with polyhydroxybutyrate: the effect on thermal, mechanical, and biodegradation properties. Adv. Polym. Technol. 30, 67–79. doi:10.1002/adv.20235.

[135] Arrieta, M. P., Samper, M. D., Aldas, M., and López, J. (2017). On the use of PLA-PHB blends for sustainable food packaging applications. Materials (Basel) 10, 1–26. doi:10.3390/ ma10091008.

[136] Plackett, D., and Siró, I. (2011). Polyhydroxyalkanoates (PHAs) for food packaging. In: Multifunctional and nanoreinforced polymers for food packaging (pp. 498–526). Elsevier. doi:10.1533/9780857092786.4.498.

[137] Arrieta, M. P., del M. Castro-López, M, Rayón, E, Barral-Losada, L. F, López-Vilariño, J. M, López, J., and González-Rodríguez, M. V. (2014). Plasticized poly(lactic acid)–poly (hydroxybutyrate) (PLA–PHB) blends incorporated with catechin intended for active food-packaging applications. J. Agric. Food Chem. 62, 10170–10180. doi:10.1021/jf5029812.

[138] Solaiman, D. K. Y., Ashby, R. D., Zerkowski, J. A., Krishnama, A., and Vasanthan, N. (2015). Control-release of antimicrobial sophorolipid employing different biopolymer matrices. Biocatal. Agric. Biotechnol. 4, 342–348. doi:10.1016/j.bcab.2015.06.006.

[139] Hankermeyer, C. R., and Tjeerdema, R. S. (1999). Polyhydroxybutyrate: plastic made and degraded by microorganisms. Rev. Environ. Contam. Toxicol. 159, 1–24. http://www.ncbi. nlm.nih.gov/pubmed/9921137.

[140] Bugnicourt, E., Cinelli, P., Lazzeri, A., and Alvarez, V. (2014). Polyhydroxyalkanoate (PHA): review of synthesis, characteristics, processing and potential applications in packaging. Express Polym. Lett. 8, 791–808. doi:10.3144/expresspolymlett.2014.82.

[141] Xavier, J. R., Babusha, S. T., George, J., and Ramana, K. V. (2015). Material properties and antimicrobial activity of polyhydroxybutyrate (PHB) films incorporated with vanillin. Appl. Biochem. Biotechnol. 176, 1498–1510. doi:10.1007/s12010-015-1660-9.

[142] Tawakkal, I. S. M. A., Cran, M. J., Miltz, J., and Bigger, S. W. (2014). A review of poly(lactic acid)-based materials for antimicrobial packaging. J. Food Sci. 79, R1477–R1490. doi:10.1111/ 1750-3841.12534.

[143] Lin, X., Fan, X., Li, R., Li, Z., Ren, T., Ren, X., and Huang, T.-S. (2018). Preparation and characterization of PHB/PBAT-based biodegradable antibacterial hydrophobic nanofibrous membranes. Polym. Adv. Technol. 29, 481–489. doi:10.1002/pat.4137.

[144] Qin, Y., Liu, D., Wu, Y., Yuan, M., Li, L., and Yang, J. (2015). Effect of PLA/PCL/ cinnamaldehyde antimicrobial packaging on physicochemical and microbial quality of button mushroom (Agaricus bisporus). Postharvest Biol. Technol. 99, 73–79. doi:10.1016/j. postharvbio.2014.07.018.

[145] Wu, Y., Qin, Y., Yuan, M., Li, L., Chen, H., Cao, J., and Yang, J. (2014). Characterization of an antimicrobial poly(lactic acid) film prepared with poly(ε-caprolactone) and thymol for active packaging. Polym. Adv. Technol. 25, 948–954. doi:10.1002/pat.3332.

[146] Gupta, B., Revagade, N., and Hilborn, J. (2007). Poly(lactic acid) fiber: an overview. Prog. Polym. Sci. 32, 455–482. doi:10.1016/j.progpolymsci.2007.01.005.

[147] Ren, J. (2011). Biodegradable poly (lactic acid): Synthesis, Modification, Processing and Applications. Springer Science & Business Media.

[148] Luckachan, G. E., and Pillai, C. K. S. (2011). Biodegradable polymers – a review on recent trends and emerging perspectives. J. Polym. Environ. 19, 637–676. doi:10.1007/s10924-011- 0317-1.

[149] Huda, M. S., Drzal, L. T., Misra, M., and Mohanty, A. K. (2006). Wood-fiber-reinforced poly (lactic acid) composites: evaluation of the physicomechanical and morphological properties. J. Appl. Polym. Sci. 102, 4856–4869. doi:10.1002/app.24829.

[150] Bax, B., and Müssig, J. (2008). Impact and tensile properties of PLA/Cordenka and PLA/flax composites. Compos. Sci. Technol. 68, 1601–1607. doi:10.1016/j.compscitech.2008.01.004.

[151] Stocum, D. L. (1998). Regenerative biology and engineering: strategies for tissue restoration. Wound Repair Regen. 6, 276–290. doi:10.1046/j.1524-475X.1998.60404.x.

[152] Sittinger, M., Reitzel, D., Dauner, M., Hierlemann, H., Hammer, C., Kastenbauer, E., Planck, H., Burmester, G. R., and Bujia, J. (1996). Resorbable polyesters in cartilage engineering: affinity and biocompatibility of polymer fiber structures to chondrocytes. J. Biomed. Mater. Res. 33, 57–63. doi:10.1002/(SICI)1097-4636(199622)33:2<57:: AID-JBM1>3.0.CO;2-K.

[153] Hablot, E., Dharmalingam, S., Hayes, D. G., Wadsworth, L. C., Blazy, C., and Narayan, R. (2014). Effect of simulated weathering on physicochemical properties and inherent biodegradation of PLA/PHA nonwoven mulches. J. Polym. Environ. 22, 417–429. doi:10.1007/ s10924-014-0697-0.

[154] Ikada, Y., and Tsuji, H. (2000). Biodegradable polyesters for medical and ecological applications. Macromol. Rapid Commun. 21, 117–132. doi:10.1002/(SICI)1521-3927 (20000201)21:3<117::AID-MARC117>3.0.CO;2-X.

[155] Venkatraman, S. S., Tan, L. P., Joso, J. F. D., +Boey, Y. C. F., and Wang, X. (2006). Biodegradable stents with elastic memory. Biomaterials 27, 1573–1578. doi:10.1016/j. biomaterials.2005.09.002.

[156] Meng, B., Wang, J., Zhu, N., Meng, Q. Y., Cui, F. Z., and Xu, Y. X. (2006). Study of biodegradable and self-expandable PLLA helical biliary stent in vivo and in vitro. J. Mater. Sci. Mater. Med. 17, 611–617. doi:10.1007/s10856-006-9223-9.

[157] Parker, T., Dave, V., and Falotico, R. (2010). Polymers for drug eluting stents. Curr. Pharm. Des. 16, 3978–3988. doi:10.2174/138161210794454897.

[158] Bose, S., Vahabzadeh, S., and Bandyopadhyay, A. (2013). Bone tissue engineering using 3D printing. Mater. Today 16, 496–504. doi:10.1016/j.mattod.2013.11.017.

[159] Nakazawa, G., Finn, A. V., Kolodgie, F. D., and Virmani, R. (2009). A review of current devices and a look at new technology: drug-eluting stents. Expert Rev. Med. Dev. 6, 33–42. doi:10.1586/17434440.6.1.33.

[160] O'Brien, F. J. (2011). Biomaterials & scaffolds for tissue engineering. Mater. Today. 14, 88–95. doi:10.1016/S1369-7021(11)70058-X.

[161] Alves, E. G. L., de F. Rezende, C. M, Serakides, R., de M. Pereira, M., and Rosado, I. R. (2011). Orthopedic implant of a polyhydroxybutyrate (PHB) and hydroxyapatite composite in cats. J. Feline Med. Surg. 13, 546–552. doi:10.1016/j.jfms.2011.03.002.

[162] Meischel, M., Eichler, J., Martinelli, E., Karr, U., Weigel, J., Schmöller, G., Tschegg, E. K., Fischerauer, S., Weinberg, A. M., and Stanzl-Tschegg, S. E. (2016). Adhesive strength of bone-implant interfaces and in-vivo degradation of PHB composites for load-bearing applications. J. Mech. Behav. Biomed. Mater. 53, 104–118. doi:10.1016/j. jmbbm.2015.08.004.

[163] Senatov, F. S., Zadorozhnyy, M. Y., Niaza, K. V., Medvedev, V. V., Kaloshkin, S. D., Anisimova, N. Y., Kiselevskiy, M. V., and Yang, K. C. (2017). Shape memory effect in 3D-printed scaffolds for self-fitting implants. Eur. Polym. J. 93, 222–231. doi:10.1016/j. eurpolymj.2017.06.011.

[164] Tamai, H., Igaki, K., Kyo, E., Kosuga, K., Kawashima, A., Matsui, S., Komori, H., Tsuji, T., Motohara, S., and Uehata, H. (2000). Initial and 6-month results of biodegradable poly-l-lactic acid coronary stents in humans. Circulation. 102, 399–404. doi:10.1161/01. CIR.102.4.399.

[165] Xu, J., and Song, J. (2015). Polylactic acid (PLA)-based shape-memory materials for biomedical applications. Shape Mem. Polym. Biomed. Appl. 197–217. doi:10.1016/B978-0-85709-698-2.00010-6.

[166] Madhavan Nampoothiri, K., Nair, N. R., and John, R. P. (2010). An overview of the recent developments in polylactide (PLA) research. Bioresour. Technol. 101, 8493–8501. doi:10.1016/j.biortech.2010.05.092.

[167] Lim, J. Y., Kim, S. H., Lim, S., and Kim, Y. H. (2001). Improvement of flexural strengths of poly (L-lactic acid) by solid-state extrusion. Macromol. Chem. Phys. 202, 2447–2453. doi:10.1002/1521-3935(20010701)202:11<2447::AID-MACP2447>3.0.CO;2-6.

[168] Lim, J. Y., Kim, S. H., Lim, S., and Kim, Y. H. (2003). Improvement of flexural strengths of poly (L-lactic acid) by solid-state extrusion, extrusion through rectangular die. Macromol. Mater. Eng. 288, 50–57. doi:10.1002/mame.200290033.

[169] Yang, C. S., Wu, H. C., Sun, J. S., Hsiao, H. M., and Wang, T. W. (2013). Thermo-induced shape-memory PEG-PCL copolymer as a dual-drug-eluting biodegradable stent. ACS Appl. Mater. Interfaces. 5, 10985–10994. doi:10.1021/am4032295.

[170] Zhang, Y., Ouyang, H., Lim, C. T., Ramakrishna, S., and Huang, Z.-M. (2005). Electrospinning of gelatin fibers and gelatin/PCL composite fibrous scaffolds. J. Biomed. Mater. Res. 72B, 156–165. doi:10.1002/jbm.b.30128.

[171] Li, X., Xie, J., Lipner, J., Yuan, X., Thomopoulos, S., and Xia, Y. (2009). Nanofiber scaffolds with gradations in mineral content for mimicking the tendon-to-bone insertion site. Nano Lett. 9, 2763–2768. doi:10.1021/nl901582f.

[172] Kunduru, K. R., Basu, A., and Domb, A. J. (2016). Biodegradable polymers: medical applications: In: Encyclopedia of Polymer Science and Technology (pp. 1–22) John Wiley & Sons, Inc., Hoboken, NJ. doi:10.1002/0471440264.pst027.pub2.

[173] Heller, J., and Himmelstein, K. J. (1985). Poly(ortho ester) biodegradable polymer systems. Methods Enzymol, 112, 422–436. http://www.ncbi.nlm.nih.gov/pubmed/3930918.

[174] Park, H., Temenoff, J. S., and Mikos, A. G. (n.d.). Biodegradable orthopedic implants. Eng. Funct. Skelet. Tissues. 55–68. doi:10.1007/978-1-84628-366-6_4.

[175] Hans, M., and Lowman, A. (2002). Biodegradable nanoparticles for drug delivery and targeting. Curr. Opin. Solid State Mater. Sci. 6, 319–327. doi:10.1016/S1359-0286(02)00117-1.

[176] Panyam, J., and Labhasetwar, V. (2004). Targeting intracellular targets. Curr. Drug Deliv. 1, 235–247. doi:10.2174/1567201043334768.

[177] Kwon, H.-Y., Lee, J.-Y., Choi, S.-W., Jang, Y., and Kim, J.-H. (2001). Preparation of PLGA nanoparticles containing estrogen by emulsification–diffusion method. Colloids Surf. A 182, 123–130. doi:10.1016/S0927-7757(00)00825-6.

[178] Dong, Y., and Feng, S.-S. (2004). Methoxy poly(ethylene glycol)-poly(lactide) (MPEG-PLA) nanoparticles for controlled delivery of anticancer drugs. Biomaterials 25, 2843–2849. doi:10.1016/j.biomaterials.2003.09.055.

[179] Carino, G. P., Jacob, J. S., and Mathiowitz, E. (2000). Nanosphere based oral insulin delivery. J. Control. Release. 65, 261–269. doi:10.1016/S0168-3659(99)00247-3.

[180] Yoo, H. S., Lee, K. H., Oh, J. E., and Park, T. G. (2000). In vitro and in vivo anti-tumor activities of nanoparticles based on doxorubicin–PLGA conjugates. J. Control. Release. 68, 419–431. doi:10.1016/S0168-3659(00)00280-7.

[181] Zhou, J., Romero, G., Rojas, E., Moya, S., Ma, L., and Gao, C. (2010). Folic acid modified poly (lactide- co-glycolide) nanoparticles, layer-by-layer surface engineered for targeted delivery. Macromol. Chem. Phys. 211, 404–411. doi:10.1002/macp.200900514.

[182] Lima, K. M., and Rodrigues Júnior, J. M. (1999). Poly-DL-lactide-co-glycolide microspheres as a controlled release antigen delivery system. Braz. J. Med. Biol. Res. 32, 171–180. doi:10.1590/S0100-879X1999000200005.

[183] Eenink, M. J. D., Feijen, J., Olijslager, J., Albers, J. H. M., Rieke, J. C., and Greidanus, P. J. (1987). Biodegradable hollow fibres for the controlled release of hormones. J. Control. Release. 6, 225–247. doi:10.1016/0168-3659(87)90079-4.

[184] Ghadi, R., Muntimadugu, E., Domb, A. J., Khan, W., and Zhang, X. (2017). Synthetic biodegradable medical polymer. In: Science and Principles of Biodegradable and Bioresorbable Medical Polymers. Elsevier. doi:10.1016/B978-0-08-100372-5.00005-2.

[185] Laurencin, C.T., Gerhart, T., Witschger, P., Satcher, R., Domb, A., Rosenberg, A.E., Hanff, P., Edsberg, L., Hayes, W., and Langer, R., Bioerodible polyanhydrides for antibiotic drug delivery: In vivo osteomyelitis treatment in a rat model system, J. Orthop. Res. 11 (1993) 256–262. doi:10.1002/jor.1100110213.

[186] Hochberg, J., Meyer, K. M., and Marion, M. D. (2009). Suture choice and other methods of skin closure. Surg. Clin. North Am. 89, 627–641. doi:10.1016/j.suc.2009.03.001.

[187] Reis, R. L., and San Román, J. (2004). Biodegradable Systems in Tissue Engineering and Regenerative Medicine. CRC Press.

[188] Han, F. Y., Thurecht, K. J., Whittaker, A. K., and Smith, M. T. (2016). Bioerodable PLGA-based microparticles for producing sustained-release drug formulations and strategies for improving drug loading. Front. Pharmacol. 7. doi:10.3389/fphar.2016.00185.

[189] Chen, Y., Geever, L. M., Killion, J. A., Lyons, J. G., Higginbotham, C. L., and Devine, D. M. (2016). Review of multifarious applications of poly (lactic acid). Polym. Plast. Technol. Eng. 55, 1057–1075. doi:10.1080/03602559.2015.1132465.

[190] Jing, X., Mi, H. Y., Huang, H. X., and Turng, L. S. (2016). Shape memory thermoplastic polyurethane (TPU)/poly(ε-caprolactone) (PCL) blends as self-knotting sutures. J. Mech. Behav. Biomed. Mater. 64, 94–103. doi:10.1016/j.jmbbm.2016.07.023.

[191] Lendlein, A., and Langer, R. (2002). Biodegradable, elastic shape-memory polymers for potential biomedical applications. Science 296, 1673–1676. doi:10.1126/science.1066102.

[192] Kellomäki, M., Niiranen, H., Puumanen, K., Ashammakhi, N., Waris, T., and Törmälä, P. (2000). Bioabsorbable scaffolds for guided bone regeneration and generation. Biomaterials 21, 2495–2505. doi:10.1016/S0142-9612(00)00117-4.

[193] Ulery, B. D., Nair, L. S., and Laurencin, C. T. (2011). Biomedical applications of biodegradable polymers. J. Polym. Sci. Part B Polym. Phys. 49, 832–864. doi:10.1002/polb.22259.

Valentina Marturano, Veronica Ambrogi, Iuliana Cota,
Pierfrancesco Cerruti

8 Biobased functional additives for polymers

Abstract: This chapter reviews the most prominent scientific approaches devoted to the development of biobased functional polymer additives, including their chemical composition and mechanism of action. Additionally, a resourceful list of already commercially available products is included. The additives are sorted out in two classes, according to their functionality:

- Protective agents: antioxidants, light stabilizers, UV absorbers, flame retardants, heat stabilizers and acid scavengers;
- Properties modifiers: impact modifiers, plasticisers, compatibilizers, coupling agents.

These novel products are expected to increasingly enter the market, provided that they ensure the same performances of their oil-based counterparts, while representing a safer and more sustainable option. The chapter is intended as a resource for researchers and practitioners working on sustainable polymer formulations and their application.

Keywords: biobased additives, antioxidants, stabilizers, plasticizers, compatibilizers, flame retardants

8.1 Introduction

Since the 1950s, when Karl Ziegler and Giulio Natta first synthesized polyethylene and polypropylene, respectively, polymers have accompanied the technological progress of humankind in all sectors [1]: from commodity household appliance and packaging to automotive, electronics, and pharmaceutical industries, and even design and construction. Even though pristine polymeric material were a good alternative to traditional materials for their lightness, low cost, and easy processability, they often showed poor properties and resulted in commercial failure [2]. The design of plastic formulations, which include additives, make these materials suitable for multiple

Valentina Marturano, Pierfrancesco Cerruti, Institute for Polymers, Composites and Biomaterials (IPCB-CNR), Pozzuoli, Italy
Veronica Ambrogi, Department of Chemical, Materials and Production Engineering (DICMAPI), University of Naples Federico II, Naples, Italy
Iuliana Cota, Université de Rennes I, "Institut des Sciences Chimiques de Rennes", Centre of Catalysis and Gree Chemistry – Team "Organometallics: Materials and Catalysis", Rennes Cedex, France

https://doi.org/10.1515/9783110590586-008

commodity and specialty applications [3]. An additive can be defined as a substance which is incorporated into plastics to achieve a technical effect in the finished product and is intended to be an essential part of the finished article. Such additives can have an effect on the artefact bulk properties of polymer products, or their surface. Pristine polyolefins, for example, are very vulnerable to thermo-oxidative degradation [4] and without the addition of protective additives they would not be the most widely employed commercial material. Rising concerns on plastic pollution and material safety and health standards have encouraged the plastic industry to increase control on the quality and safety of the additives as well as to increase research and development of greener alternatives to oil-based products [5]. The three key issues concerning functional additives involve the use of halogen-based products (widely used in flame-retardant formulations), heavy metals, and plasticizers. Infamously, in 2009 Bisphenol A, one of the most performing precursors for the production of plastics was found in many commodity products even in food contact and was banned by the Canadian government because of its action as an endocrine disruptor [6].

Another societal issue involving plastic use and regulation focuses on end-of-life options for plastic items, especially single-use plastics that have higher probability to be disposed in the terrestrial and marine environment as litter [7]. The chemical nature and composition of additives become highly important under these circumstances, because as the plastic material is recycled, incinerated, disposed in the waste, or littered, the contained additives can leach, pollute, or release dangerous combustible byproduct in the atmosphere [8]. For this reason, several research and development teams are studying more sustainable alternatives to the currently functional polymer additives, with applications in even wider areas than those initially envisaged.

The aim of this chapter is to produce a resourceful list of common commercially available biobased functional additives. Moreover, it provides scientific knowledge on their chemical composition and mechanism of action. Additives are sorted out in two categories considering their functionality:

- Protective agents: antioxidants (AO), light stabilizers, UV absorbers, flame retardants (FR), heat stabilizers, and acid scavengers.
- Properties modifiers: impact modifiers, plasticizers, compatibilizers, and coupling agents.

8.2 Protective additives

Plastic products are durable, light, and cheap materials. However, the first polymers synthesized were considered very unsatisfactory when compared to other commercial materials such as metals and ceramics [9]. The exposure of a pristine polymer to a natural environment (sunlight, temperature variation, or humidity) triggers polymer degradation – often called weathering – ultimately resulting in discoloration and mechanical failure [10]. Plastic materials need protection also from fire and acid environment. A schematization of common factors that can endanger polymer manufacturers in their life cycle (from processing to disposal) is reported in Figure 8.1. In this paragraph, the structure and the action mechanism of different biobased additives for polymer stabilization are analyzed.

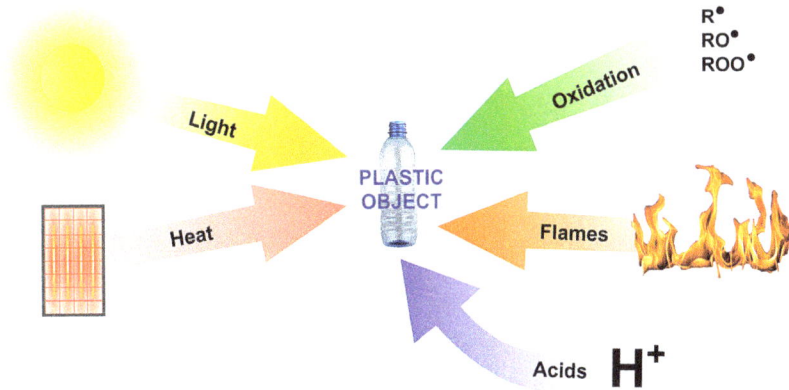

Figure 8.1: Common threats in plastic life cycle.

8.2.1 Antioxidants

In organic materials, oxygen absorption and aging are strictly linked, as discovered for the first time by Bollard et al in 1949 [11].

Oxygen and sunlight, primarily the UV wavelengths, are the principal degrading agents for hydrocarbon polymers during outdoor weathering [12] since they are able to initiate an autoxidation reaction [13]. As shown in Figure 8.2(a), hydrocarbon compounds react with molecular oxygen forming oxidation products, mainly peroxy radicals ($ROO^\bullet$), which further react with organic materials leading to hydroperoxides ($ROOH$).

In both thermal and photo oxidation hydroperoxides are identified as the main branching agents of the autoxidation reaction. Hydroperoxides and their decomposition products are therefore responsible to a large extent for the changes in molecular structure and the overall molar mass decrease in the polymer matrix, resulting in the loss of mechanical properties (e.g., impact, flexure, tensile, and elongation) and a variation in the physical properties of the polymer surface (e.g., loss of gloss, reduced transparency, cracking, and yellowing) [14].

The chemical compounds able to protect organic materials from oxidative processes are called AO. The chemical structures of the most common AO are reported in Figure 8.2(c–e). Living organisms can produce AO that scavenge free

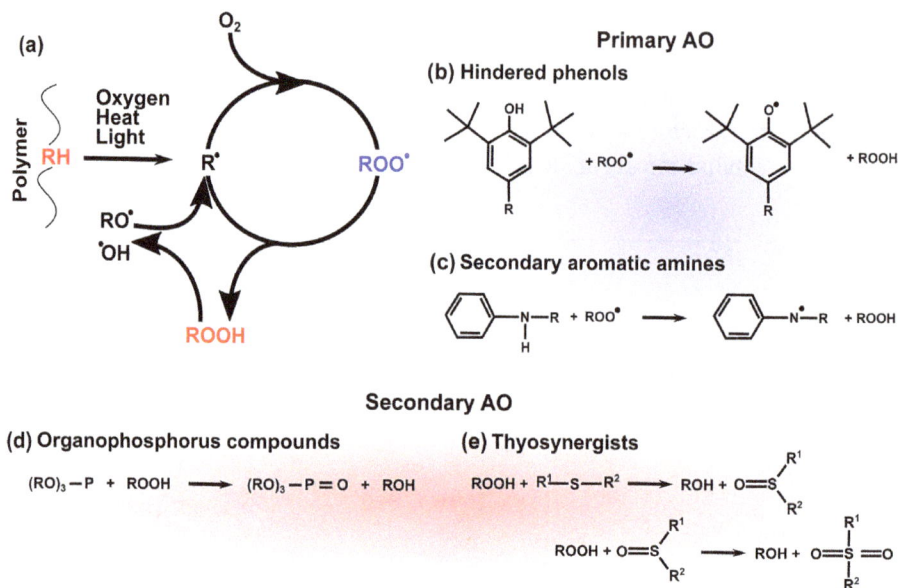

Figure 8.2: Schematization of the autoxidation cycle in polymer matrixes: (a) action mechanism of hindered phenols (b) and secondary amines (c) as primary antioxidants, and organophosphorus compounds (d) and thiosynergists (e) as secondary antioxidants.

radicals protecting cell material. A few of the most renowned natural AO are phenols, flavonoids, anthocyanins, stilbenes, carotenes, and lycopene, among others [15]. Currently, many natural AO are being tested as alternatives to the commercially available oil-based products [16]. However, few of them are available as a viable commercial option.

AO are generally classified into primary and secondary, accordingly to their protection mechanism [10]. Primary AO (Figure 8.2(b, c)) are chain-breaking compounds (chain terminators and chain scavengers) that are able to scavenge low molecular as well as polymeric radicals by chain-breaking electron-donor mechanism. Secondary AO (Figure 8.2(d, e)) are known as peroxide decomposers as they are able to decompose hydroperoxy groups (HOO–) present in polymers, converting them in less reactive substances.

8.2.1.1 Primary antioxidants

In these compounds, the inhibition of the oxidative process occurs via chain terminating reactions. Primary AO, such as hindered phenols and secondary aromatic amines, have reactive OH or NH groups which are able to transfer a proton to the chain propagating free radical species, as depicted in Figure 8.2(b, c). [17]. The resulting radical is therefore stabilized and unable to further attack the protons present in the polymer chain. Sterically hindered phenols are the most widely employed phenolic stabilizers because of their high performances during both processing and long-term aging. Most importantly, the phenoxy radicals generated are very stable due to their ability to build numerous mesomeric forms [18, 19].

Amines also act as primary AO via a deactivation reaction of peroxy radicals displayed in Figure 8.2(c). Secondary aromatic amines are acknowledged as the most efficient hydrogen donors with a much higher activity than hindered phenols, because of less steric hindrance. Aromatic amines, however, possess a discoloration effect especially when exposed to light or combustion gases [20]. In Table 8.1 two commercial biobased AO are listed, both having an amine-based composition.

Table 8.1: Commercially available biobased primary antioxidants.

Commercial name	Producer	Composition	Derived from	Used in	Food contact
Revonox® 420 V	Chitec Technology	Hydroxylamine	Palm oil	Polyolefins	Yes
GENOX® EP stabilizer	SI Group	Blend of amines	Rapeseed oil	Polyolefins	Yes

8.2.1.2 Secondary antioxidants

Secondary AO are commonly referred to as hydroperoxide decomposers because they are able to decompose hydroperoxide into nonradical, nonreactive, and thermally stable products, thus preventing the split of hydroperoxides into extremely reactive alkoxy and hydroxyl radicals. They are often used in combination with primary AO to yield synergistic stabilization effects [21].

Organophosphorus compounds decompose peroxides and hydroperoxides according to Figure 8.2(d). Although they are particularly efficient at high temperatures during polymer processing, they are normally used in combination with a primary AO [22]. When accounting for sustainability and safety, organophosphorus compounds have many drawbacks. Due to their sensitivity to water, some compounds can hydrolyse, producing acidic species and are sometimes labelled as toxic [23]. Another type of secondary AO are sulfur-based hydroperoxide decomposers, known as thiosynergists, that act according to the schematization proposed in Figure 8.2(e) [24].

8.2.2 Photostabilizers

The initiation of the oxidation of the polymeric chain (Figure 8.2(a)) can occur also by means of a light photon; in this case, the process leading to properties loss and discoloration is called photodegradation [25]. Photo oxidative degradation takes place at the surface, as the intensity of the UV fraction of sun radiation is maximum at the surface and has low penetration efficiency, when compared to less energetic wavelengths, such as infrared radiation [26].

Photostabilizers are a family of compounds added to polymeric materials to prevent photochemical destructive process and the reactions caused by exposure to light [27]. Photostabilizers can be classified into several main classes: UV absorbers, quenchers, hindered amine light stabilizers (HALS), and UV screeners.

8.2.2.1 UV absorbers

UV absorbers need to interfere with the first step of the photooxidation process by absorbing the harmful UV radiation (300–400 nm) before it reaches the photo active site in the polymer backbone. A UV absorber must be light stable, because otherwise it would be destroyed during the stabilizing reactions [28]. The most common mechanism for energy dissipation is the conversion of harmful UV radiation into harmless IR radiation or heat that is in turn dissipated in the polymer. Due to their filtrating action, many aromatic compounds possess a photostabilizing effect [29, 30]. Hydroxyaromatic compounds, such as hydroxybenzophenones, were originally designed to absorb wavelengths between 290 and 400 nm in the region where polymer systems are the most vulnerable to UV radiation. A schematization of the mechanism of energy dissipation of hydroxyaromatic compounds is reported in Figure 8.3(a), where in avobenzones the reversible formation of a six membered hydrogen bonded ring occurs. The two tautomers, in thermodynamic equilibrium, allow an easy deactivation of the excited state induced by the light absorption.

8.2.2.2 Quenchers

These compounds deactivate the singlet/triplet excited states of chromophore moieties present in the polymer [31, 32]. Effective quenchers are able to dissipate the accumulated energy without the production of harmful by-products such as free radicals. [33, 34]. Figure 8.3(c) represents an exemplification of the quenching reaction involving metal chelates. Quenchers can be distinguished by their chemical nature. Metal complexes are often considered the most performing, especially nickel chelates which are very effective quenchers of the triplet state of carbonyl groups in polyolefins [35].

(a)

(b)

(c)

$$Ch, O_2 \xrightarrow{hv} Ch^*, O^1{}_2 \xrightarrow{Q} Ch, O_2 + \boxed{Q^*}$$

$$\boxed{Q^* \longrightarrow Q + heat}$$

Q=quencher R$_2$N—⟨S...S⟩Ni⟨S...S⟩—NR$_2$

Ch=chromophore (dyes, pigments, carbonyl groups)

(d)

UV radiation

Coatings

Incorporated pigments

Polymer matrix

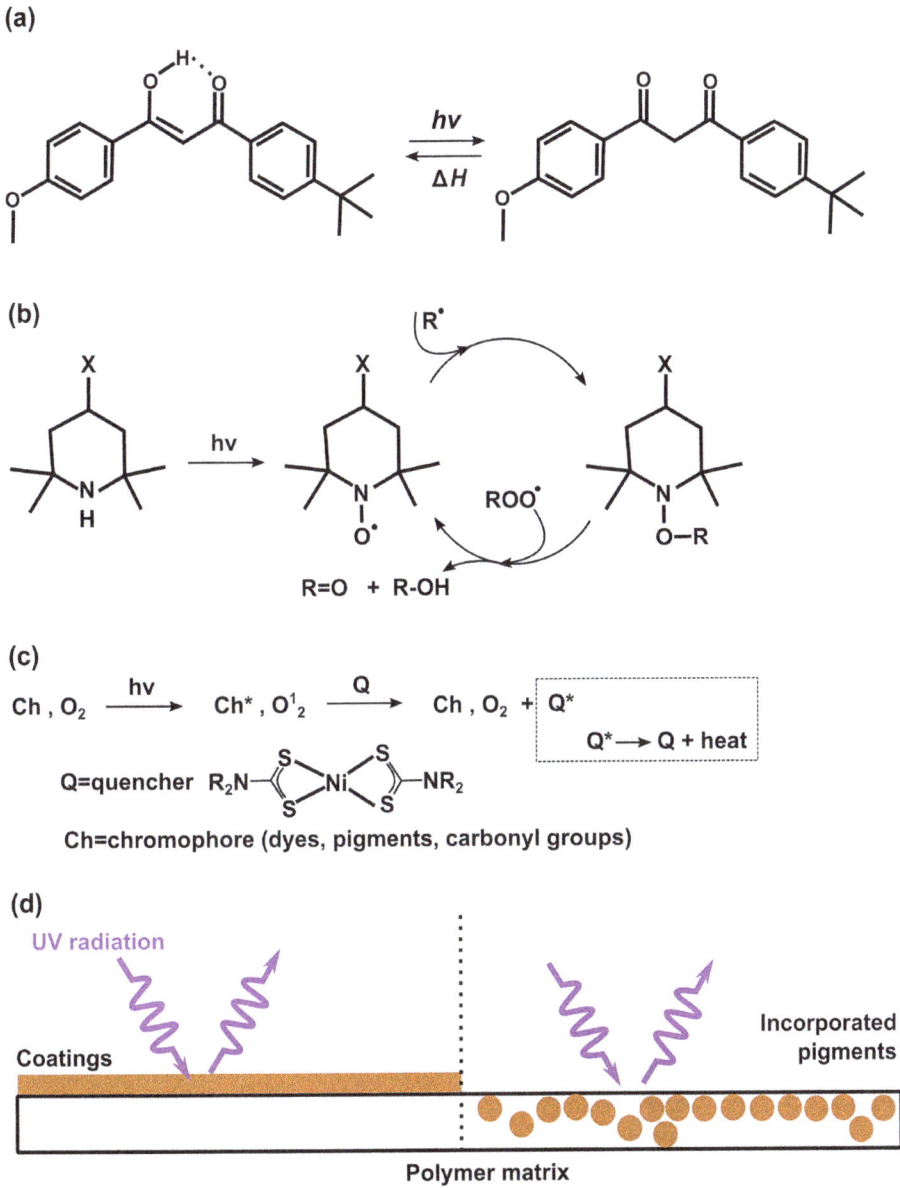

Figure 8.3: Schematization of photostabilization mechanisms via avobenzone light stabilizers (a), hindered piperidine light stabilizers (b), metal-based quenchers (c), and coating and pigments as light (UV) screeners (d).

8.2.2.3 Hindered amine light stabilizers

HALS are generally derivatives of tetramethyl piperidine. During UV irradiation, in presence of oxygen and free radicals, piperidine forms piperidinoxy radicals that in turn deactivate other harmful radicals involved in the reaction (Figure 8.3(c))[36]. HALS are regarded as the most efficient UV stabilizers for a wide range of plastics even in specialty applications, such as automotive industry [37]. Both biobased HALS and metal-based quenchers are commercially available as shown in Table 8.2.

Table 8.2: Commercially available quenchers for polymers photostabilization.

Commercial name	Composition	Producer	Composition	Derived from
CAPLIG 770	HALS	Nanjing Capatue Chemical	Bis (2, 2, 6, 6-tetramethyl-4-piperidyl) sebacate	Plant-based sebacate
Akcrostab™ LT-4803	Metal-based	Valtris	Mixed metal epoxidized soybean oil blend	Soybean oil

8.2.2.4 UV screeners

The action of these additives does not involve their chemical structure. UV screeners are simply materials that can reflect light away from the polymeric object hindering its damaging effect. Coatings and pigments, as shown in Figure 8.3(d), are the two most widely used technologies to screen plastic items [38] and confine the photooxidative mechanisms to their surface [39].

8.2.3 Heat stabilizers

During processing, polymers undergo both mechanical deformation and thermal stresses. Heat stabilizers are added to the polymeric formulation in order to obstruct the heat-induced oxidation of the processing step, but they are also useful during application. Heat stabilizers attack chemical compound derived from oxidation, therefore, stopping the oxidative process [40]. Heat stabilizers can be classified according to their chemical composition: (i) metallic salts, where two or more metals (barium, cadmium, lead, or zinc) are synergistically used together, (ii) organometallic compounds (mainly tin-based), and (iii) nonmetallic organic stabilizers.

To improve safety and sustainability of these much-needed product for plastic formulation, a reduction in the use of metal-based compounds, especially lead, has been promoted since the 1990s [41]. In Table 8.3 several biobased options are listed,

Table 8.3: Commercially available biobased heat stabilizers.

Class	Commercial name	Producer	Composition	Derived from	Used in
Barium/ zinc	Ligastab BAL	Peter Greven	Barium salt of a technical lauric acid	Lauric acid-based	PVC
Calcium/ zinc	Ligastar CA 12 OXY	Peter Greven	Calcium salt of a 12-hydroxy stearic acid	Stearic acid	PVC
	Liga Calcium behenate	Peter Greven	Calcium salt of a technical behenic acid	Behenic acid	PVC
	Baerostab NT 170 PS	Baerlocher	2-EHA-free epoxidized soya bean oil-based calcium–zinc complex	Soya bean-oil	PVC Food approved
Organic	Baerostab LSA	Baerlocher	Epoxidized soya bean oil	Soya bean oil	Rigid PVC
	Baerostab LSU	Baerlocher	Epoxy octyl stearate	Fatty acid	Rigid PVC

including barium and calcium/zinc salts of biobased lauric, stearic, and behenic acids as well as organic heat stabilizers based on soybean oil.

8.2.4 Flame retardants

Fire constitutes a serious threat to all organic (i.e., carbon-based) materials, but while wood has been more or less harmlessly burnt for thousands of years to produce energy, the combustion of plastics items leads to the release of low-molecular-weight fragments in the atmosphere containing potentially harmful elements such as sulfur, fluorine, and chlorine [42]. The action of FR agents consists of a chemical or physical interference with the combustion process [43]. Many commercially available FR belong to three main chemical classes: halogen-based, phosphorus-based, and metal hydrate FR. Some FR belonging to the halogen-based class, especially brominated, are considered harmful for the environment [44] and have been banned by some countries. Among them, polybrominated biphenyl and polybrominated diphenyl ethers are no longer manufactured. Biobased FR constitutes a safer and greener option, and many biobased compounds (e.g., cellulose, starch, and lignin) have an intrinsic ability to produce a char layer during thermal decomposition [45]. However, many biobased products have been never introduced to the market because they are not able to perform appropriately. In Table 8.4 some commercially available biobased FR are listed. For example, Kronitex® CDP is known among all biobased products as

Table 8.4: Biobased FR introduced in the global additives market.

Commercial name	Producer	Composition	Used in
Polycard XFN™ 53	Composite Technical Services	Renewable aromatic multifunctional polyol	Biobased rigid and flexible urethane prepolymers.
Kronitex® CDP	Great Lakes (Chemtura Group)	Naturally derived cresyl diphenyl phosphate	Natural and synthetic rubber and PVC
Charmor™ PM40 Care	Perstorp	Polyhydric alcohol containing four primary hydroxyl groups	

one of the most performing phosphorus-based FR with char forming agents used in natural and synthetic rubber.

8.2.5 Acid scavengers

Polymers are used in a wide variety of applications, therefore they need to be resistant to different types of aggressive environments [46, 47]. One important feature to provide stable polymer formulations is the ability to effectively neutralize acidity. Commonly used acid scavengers are metal stearates, usually employed in polyolefins. An increasing number of vegetable-based acid scavengers, mostly calcium and magnesium stearates, are now commercially available, as reported in Table 8.5. Interestingly, many acid scavengers behave as well as the property modifiers, influencing crystallization and processing behavior [48]. In fact, for all products listed in Table 8.5, their manufacturers reported a side effect as lubricant or processing aid. Sun Ace, for example, produces vegetable-based fatty acid and metal stearates (calcium, magnesium, zinc, and potassium) in liquid and powder formulations, which act as acid scavengers as well as lubricant, mold release agents, extenders, and water-repellent agents.

Table 8.5: Commercially available biobased acid scavengers.

Commercial name	Producer	Composition	Other properties	Used in
Ceasit AV Veg	Baerlocher	Vegetable-based Calcium stearate grade	Lubricant, processing aid	Polyolefins and polybutylene terephthalate
SAK- DCS	Sun Ace	Dispersible calcium stearate	Lubricant and mold release agent	Plastic and rubber
SAK- DMS	Sun Ace	Dispersible magnesium stearate	Lubricant and mold release agent,	Plastic and rubber

8.3 Modification of polymer properties and compatibilization

In real-life applications, pristine polymers are not expected to exhibit good perform-ances because they typically do not possess good mechanical properties. However, once protective properties are imparted to the polymer, it is possible to improve its mechanical performance by either increasing flexibility, processability, and disten-sibility (plasticizers) or imparting impact resistance (impact modifiers). A further improvement can be achieved by using additives that can maximize the compatibil-ity of the polymer matrix with other polymers or fillers (compatibilizers, coupling agents, and adhesion promoters) (Figure 8.4) [49].

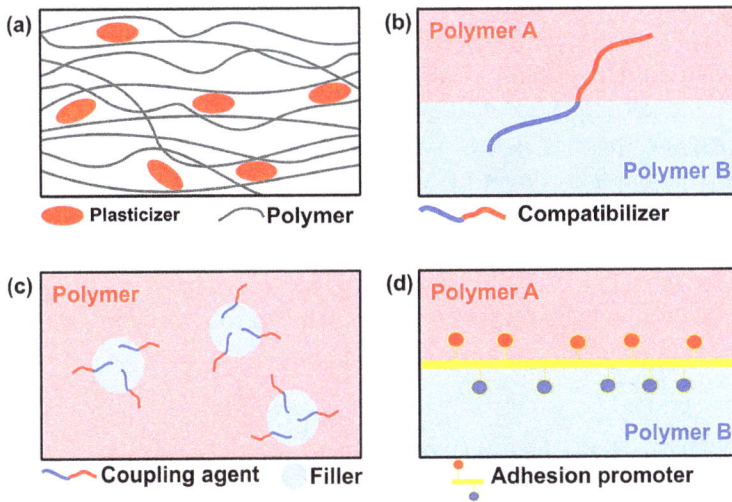

Figure 8.4: Modification of polymer properties by addition of plasticizers (a), compatibilizers (b), coupling agents (c), and adhesion promoters (d).

8.3.1 Plasticizers

Without plasticizers, polymer industry would not have spread as much as it has in the last 70 years, because pristine polymers lack in flexibility, extensibility, and pro-cessability. Plasticizers are an essential tool for the performance improvement of both traditional and biobased polymers. In fact, they increase flow and thermoplas-ticity of plastic materials by decreasing the viscosity of polymer melts, the glass tran-sition temperature (T_g), the melting temperature (T_m), and the elasticity modulus of finished products [50]. As depicted in Figure 8.4(a), the flexibility is imparted by to the macromolecular chains with the addition of smaller plasticizing molecules, which increase the chain mobility enhancing their mechanical properties, lowering the brittle-to-tough transition of the system, and extending the temperature range for its rubbery or viscoelastic state behavior [51].

Plasticizers for PVC are usually divided according to the chemical nature of their nonpolar functional group [52], which can be either a phthalic acid ester, having polar groups attached to aromatic rings, or a polar aliphatic group. Concerns have been raised regarding the use of phthalates because they are not chemically bound to the polymer matrix and are therefore able to leach in indoor or outdoor atmosphere, thus migrating to food products or to other items [52, 53]. Phthalates are broadly classified as endocrine disruptors [54], therefore, for health and environmental reasons, the market has been pushed to search for biobased alternatives to traditional petrol-based products [55]. In Table 8.5, the classification of biobased products is presented on the basis of their chemical structure. Therefore adipates, benzoates, citrates, epoxidized oils, polyesters, and sebacates are listed. Being natural/renewable sourced, biobased plasticizers are sometimes easily approved for food contact and medical applications. This category of plasticizers can be easily incorporated in toys and teething products for infants. Some of them have also found use in wire insulation and jacketing, household and consumer goods, flooring, carpet backing, and other building and construction end use applications [56]. Vegetable oil derivatives are the most widely used natural product-type plasticizers. Products consisting of triglyceride esters of unsaturated fatty acids (e.g., soybean oil and linseed oil) – in which the double bonds in the fatty acid residues are epoxidized – have been commercial products for decades (Table 8.6) [57].

Table 8.6: Classification of commercially available biobased plasticizers based on their chemical composition.

Class	Commercial name	Producer	Composition	Used in
Adipates	Syncroflex™ 3114	Croda	Polyadipate ester	PVC and rubber
	Syncroflex™ 3142	Croda	Polyazelate ester	PVC and rubber
	EDENOL® 111	Emery Oleochemicals	Adipate ester	synthetic rubbers.
	EDENOL® 1200	Emery Oleochemicals	Adipic acid-based polymeric plasticizer	PVC, PLA, and synthetic rubbers.

Table 8.6 (continued)

Class	Commercial name	Producer	Composition	Used in
Benzoates	Uniplex 280CG	Lanxess	Sucrose benzoate	PMMA, thermosets MF and UF resins, cellulosic polymers, PVA, and PVC
	VELSIFLEX™ 328	Velsicol	Blend of diethylene glycol dibenzoate and dipropylene glycol dibenzoate	Flexible PVC
Citrates	Proviplast® 2624	Proviron	Tributyl 2-acetylcitrate	Cellulosic polymers and PVC
	HALLGREEN® R-C discontinued	Hallstar	Tributyl citrate	Cellulosic polymers PHA, PLA, polyester resin, and PVC
	CITROFOL® AI	Jungbunzlauer	Triethyl citrate	PVC, used in food contact (FDA approved)
	OXBLUE® ATBC	Oxea	Acetyl tributyl citrate	PVC and cellulosic resins
	Citroflex® 4	Vertellus Specialties	Tri-*n*-butyl citrate	PVC, PLA, and cellulosic resins
Epoxidised oils	ADK CIZER® O-130P	Adeka	Epoxidized soybean oil	PVC, PS, and ABS
	Epoxol® 7-4	American Chemical Service	Epoxidized soybean oil	PVC
	Paraplex® G-60	Hallstar	HMW epoxidized soybean oil	Chlorinated rubber and PVC
	Plasthall® ELO	Hallstar	Epoxidized linseed oil	Chlorinated rubber and PVC
Polyesthers	Radiamuls® 2130	Oleon (Avril Group)	Acetylated monoglyceride	PVC
	Radia® 7051	Oleon (Avril Group)	Butyl stearate	PVC, TPU, ABS, and PS
	Syncroflex™ 3114	Croda	Polyadipate ester	PVC
Sebacates	Proviplast® 1944	Proviron	Dibutyl sebacate	PVC

Table 8.6 (continued)

Class	Commercial name	Producer	Composition	Used in
	Oleris® Dimethyl Sebacate	Arkema	Dimethyl sebacate	Cellulosic polymers
	Levulinic ketals	GFBiochemicals	Ketalized levulinic acid esters	PVC, flexible or Rigid and PUR

8.3.2 Impact modifiers

We often refer to plastics as resistant materials, to the point that polymers are occasionally themselves used as toughening additives. However, many commodity polymers, such as polystyrene, polypropylene, polyvinyl chloride, and others, often show a brittle impact behavior at room temperature. This characteristic, attributed to the fact that their glass transition is higher or close to room temperature, can easily be modified by adding to the polymer formulation an impact modifier additive. These materials are generally added in the form of particles with good size distribution and homogeneous dispersion, with a low glass transition temperature that is able to guarantee elastomeric properties at room temperature [58]. Commonly used rubber modifiers are butadiene-based graft copolymers, butadiene-based block copolymers, ethylene-based rubbers, polyurethanes, and Acrylonitrile Butadiene Styrene – Nitrile Butadiene Rubber (ABS-NBR) core–shell graft copolymers [2]. As shown in Table 8.7, the commercial options involving biobased impact modifiers are currently rather limited.

Table 8.7: Commercially available biobased impact modifiers.

Commercial name	Producer	composition	Used in	Concentration
HD-L01	Polyvel	Polylactic acid	Polylactic acid	5–10%
Terratek® Flex	Green Dot Bioplastics	Starch-based compostable elastomer	Polylactic acid	10–30%

8.3.3 Compatibilizers, coupling agents, and adhesives

Polymers are rarely found in their pristine form, more often they are mixed with additives and fillers or matched to other polymers or materials to combine their excellent properties to form a higher performance finite product. Polymers with different chemical structures are often hardly miscible and the formation or a homogeneous blend is therefore thermodynamically hindered [59]. In this case, taking into account a two-component blend, the most abundant polymer constitutes the continuous matrix, while the second component is dispersed. Compatibilizers are added to the formulation because of the low adhesion between the continuous and dispersed phases. These chemicals are macromolecular species with a two block structures with each side having a maximized compatibility with either the continuous or the dispersed phase, as represented in Figure 8.4(b). [60, 61]. Although the potential of the biopolymers market is undeniable, these green materials are still more expensive than commodity polymers with much inferior properties. For this reason, mixing and compatibilizing two biopolymers is a positive strategy to obtain viable properties [62]. Moreover, compatibilization plays an important role in the recycling of waste polymers, particularly those used in packaging applications [63]. In a similar way, the compatibility between polymeric matrix and the filler (particles or fibers) is promoted by coupling agents as shown in Figure 8.4(c). Silane coupling agents have been most widely used in commercial applications involving traditional reinforcement not only such as glass fibers [64] but also natural fibers thanks to their ability to couple polar fillers with most nonpolar polymeric matrices [65, 66]. Research efforts aimed to the replacement of commercial compatibilizers, mostly based on oil-based copolymers, with modified vegetable oils as well as copolymers synthesized using biobased comonomers [67–69] are ongoing. Research and development in adhesion promoters and adhesives also arose to accommodate the needs of the current packaging market. In fact, in multilayer films, two polymers are often coupled and sometimes other materials are employed, such as paper or aluminum. [70, 71]. Unfortunately, the market of compatibilizers, coupling agents, and adhesives still lacks in number of biobased options (Table 8.8).

Table 8.8: Commercially available biobased adhesives.

Class	Commercial name	Producer	Composition	Used in
Biobased PE	Renew	Yparex®	Biobased LDPE and LLDPE	Tie layer in multilayer flexible films
Acrylics	S500	BioTAK® Adhesives	Acrylic	Laminating adhesive

8.4 Conclusions

Thanks to public awareness regarding plastic pollution, in many countries industries and politics synergically aspire to a more sustainable and safe approach to plastic use. The focus is mainly directed to biopolymers and biobased materials. In parallel, concerns on safety and regulation of plastic additives are constantly pushing toward the investigation of biobased additives. These novel products are able to enter the market provided that they ensure high performances of their oil-based alternatives while resulting as a safer and more sustainable option. Though ongoing research does not always travel in parallel with industrial development, the progress toward 100% biobased formulations seems inevitable and research efforts are thriving.

References

[1] Hutley, T. J., and Ouederni, M. (2016). Polyolefins – the history and economic impact. In: Polyolefin Compounds and Materials (pp. 13–50). Springer, Cham.
[2] Marturano, V., Cerruti, P., and Ambrogi, V. (2017). Polymer additives. Phys. Sci. Rev. 2(6).
[3] Plastic Additives, An A-Z reference. G. Prithchard.
[4] Bockhorn, H., Hornung, A., Hornung, U., and Schawaller, D. (1999). Kinetic study on the thermal degradation of polypropylene and polyethylene. J. Anal. Appl. Pyrolysis 48(2), 93–109.
[5] Thompson, R. C., Moore, C. J., Vom Saal, F. S., and Swan, S. H. (2009). Plastics, the environment and human health: current consensus and future trends. Philos. Trans. R. Soc. B: Biol. Sci. 364(1526), 2153–2166.
[6] Vogel, S. A. (2009). The politics of plastics: the making and unmaking of bisphenol a "safety". Am. J. Public Health, 99(S3), S559–S566.
[7] Depledge, M. H., Galgani, F., Panti, C., Caliani, I., Casini, S., and Fossi, M. C. (2013). Plastic litter in the sea. Mar. Environ. Res. 92, 279–281.
[8] Geyer, R., Jambeck, J. R., and Law, K. L. (2017). Production, use, and fate of all plastics ever made. Sci. Adv. 3(7), e1700782.
[9] Pritchard, G. (2012). Plastics Additives: An AZ Reference (Vol. 1). Springer Science & Business Media.
[10] Ambrogi, V., Carfagna, C., Cerruti, P., and Marturano, V. (2017). Additives in polymers. In: Modification of Polymer Properties (pp. 87–108). William Andrew Publishing.
[11] Bolland, J. L. (1949). Kinetics of olefin oxidation. Q. Rev. Chem. Soc. 3, 1–21.
[12] Feldman, D. (2002). Polymer weathering: photo-oxidation. J. Polym. Environ. 10(4), 163–173.
[13] Crounse, J. D. (2013). Autoxidation of organic compounds in the atmosphere. J. Phys. Chem. Lett. 4, 3513–3520.
[14] Singh, B., and Sharma, N. (2008). Mechanistic implications of plastic degradation. Polym. Degrad. Stab. 93(3), 561–584.
[15] Xu, D. P., Li, Y., Meng, X., Zhou, T., Zhou, Y., Zheng, J., et al. (2017). Natural antioxidants in foods and medicinal plants: extraction, assessment and resources. Int. J. Mol. Sci. 18(1), 96.

[16] Kirschweng, B., Tátraaljai, D., Földes, E., and Pukánszky, B. (2017). Natural antioxidants as stabilizers for polymers. Polym. Degrad. Stab. 145, 25–40.

[17] Pospisil, J. (1993). Chemical and photochemical behaviour of phenolic antioxidants in polymer stabilization – a state of the art report. Polym. Degrad. Stab. Pt I 39, 103–115, Part II 40, 217–232.

[18] Grossman, R. F., and Lutz, J. T., Jr (Eds.). (2000). Polymer Modifiers and Additives. CRC Press.

[19] Ingold, K. U., and Pratt, D. A. (2014). Advances in radical-trapping antioxidant chemistry in the twenty-first century: a kinetics and mechanisms perspective. Chem. Rev. 114(18), 9022–9046.

[20] Pocius, A. V., and Dillard, D. A. (Eds.). (2012). Adhesion Science and Engineering: Surfaces, Chemistry and Applications. Elsevier.

[21] Shanina, E. L., Zaikov, G. E., Fazlieva, L. K., Bukharov, S. V., and Mukmeneva, N. A. (2002). Influence of synergists: the influence of hydroperoxide decomposers on phenolic inhibitors consumption in oxidized polypropylene, J. Appl. Polym. Sci. 85, 2239–2243.

[22] Schwetlick, K., Pionteck, J., König, T., and Habicher, W. D. (1987). Organophosphorus antioxidants – VIII. Kinetics and mechanism of the reaction of organic phosphites with peroxyl radicals. Eur. Polym. J. 23(5), 383–388.

[23] Kumar, S. V., Fareedullah, M. D., Sudhakar, Y., Venkateswarlu, B., and Kumar, E. A. (2010). Current review on organophosphorus poisoning. Arch. Appl. Sci. Res. 2(4), 199–215.

[24] Beißmann, S., Grabmayer, K., Wallner, G., Nitsche, D., and Buchberger, W. (2014). Analytical evaluation of the performance of stabilization systems for polyolefinic materials. Part II: interactions between hindered amine light stabilizers and thiosynergists. Polym. Degrad. Stab. 110, 509–517.

[25] Emad, Y., and Raghad, H. (2013). Photodegradation and Photostabilization of Polymers, Especially Polystyrene: Review. Springerplus, 2, 398. doi:10.1186/2193-1801-2-398.

[26] Rabek, J. F. (1996). Practical aspects of polymer photodegradation. In: Photodegradation of Polymers (pp. 161–191). Springer, Berlin, Heidelberg.

[27] Rabek, J. F. (2012). Polymer Photodegradation: Mechanisms and Experimental Methods. Springer Science & Business Media.

[28] Aloui, F., Ahajji, A., Irmouli, Y., George, B., Charrier, B., and Merlin, André. (2007). Inorganic UV absorbers for the photostabilisation of wood-clearcoating systems: Comparison with organic UV absorbers. Appl. Surf. Sci. 253. 3737–3745.

[29] Rabek, J. F., and Ranby, B. (1975). Photodegradation, Photooxidation and Photostabilization of Polymers. John Wiley, New York.

[30] Heller, H. J. (1969). Photochemistry of dyed and pigmented polymers. Eur. Polym. J. Suppl., 99–105.

[31] Wiles, D. M. (1979). Photostabilisation mechanisms in polymers: a review. Pure Appl. Chem. 50, 291.

[32] Allen, N. S., and Edge, M. (1992). Fundamentals of Polymer Degradation and Stabilization. Elsevier Applied Science, London.

[33] Guillory, J. P., and Cook, C. F. (1973). Energy transfer processes involving ultraviolet stabilizers. Quenching of excited states of ketones. J. Am. Chem. Soc. 95(15), 4885–4891.

[34] Heller, H. J. (1969). Protection of polymers against light irradiation. Eur. Polym. J. 5, 105–132.

[35] Lala, D., and Rabek, J. (1980). Polymer photodegradation: mechanisms and experimental methods. Polym. Degrad. Stab. 3, 383–391.

[36] Yousif, E., Salimon, J., and Salih, N. (2011). Photostability of Poly (Vinyl Chloride). VDM Verlag: Dr. Muller Publisher.

[37] Zareanshahraki, F., Davenport, A., Cramer, N., Seubert, C., Lee, E., and Cassoli, M. (2020). Durability Study of Automotive Additive Manufactured Specimens (No. 2020-01-0957). SAE Technical Paper.

[38] Pickett, J. E., and Moore, J. E. (1996). Photostability of UV Screeners in Polymers and Coatings.

[39] Yousif, E. (2012). Photostabilization of Thermoplastic Polymers. Lambert Academic Publishing, Germany.

[40] Heat stabilizers. (1999). Plast. Addit. Compd. 1(4), 24–29.

[41] Prunier, J., Maurice, L., Perez, E., Gigault, J., Wickmann, A. C. P., Davranche, M., and Ter Halle, A. (2019). Trace metals in polyethylene debris from the North Atlantic subtropical gyre. Environ. Pollut. 245, 371–379.

[42] Morgan, A. B., and Gilman, J. W. (2013). An overview of flame retardancy of polymeric materials: application, technology, and future directions. Fire Mater. 37, 259–279.

[43] Morgan, A. B., and Gilman, J. W. (2013). An overview of flame retardancy of polymeric materials: application, technology, and future directions. Fire Mater. 37, 259–279.

[44] de Wit, C. A. (2002). An overview of brominated flame retardants in the environment. Chemosphere 46(5), 583–624.

[45] Costes, L., Laoutid, F., Brohez, S., and Dubois, P. (2017). Biobased flame retardants: when nature meets fire protection. Mater. Sci. Eng.: R: Rep. 117, 1–25.

[46] Munaro, M., and Akcelrud, L. (2008). Polyethylene blends: a correlation study between morphology and environmental resistance. Polym. Degrad. Stab. 93(1), 43–49.

[47] Ribeiro, M. C. S., Tavares, C. M. L., and Ferreira, A. J. M. (2002). Chemical resistance of epoxy and polyester polymer concrete to acids and salts. J. Polym. Eng. 22(1), 27–44.

[48] Thürmer, A. (1998). Acid scavengers for polyolefins. In: Plastics Additives (pp. 43–48). Springer, Dordrecht.

[49] Kutz, M. (Ed.). (2016). Applied Plastics Engineering Handbook: Processing, Materials, and Applications. William Andrew.

[50] Chanda, M., and Roy, S. K. (Eds). (2007). Plastics Technology Handbook, Fourth Edition. CRC Press, Boca Raton.

[51] Štěpek, J., and Daoust, H. Additives for plastics. Volume 5 of the series Polymers pp. 7–33. Plasticizers.

[52] Heudorfa, U., Mersch-Sundermannb, V., and Angererc, J. (2007). Phthalates: toxicology and exposure. Int. J. Hyg. Environ. Health. 210(5), 623–634.

[53] Analysis of phthalate migration from plastic containers to packaged cooking oil and mineral water. Qian, Xu, Xueyan Yin, Min Wang, Haifeng Wang, Niping Zhang, Yanyan Shen, Shi Xu, Ling Zhang, and Zhongze Gu. J. Agric. Food Chem., 2010, 58 (21), 11311–11317. DOI: 10.1021/jf102821h

[54] Zamkowska, D., Karwacka, A., Jurewicz, J., and Radwan, M. (2018). Environmental exposure to non-persistent endocrine disrupting chemicals and semen quality: an overview of the current epidemiological evidence. Int. J. Occup. Environ. Health. 31(4), 377–414.

[55] Vieira, M. G. A., da Silva, M. A., dos Santos, L. O., and Beppu, M. M. (2011). Natural-based plasticizers and biopolymer films: a review. Eur. Polym. J., 47(3), 254–263.

[56] Xie, Z., Chen, Y., Wang, C., Liu, Y., Chu, F., and Jin, L. (2014). Effects of biobased plasticizers on mechanical and thermal properties of PVC/wood flour composites. BioResources, 9(4), 7389–7402.

[57] Bocqué, M., Voirin, C., Lapinte, V., Caillol, S., and Robin, J. J. (2016). Petro-based and bio-based plasticizers: chemical structures to plasticizing properties. J. Polym. Sci. Pt A: Pol. Chem., 54(1), 11–33.

[58] Nazari, D., Bahri-Laleh, N., Nekoomanesh-Haghighi, M., Jalilian, S. M., Rezaie, R., and Mirmohammadi, S. A. (2018). New high impact polystyrene: Use of poly (1-hexene) and poly (1-hexene-co-hexadiene) as impact modifiers. Polym. Advan. Technol., 29(6), 1603–1612.

[59] Nesterov, A. E., and Lipatov, Y. S. (1998). Thermodynamics of Polymer Blends (Vol. 1). CRC Press.

[60] Manson, J. A. (2012). Polymer Blends and Composites. Springer Science & Business Media.

[61] Compatibilization of Polymer Blends, L. A. Utracki.

[62] Imre, B., and Pukánszky, B. (2013). Compatibilization in biobased and biodegradable polymer blends. Eur. Polym. J., 49(6), 1215–1233.

[63] Maris, J., Bourdon, S., Brossard, J. M., Cauret, L., Fontaine, L., and Montembault, V. (2018). Mechanical recycling: compatibilization of mixed thermoplastic wastes. Polym. Degrad. Stabil., 147, 245–266.

[64] Park, S. J., and Jin, J. S. (2001). Effect of silane coupling agent on interphase and performance of glass fibers/unsaturated polyester composites. J. Colloid. Interf. Sci., 242(1), 174–179.

[65] Xie, Y., Hill, C. A., Xiao, Z., Militz, H., and Mai, C. (2010). Silane coupling agents used for natural fiber/polymer composites: a review. Compos. Pt A. Appl. Sci. Manuf., 41(7), 806–819.

[66] Lu, J. Z., Wu, Q., and McNabb, H. S. (2007). Chemical coupling in wood fiber and polymer composites: a review of coupling agents and treatments. Wood Fiber Sci., 32(1), 88–104.

[67] Ignaczak, W., Sobolewski, P., and El Fray, M. (2019) Bio-Based PBT–DLA Copolyester as an alternative compatibilizer of PP/PBT blends. Polymers, 11, 1421.

[68] Liminana, P., Garcia-Sanoguera, D., Quiles-Carrillo, L., Balart, R., and Montanes, N. (2019). Optimization of maleinized linseed oil loading as a biobased compatibilizer in poly (butylene succinate) composites with almond shell flour. Materials, 12(5), 685.

[69] Turco, R., Ortega-Toro, R., Tesser, R., Mallardo, S., Collazo-Bigliardi, S., Chiralt Boix, A., Santagata, G. (2019). Poly(lactic acid)/thermoplastic starch films: effect of cardoon seed epoxidized oil on their chemicophysical, mechanical, and barrier properties. Coatings, 9(9), 574.

[70] Morris, B. A. (2016). The Science and Technology of Flexible Packaging: Multilayer Films from Resin and Process to End Use. William Andrew.

[71] Tena, M. T. (2019). Adhesives in Food Packaging. Food Contact Mater. Anal: Mass Spectrom. Tech., 10, 82.

Sarai Agustin-Salazar, Pierfrancesco Cerruti, Gennaro Scarinzi

9 Biobased structural additives for polymers

Abstract: Biobased structural additives, in the last few years, have gained more and more interest for their positive environmental profile and their specific properties. Due to their natural origin, biodegradability and large availability they represent a valuable alternative to traditional synthetic fibers in composite formulation. In this chapter, an overview on the applications of natural fillers is reported. The first part of the work concerns the general features of composites and biobased fillers along with the main characterization techniques; a short discussion of chemical and physical modification methods of natural fibers is also provided. The second part of this contribution is focused on the description of some lignocellulosic and animal origin fillers as well as the analysis of their reinforcing role in biocomposites. Finally, the biodegradation behavior of composites based on natural fibers is also examined.

Keywords: lignocellulosic materials, biocomposites, polymer composites, sustainability, animal origin fibers, biofillers, biowastes, biomass, byproducts

9.1 Introduction

Since the beginning of the plastic age, additives have been included in polymer formulations in order to modify and improve their properties and widen their application field. Antioxidants, heat stabilizers, and plasticizers are largely used during processing with the aim of enhancing polymer thermal resistance and improve their rheological properties. Light and ultraviolet stabilizers are frequently combined with polymeric materials in order to extend their life cycle in outdoor applications. Flame retardants are often requested when fire hazards are foreseen. Most of these products are usually added in relatively small amounts to affect a specific property of the material, without altering its intrinsic arrangement; in this instance, they can be categorized as nonstructural additives. However, a great number of polymer applications involve mechanical strength along with dimensional and morphological stability. In order to meet these requirements, polymer systems are often modified by the addition of reinforcing agents or fillers. These products usually retain their original shape and physical form, producing a separate, disperse phase

Sarai Agustin-Salazar, Departamento de Investigaciones Científicas y Tecnológicas de la Universidad de Sonora, Hermosillo, Sonora, México; Institute for Polymers, Composites and Biomaterials (IPCB-CNR), Pozzuoli, Italy
Pierfrancesco Cerruti, Gennaro Scarinzi, Institute for Polymers, Composites and Biomaterials (IPCB-CNR), Pozzuoli, Italy

https://doi.org/10.1515/9783110590586-009

inside the polymer matrix; in this way, they alter its pristine structure and can be termed as structural additives.

The heterophasic system obtained by including the reinforcing phase in the polymer matrix is defined as a composite [1–4]. Polymer composites are commonly used due to their easy fabrication, lightweight, low cost, and excellent insulation properties [5, 6]. They are of outmost importance in modern world, and they have found a large number of applications as packaging materials, automotive parts, furniture, and construction components [6–8]. Polymer composites, compared to traditional materials, show, some remarkable advantages, such as higher chemical and environmental resistance, milder processing conditions, lower manufacturing costs, and lower density. These properties make them an economically viable alternative to metals, ceramics, and other engineering materials [1, 5, 9].

Traditional composites are made of fossil-based nonbiodegradable polymers as matrices, and inorganic fillers or man-made fibers (e.g., fiberglass, carbon fibers, and boron fiber) as reinforcing agents. These characteristics do not comply with the current environmental concerns related to sustainability, recyclability, and biodegradability [6, 10–23]. In this regard, the use of biomass-derived structural additives in composite formulations has gained more and more interest.

This chapter deals with the application of naturally occurring fillers in thermoplastic and thermoset polymer composites. General concepts about polymer composites are first presented, then the use of biobased fillers including their chemical modification methods along with the main characterization approaches are recalled. Finally, several examples of lignocellulose as well as animal origin fillers used as reinforcing agents in biocomposites are reviewed and discussed.

9.2 Polymer composites

Composites are multiphasic materials constituted by two or more components. Usually, we distinguish a continuous phase called matrix and a second constituent, denoted as filler or charge. Composite matrices are often of polymeric nature, and their role is to confer processability and workability to the material, provide a means for the dispersion of the second component, allow the transfer of the mechanical load to the charge, and protect the latter from environmental agents. Filler function is to modify mechanical and physical properties of the matrix such as stiffness, strength, thermal and electrical conductivity, and gas and water permeability. A third component of several composite formulations are coupling agents. They are often used in order to compensate the mismatch between filler and matrix. Indeed, most composite systems are based on apolar or slightly polar organic polymers and polar inorganic charges. These systems are characterized by a low adhesion at their interface that affects mechanical performances, water absorption, and gas permeability with deleterious effects on material properties. The addition of other components, in a relatively small quantity, is aimed also to modify the rheological properties of the matrix, enhance thermal stability, and provide the material with antioxidant properties [6, 9, 24, 25].

Composite materials can be classified in three main groups: traditional composites (polymer matrix + inorganic filler), biocomposites (polymer matrix + biobased filler, specially from agrifood wastes), and hybrid composites (polymer matrix + organic filler + inorganic filler). Biocomposites are composite materials based on biobased structural additives dispersed in a polymer matrix. With respect to traditional composites, they show lower preparation costs and reduced environmental impact. In this class of materials, the substitution of conventional fillers (metals, synthetic fibers, clays, etc.) with biobased products [6, 9, 26] has allowed to expand the application field of composites (packaging, automotive industry, furniture, construction, fabrication of electronic and medical devices, etc.) [5–7, 27–29].

Hybrid composites are multicomponent systems containing two or more different types of fillers. They are usually prepared by combining charges of different nature and properties, for example, organic and inorganic particles or fibers, in order to reduce the drawbacks of one filling with respect to the other, and to tune the properties of the final material. Typical filler systems for hybrid composites are glass-jute, glass-bagasse, and glass-flax. They find use in wall partitions, packaging materials, automobile interior parts like inner fenders and dashboard, lamp shades, floor tiles, and so on [9, 30–32].

The production of composite materials is determined by the origin of the biobased additives and by some parameters like damage during processing, wettability, weak bonding between matrix and filler, and high moisture absorption. The addressed solutions have improved the type and strength of interfacial interactions, and the performance of the resulting material [6, 23].

9.3 Biobased fillers in polymer composites

Compared to traditional reinforcing fillers, biobased fillers, and reinforcements confer multiple benefits to composites, including the decrease of the carbon footprint of the material and the lessening of the production costs. These issues are much more relevant if wastes and byproducts from agriculture or agrofood industry are considered as sources of additives; the recovery of these cheap products or residues strongly reduces the amount of wastes to be discarded in landfills and also helps the economic balance of the composite production [6, 8, 10, 33–36].

Crop biowastes, such as vegetable fiber, are considered as a low-cost reinforcing phase in composite formulation based on thermoplastics of general use, that is, used to enhance their mechanical properties [5, 6, 37–40]. Biobased structural materials can be manufactured through mixing the biobased reinforcing filler with the polymer matrix, followed by processing under high-pressure and high-temperature conditions. The resulting composite materials may be transformed by extrusion, injection, casting, or thermoforming methods, and they can be used in structural and nonstructural applications [3, 5, 6, 23]. Depending on its amount, shape, and size, the biobased component can be used as a particulate filler or an anisotropic fibrous reinforcement [6, 41–45]. There are three sources of natural fibers: animal, plant, and mineral [8, 34, 46–48]. Additionally, fibers can basically exist in two physical forms: short fiber or filaments. Those shapes can be visible at micro- and macro-scale. Depending on the shape, the fiber can present different properties and composition, and consequently can be employed for different applications [6, 41–45].

The use of biobased structural additives into polymer composites has shown good performance in life cycle assessments and has also provided important advantages in relation to toxicity, emission of effluents, energy consumption, abundance of disposal options, cost, and biodegradability. Furthermore, improvements in mechanical performance, such as tensile strength and elastic modulus, comparable to inorganic fillers, were recorded as well [10, 16, 45, 49, 50]. Applications, sustainability, competitive edge, and good performance are the most important factors determining the commercial success of biobased reinforcement materials [6, 10, 51, 52].

Notwithstanding the many advantages of these materials with respect to their synthetic counterparts, they also show a number of drawbacks. First, the application field is often limited by their lower mechanical and physical performance [6, 10, 33, 34, 53, 54]. Moreover, the hydrophilic nature of most biobased materials can cause poor interfacial adhesion between the structural additive and the hydrophobic polymer matrix that produces a decrease in the mechanical properties of the composite. Nonetheless, the high hydrophilicity of the biomass-derived products can increase the water absorption of the product, favoring biological attack and biodegradation. The reported drawbacks can be faced by adequate physical and chemical modifications of both matrix and fillers [6, 8, 9, 35, 55].

9.4 Chemical and physical treatments in biobased additives

The high polarity of the biomass-derived fillers can be limited by coating methods or hydrothermal treatments, as well as other physical and chemical processes. On the other hand, the scarce polymer–biofiller interaction can be improved by reactive functionalization of the matrix, the use of polymer blends as the continuous phase, the addition of compatibilizers or coupling agents, and other additives (clays, ashes, etc.). These treatments not only affect interfacial interactions between the components, but they also improve the dispersion and orientation of the additives inside the polymer matrix [6, 17, 35, 56–58].

Chemical treatments in biobased materials are a very useful way to improve the compatibility with the polymer matrix. This effect can be obtained by reducing the hydrophilic feature of the natural fiber components by removing the hydroxyl groups (OH) [23, 59] or through the generation of reactive functional groups on fiber surface. Chemical modification can be conducted by silane treatments, graft copolymerization, cyanate treatment, impregnation of fibers, alkali swelling, and substitution reactions [1, 10, 23].

Physical methods are used to alter the structural properties of the fibers or induce changes on the surface of the materials. The aim is to modify the surface properties of the fibers in order to generate a physical interlocking between matrix and filler or to affect its processability and mechanical properties. Physical modifications include sputtering, corona discharge, low-temperature plasma, calendaring, stretching, thermal treatment, the production of hybrid yarns, or mechanical surface microfibrillation [10, 23].

Bodîrlău et al. [59] modified softwood disks by using succinic anhydride. The chemical treatment produced a decrease of the hydrophilic character of the material. Tripathi et al. [9] reported the treatment of jute and bagasse fibers with NaOH. The alkali extraction was performed in order to remove the structural amorphous components such as hemicelluloses and lignin, and to clean the biomass surface by solubilizing impurities, waxy substances, and natural oils. The authors detected an increase in the number of free hydroxyl groups in the fiber surface and an improvement of adhesion with the polymer matrix in epoxy-based composites. Ramesh et al. [60] applied the alkaline method to modify the surface of kenaf fibers introduced as filler in polylactic acid (PLA) hybrid composites. The same method was adopted for the modification of hemp fibers used as reinforcing agents in polybenzoxazine hybrids [7, 61] and polypropylene (PP) composites [14]; in both cases, an enhancement in mechanical and thermal properties of the matrix was recorded. Labidi et al. [23] studied the effects of stearic acid and potassium permanganate in the chemical treatment of alfa plant fibers. The authors detected an enhancement of mechanical properties and melt flow index of the composites.

9.5 Techniques for the characterization of composite materials

Composite characterization is of fundamental importance to investigate the effect of the natural filler on the performance of the material, and to tailor its properties in order to meet its end-use requirements. The characterization techniques of composites are essentially those of classical polymeric materials. A few techniques of particular importance in this field are described further.

9.5.1 Mechanical properties

The study of mechanical properties is the principal method in biocomposite characterization as it provides useful information on the application field of the materials. The mechanical behavior of composites depends on the properties of the polymer matrix, nature, content, dimension and shape of the filler [3, 7, 35, 55], as well as the physical and chemical interactions between the components [6, 8, 61].

Mechanical properties of polymer composites are strongly affected by the interfacial adhesion between filler and matrix [62]. If the interaction among the two components is poor, when an external load is applied to the material, filler particles easily debond from the continuous phase producing voids and the macroscopic failure of the sample [62]. In biocomposites, due to the difference in polarity of most polymeric matrices and biomass-based fillers, interfacial adhesion is usually scarce. Several methods, which involve both physical and chemical techniques, have been proposed in order to modify this parameter.

Yuan et al. [63] reported the study of PP composites charged with plasma-treated wood fibers. Samples prepared with the treated filler exhibited an improvement of tensile strength and modulus. Based on scanning electron microscopy (SEM) analysis, this effect was attributed to the increase in surface roughness of the fiber generated by the plasma application. This change in morphology allowed an enhancement in adhesion between filler and matrix by mechanical interlocking effects.

The use of functional polymers is another method often used to modify interfacial adhesion in polymer composites. They are usually constituted by polymers, with the same chemical structure of the matrix, bearing functional groups able to interact with the filler by physical or chemical linkages [62]. Maleated polypropylene (MAPP) has been reported as a coupling agent in PP/wood composites [64–66] and in PP/sugarcane bagasse composites [67]. The mechanical characterization of the wood charged composites showed that the use of MAPP does not affect the stiffness of the samples but strongly influences the ultimate properties producing remarkable improvements in tensile and impact strength [65]. These results, based on SEM analysis of fractured surfaces, were attributed to the good adhesion between filler and matrix. This conclusion was further confirmed by the micromechanical characterization of the samples [64]. MAPP was also successfully used in PP/sugarcane bagasse composites [67]; similarly to PP/wood composites, the introduction of a small quantity of functional polymer produced an improvement in tensile and impact properties. Even in this case, the coupling agent affected the micromechanics involved in the mechanical response of the material. The use of PP–maleic anhydride copolymers as a modifying agent in PP/ cellulose fiber composites was reported by Felix et al. [68]. The copolymer was first applied to the cellulose fiber, which, in a second step, was compounded with the PP matrix. The authors claim that the use of the functional polymer produced a general improvement in composite mechanical properties, a better dispersion of the filler throughout the matrix, and an enhancement in fiber–polymer interface adhesion.

Low-molecular-weight coupling agents have also been applied to biocomposite formulations with the aim to modify the interphase interaction and improve their mechanical properties. Faludi et al. [69] revealed the use of aromatic maleimides in PLA/wood composites. The authors tentatively proposed that their coupling action was based on a Michael-type addition of the maleimide double bonds with hydroxyl functionalities from biomass and terminal moieties of PLA. The tested molecules proved to be effective coupling agents; they produced an enhancement in tensile properties of the composites with a large effect on strength. The coupling effect was confirmed through the morphological analysis of the fractured surfaces. SEM characterization showed that, in presence of maleimide modifiers, particle debonding was limited and the main deformation process was fracture of the filler (Figure 9.1). The effect of isocyanates on the mechanical properties of PVC/wood fiber composites was investigated by Kokta et al. [70]. The same authors also studied the role of isocyanates and vinyl silane coupling agents on LDPE reinforced with aspen fiber, wood flour, and cellulose flour [71]. The treatment of the fillers with the coupling agents produced scarce effects on tensile modulus but improved strength. The effects on tensile properties were comparable to mica and glass fiber. Organosilane compounds were reported as coupling agents in LDPE/henequen fiber composites [72]. Tensile characterization of the prepared samples showed that their mechanical properties depend on the amount of silane deposited on the filler: a maximum tensile strength was recorded for an optimum modifier concentration, and then, at higher contents, a decrease was detected. The prepared specimens were also tested through flexural tests but, in this case, no remarkable effect on elastic modulus was recorded. Silane treatment also affected the failure mode of the tested samples: SEM analysis of the fractured surfaces showed that by increasing the amount of coupling agent, the failure mode of the composites changed from interfacial failure to matrix failure.

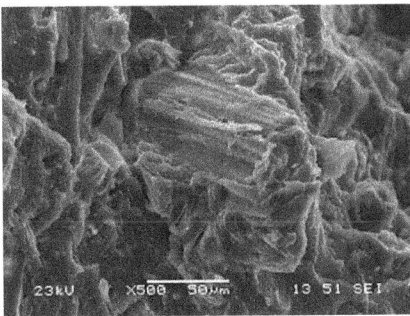

Figure 9.1: SEM micrograph of the fracture surface of a PLA/wood composite (30 vol%) [69].

Alkali treatment of natural fibers is a simple and valuable method to modify their surface properties and mechanical performance [9, 73–75]. This approach was applied to alfa fibers used as reinforcement in PP-based composites [23]. Compared to the PP/raw fiber system, composites charged with NaOH extracted fiber exhibited improvements in Young's modulus and tensile strength. The effect was related to the removal of amorphous noncellulosic components and breakdown of fiber bundles on behalf of alkali treatment. The same modification method was used in epoxy hybrid composites charged with jute, bagasse, and glass fibers [9]. Similarly to the PP-based system described earlier, the alkali extraction of jute and bagasse biofillers led to a tensile strength increase. Also, in this case, the removal of amorphous components of the fibers along with natural oils, waxes, and impurities was invoked to explain the results. In addition, other effects such as increase of the number of hydroxyl groups on fiber surface and the consequent improvement of fiber–matrix adhesion were also envisaged. However, it was also shown that the action of the alkali treatment on natural fibers sometimes cannot be foreseen and it often depends on the nature of the biomass. Indeed, Oladele et al. [55], in their tribological and mechanical study on polyester thermoset/cow hair composites, showed that by alkali treating the animal fiber a reduction in fiber tensile properties was recorded. This decay in mechanical properties was also verified in the corresponding composites, as at high filler loading (15–20 wt%) a reduction in tensile and flexural modulus was detected. In order to explain this worsening outcome, the authors claim that the alkali treatment of the biomass, in addition to the removal of lipids and other components, also leads to a damage of the fiber as testified by the SEM analysis. Alkali treatment of biofillers was also performed in a more sustainable way and in milder conditions. Fiore et al. [76] reported the extraction of sisal fiber with a sodium bicarbonate aqueous solution at room temperature. Fiber characterization showed the removal of hemicelluloses and partial eradication of lignin along with a remarkable improvement of its tensile properties. The modified fiber was used as reinforcement in an epoxy-based resin. Composite characterization, carried out by flexural tests, revealed an enhancement in the main mechanical parameters.

9.5.2 Thermal properties

In composite characterization, thermal properties are of key importance to establish the processing conditions and the stability of the material, which in turn dictate the application field of the final product [6, 22, 59, 61, 77–80]. The most used thermal techniques are differential scanning calorimetry (DSC) and thermogravimetric analysis (TGA). DSC characterization enables to detect the temperature regions where the material (mainly polymers) undergoes transitions from a hard glassy state to a soft rubbery condition, by measuring the amount of heat required to increase the temperature of a sample and a reference. On the other hand, TGA measures the changes in the weight of a heated sample as a function of temperature or time. TGA-relevant parameters are the temperature, the rate of temperature change, the purging gas, and its flow rate and the pressure. By measuring the mass change, TGA can be used to correlate the thermal decomposition of any material to its thermal stability. Typically, solids or low volatility liquids are tested [79, 80].

TGA has been widely used to investigate the thermal behavior of lignocellulosic materials [6, 49, 59]. Wood and related products show a thermal degradation behavior that depends on the botanical origin of the biomass and its composition. The TGA curve of wood usually exhibits a first mass drop stage (at around 100 °C) related to the evolution of moisture [49, 80, 81], followed by further thermal degradation, which takes place as a multistep process. The latter is a complex phenomenon that involves the thermal decomposition of the main wood components and can be described in three stages: depolymerization of hemicellulose (150–350 °C), the random cleavage of the glycosidic linkage of cellulose (275–350 °C), and lignin degradation (250–550 °C). Hemicelluloses are the less thermally stable components of the lignocellulose biomass, while lignin exhibits a higher thermal stability with a decomposition temperature close to that of polymers frequently used as matrix in composite materials [49, 80–82].

As an example, Figure 9.2(a) shows the multistep thermal degradation process of *Cissus quadrangularis* root fibers, reported by Indran et al. [83]; the picture includes the thermogravimetric curve recorded in nitrogen, along with its derivative (DTG) plot. The initial weight loss is attributed to the removal of moisture from the fiber. The second minor dip is noted at around 164 °C, where pyrolysis of lignin is initiated. The third step occurs between 230 °C and 330 °C due to the thermal decomposition of the hemicellulose and glycosidic links of cellulose. The thermal decomposition of cellulose is visible as a large weight loss step in the TGA plot as well as a peak at 329 °C in the DTG curve. Thermal degradation of polysaccharides can occur by cleavage of glycosidic, C–H, C–O, and C–C bonds, dehydration, decarboxylation, and decarbonylation reactions, with formation of C–C, C = C, C–O bonds, as well as carbonyl and carboxyl groups [84]. At temperatures above 500 °C, molecules break down into a variety of low-molecular-weight products, such as CO_2, CO, H_2O, hydrocarbons, and hydrogen [83]. Finally, at about 700 °C, the almost complete pyrolysis of lignin can be observed.

Thermogravimetry can be conducted under inert (nitrogen) or oxidative (air or oxygen) atmospheres, in order to better investigate the mechanisms underlying the mass loss. In Figure 9.2(b), the thermal behavior of a lignocellulosic material [pecan nutshell (PNS)] under both atmospheres is reported. PNS is made up of about 50 wt% of lignin and almost 40 wt% holocellulose (HC), and a minor amount of protein, lipids, and phenolic compounds [49]. The thermal testing of the material can provide hints to the identification of its components and to roughly assess their percentage. Indeed, TGA and DTG curves of PNS reflect its multicomponent nature. In the graph acquired under nitrogen, a main peak at 282 °C (DTG curve) is due to decomposition of polysaccharide components and the initiation of the pyrolysis of lignin. Subsequently, a second weight loss step (360 °C) accounts for the degradation of this product. As reported by other

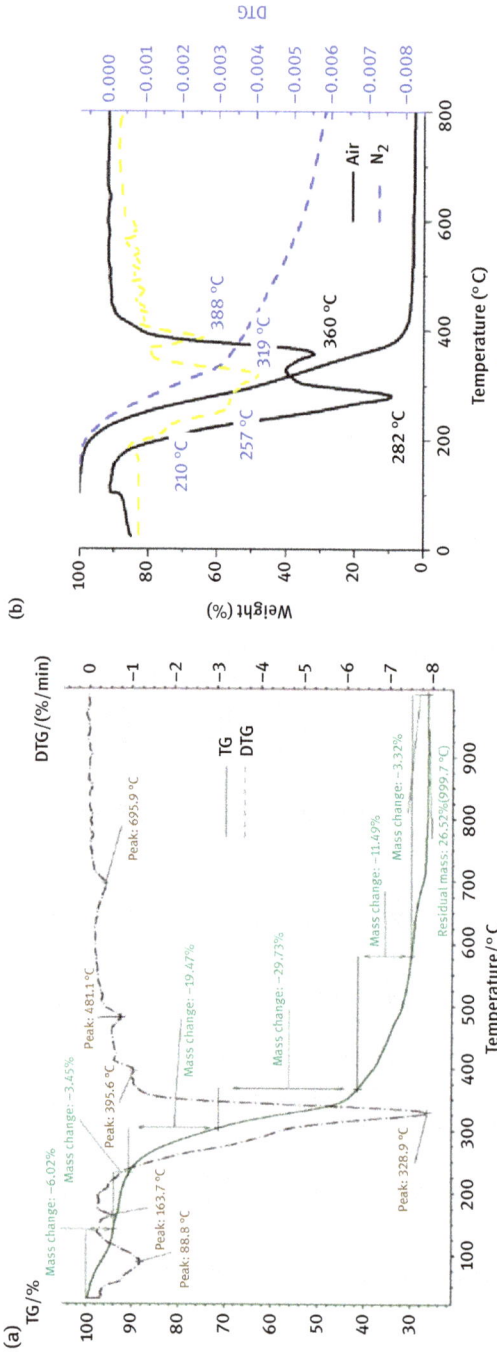

Figure 9.2: Thermal analysis of lignocellulose: (a) TGA and DTG curves of *Cissus quadrangularis* root fibers under nitrogen reported by Indran et al. [83], and (b) TGA and DTG curves of pecan nutshell (PNS) under both nitrogen and air.

authors for lignocellulosic biomasses, the whole nutshell thermal decomposition appeared to be a complex process characterized by more than one component.

TGA and DSC are also valuable methods for the characterization of biocomposites. In the following, thermal data concerning a PLA/PNS system charged with 50 wt% filler are discussed [49]. As shown in Figure 9.3(a), relative to the TGA analysis of the system, the use of PNS, as a reinforcing material in PLA, slightly decreased the onset of thermal degradation under both air and nitrogen. However, at higher temperatures, it can be noticed that the filler also affected the main weight loss step; indeed, it produces a shift of this parameter from 350 °C, of plain polymer, to around 400 °C in the composite. Moreover, from the curve recorded under nitrogen, it can be envisaged that the addition of the nutshell allows a 20% residual char at the end of measurement. As

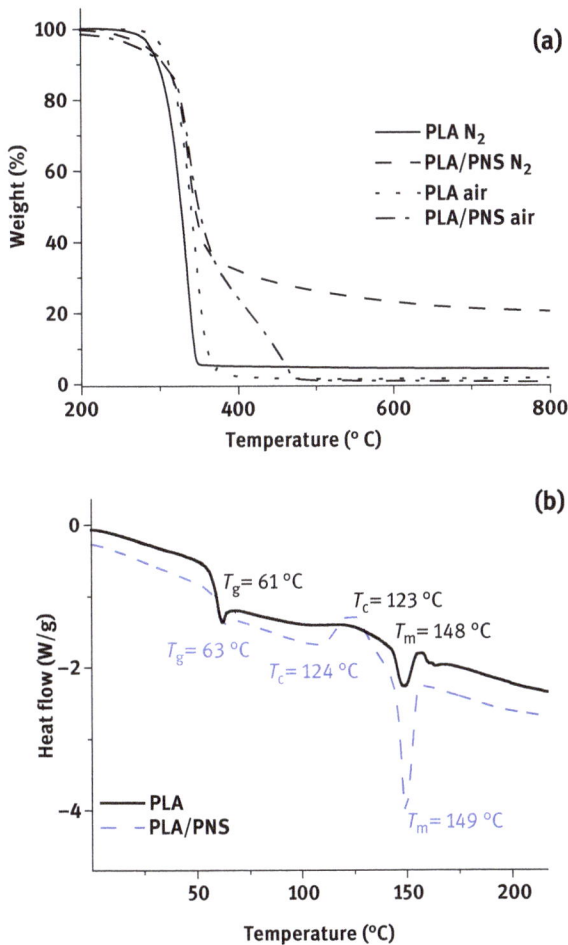

Figure 9.3: (a) TGA curves and (b) DSC curves of PLA and PLA/PNS composite (50 wt%).

evidenced in Figure 9.3(b), relative to DSC analysis, apart from impacting on thermal stability, the use of biobased fillers in composites can also influence other properties. Indeed, small but important changes in some thermal parameters such as glass transition temperature (T_g), crystallization temperature (T_c), and melting temperature (T_m) are detected in the calorimetric plot. In addition, PNS acted as a heterogeneous nucleating agent on PLA, promoting polymer crystallization. In Figure 9.3(b), this effect is revealed by the increase in intensity of the PLA/PNS cold crystallization peak centered at 123 °C, which in the plain PLA plot is scarcely detectable.

9.5.3 Water sorption properties

The study of water sorption behavior of composite materials is of remarkable importance as some moisture-dependent properties, such as fungal attack and mechanical performance, can be affected by moisture uptake [6, 23]. The immersion of biocomposites in water results in lower tensile strength, flexural strength, impact strength, and Young's modulus [23]. However, the water penetration behavior of biobased composites is highly complex. This problem arises from the different polarity of fillers and matrices. Biobased fillers are highly polar and exhibit a remarkable hydrophilicity. Conversely, polymer matrices are mostly hydrophobic materials. When the biofiller is incorporated into a polymer with hydrophobic nature, it undergoes a certain degree of protection against moisture sorption, so the diffusion of the penetrant is limited. On the other hand, the natural tendency of the hydrophilic filler to absorb water can generate an enhancement in moisture uptake of the composite. As a result, the final behavior of the material is not fully predictable.

Krause et al. [3] showed that the equilibrium water uptake in PP/wood composites depends on the source of the natural fillers. A similar effect was detected in epoxy composites charged with kenaf and bamboo fibers [8, 60]. In the last case, both fillers produced an increase in water uptake with respect to plain epoxy, but the effect was more remarkable for kenaf fibers. This finding was attributed to the different composition of the two materials. Particle size of fillers and filler loading are also important factors that can affect the water absorption process of composites. In this instance, it has been reported that large particle sizes help moisture diffusion and produce higher moisture uptake [6, 8, 9, 55].

The treatment of biofillers with coupling agents also influences water diffusion in composites. Thermal treatments of the charge, performed prior or during the processing steps, produce effects on moisture diffusion. They affect water uptake through loss of intrinsic water, thermal decomposition of organic acids, and degradation of the more thermally unstable moieties of lignin [6, 45, 85, 86]. Composites containing raw fibers can absorb more water than composites with treated fibers (alkaline extraction or chemical treatments). The higher amount of noncellulosic materials in raw fibers and a poorer adhesion between fiber and matrix are responsible for the higher water uptake [23].

9.5.4 Thermal conductivity, morphology, and electric and surface properties

Thermal conductivity and dielectric constant measurements are also reported as characterization techniques for composites. These two parameters define the potential use of the material in microelectronic applications. In this field, the interest toward natural fillers arises from their better electrical insulating properties with respect to synthetic fibers [5, 6, 87]. The nature of the fibers (crystalline and noncrystalline phases) and their distribution in the composite material influence their electrical behavior and potential applications. In bacterial cellulose composites reinforced with graphene oxide, for example, a rise in the conductivity was correlated to the increase in cellulose content. The dynamic mobility in this composite was affected by the number of hydroxyl groups in cellulose chains [88].

Surface properties of polymer composites can be evaluated through a number of characterization techniques. The most used are contact angle analysis, Fourier transform infrared (IR) spectroscopy, atomic force microscopy, SEM, and X-ray photoelectron spectroscopy [6, 15, 89]. Sullins et al. [14] used IR spectroscopy to study the effect of alkaline treatment on hemp fiber. The IR spectra of the treated biomasses show an increase of hydroxyl group concentration and a reduction of impurities content. The same characterization method was also used by Labidi et al. [23] to evidence the removal of waxes and other noncellulosic components from alfa fiber treated by steam explosion and alkaline extraction. In PLA–lignocellulosic nanocomposites, a good interfacial adhesion was evidenced by IR spectroscopy and atomic force microscopy characterization [28].

SEM is a technique often used for the morphological characterization of composites. It provides information concerning the degree of dispersion of the filler within the matrix, and the effects of chemical and mechanical treatments on the adhesion between filler and continuous phase. In addition, it is also a useful technique for the analysis of fracture surfaces obtained through mechanical testing [8, 12, 23, 77, 89, 90].

As an example, SEM micrographs of the biodegradation of soil-buried polyhydroxybutyrate (PHB) and a biocomposite formulated with PHB and *Arundo donax* (at 30 wt%) are displayed in Figure 9.4.

Bodîrlău et al. [59] also used SEM analysis to monitor the effects of accelerated weathering tests performed on wood charged composites. The investigated samples were prepared by using PP or PLA as matrices and spruce wood fibers as fillers. According to Teacà et al. [6], after the exposure to accelerated weathering conditions, the composite surfaces showed an increased roughness and formation of

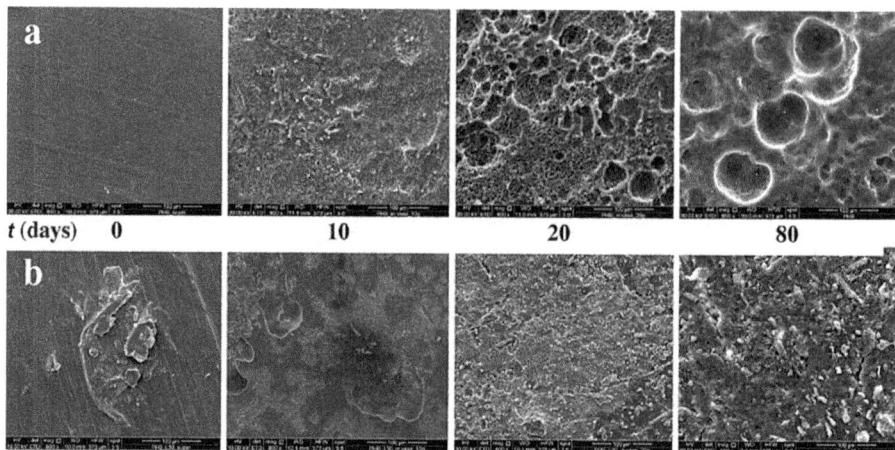

Figure 9.4: SEM micrographs of (a) PHB and (b) PHB/*Arundo donax* composite after 0, 10, 20, and 80 days of soil burial.

cracks due to the loss of degraded wood components as well as the degradation of the polymer matrix.

9.6 Lignocellulosic materials as fillers in biocomposites

Nowadays, the industrial interest toward lignocellulosic biomass is mainly focused on the production of cellulose pulp for paper manufacture. However, the growing demand for green and sustainable materials has made this feedstock more and more attractive as a renewable and largely available source of monomers, chemicals, additives, and fillers for composites [34, 45, 52, 86, 87, 91]. In fact, new uses for lignocellulose have been proposed in the production of biofuels and chemical intermediates, as well as monomers for the production and synthesis of plastic materials and finally as fillers for composites [86, 92, 93]. In this section, general information on the structural components of lignocellulose are provided, while the next paragraphs give further details on the use of biomass from specific crops as structural additives in biocomposites.

The primary components of the plant biomass are cellulose, hemicellulose, and lignin. Hemicelluloses are polysaccharides located in the cell wall of plants where they contribute to the strengthening of the lignocellulosic structure by interacting with cellulose and lignin. Hemicelluloses are synthesized by glycosyltransferases (biosynthesis of xyloglucans and mannans) located in the Golgi membranes. They are made of different sugar units such as glucose, mannose, and xylose mainly linked through β-1,4-linkages but other kinds of chemical connections are also reported [94]. These monomeric units are mostly substituted with acetic acid. Its structural heterogeneity is further complicated by branching. As a result, this class of polysaccharides is essentially amorphous [95–100].

The detailed structure of hemicelluloses and their abundance vary widely among different species and cell types [96, 101–103].

Due to their random molecular structure, hemicelluloses are amorphous polymers with low mechanical strength. In addition, as a result of their high hydroxyl content, they are remarkably hydrophilic. These features limit their industrial application. Chemical modification of hemicelluloses has been often applied in order to overcome these shortcomings. Esterification, in particular, has been performed in order to convert its hydrophilic hydroxyl groups to more hydrophobic ester moieties. This modification positively affected compatibility and adhesion with organic polymers [94]. Fundador et al. [104] reported the modification of xylan hemicellulose through esterification with acetic anhydride. The authors revealed an increase in thermal resistance, measured by TGA, compared to the parent biopolymer. Similar effects were reported by Sun et al. [105], which also recorded an increased solubility in organic solvents. These results make hemicelluloses a potential component in biocomposite formulations.

Cellulose is a high-molecular-weight linear homopolymer made of D-glucose units linked through β-1,4-glycosidic linkages. It can exist in six different polymorphic crystalline forms but only one is reported for native cellulose, that is cellulose I, an enzymatically directed structure with an all-parallel chains arrangement [23, 50, 96, 97, 99, 106–110]. Celluloses have crystalline regions due to strong hydrogen and van der Waals bonding. Breaking down this hydrogen bonding can facilitate the conversion of crystalline cellulose to amorphous cellulose and cellulose II. The latter is obtained through regeneration or mercerization, and has a lower free energy organization with antiparallel chains [111]. Cellulose is the most abundant polymer in nature with an annual production of 75 Gton. It is mostly of vegetal origin but it is also produced by some marine animals (tunicates) and a small number of bacterial species [96, 99, 112, 113]. Cellulose is characterized by a complex, tridimensional network of hydrogen bonds that, along with its semicrystalline structure, are responsible for its strength and rigidity [4]. Native cellulose is organized in a hierarchical architecture that consists of small, highly crystalline entities, called fibrils, tightly packed in microfibers. Fibrils are characterized by high crystallinity and high mechanical strength [112, 114, 115]. Cellulose fibrils can be isolated from the native material by deconstructing the hierarchical structure described earlier though chemical or mechanical methods. The recovered fibrils have shown many interesting properties and have found a great deal of applications [50, 110, 116].

Cellulose from animal sources can be obtained by various sea creatures (tunicates, e.g., *Microcosmus fulcatus*). Typically, this form of cellulose is characterized by nearly perfect, rod-like crystals. This morphological feature makes it a very interesting product for advanced applications. In addition, several strains of bacteria, such as the gram-negative bacterium *Acetobacter xylinum*, under special cultivation conditions, are able to produce cellulose. Bacterial cellulose is hemicellulose and lignin free, and highly crystalline with high polymerization degree. It has been applied in biomedical research and it is also promising as filler in composites [99, 106, 109, 110, 112, 117].

Lignin is the second most abundant polymer in nature, a complex polyphenolic component of the secondary cell walls of plants. The name of "lignin" is granted to a large family of complex aromatic polymers, intimately intertwined with hemicellulose [45, 77, 81, 118]. Lignin chemical structure is based on three aromatic units: *p*-hydrophenyl, guaiacyl, and syringyl. In plant cell wall, they are biosynthesized, respectively, from three alcohols: coumaryl, coniferyl, and sinapyl (Figure 9.5) [85, 96, 118, 119]. The free-radical polymerization of these precursors produces a heterogeneous and polydisperse polymer. As a consequence, lignin chemical structure can vary widely on the basis of its botanical origin. In plants, lignin is a component of cell wall structure and contributes to its strength. It also controls the fluid flow and brings protection against biochemical stresses [45, 120].

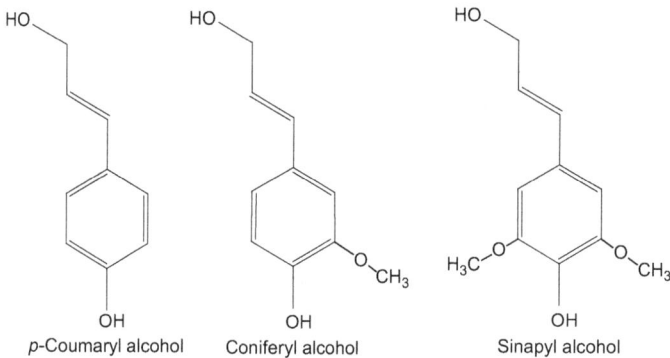

Figure 9.5: Chemical structures of the precursor alcohols of lignin.

Lignin is a biowaste from several industrial processes; pulp and paper manufacture is the main responsible of this byproduct production. From the generated lignin as side-stream from this industrial sector less than 2% is recovered for utilization. The remaining is considered a waste and usually burned for recovering energy [45, 81, 113]. Lignin structure, in addition to the plant origin of the parent biomass, also depends on the industrial process applied for its isolation.

Lignin, due to its renewable origin, large availability, versatility, and lack of toxicity, would represent an economically viable and sustainable resource for a number of industrial fields. Currently, its commercial applications, though limited, range from animal feed additives and agriculture to construction, textile, oil drilling, binders, and dispersants. In recent years, lignin has been also exploited in polymer applications, as stabilizing agents, lubricants, coatings, plasticizers, surfactants, and base for composites. In addition, owing to its high content in aromatic moieties and its remarkable charring capability, it has been proposed as a carbon source in intumescent flame-retardant systems for polymeric materials. For the same reason, it has been explored as a structural additive in polymer composites [45, 121–125].

9.6.1 Wood

Wood and its byproducts are interesting fillers for the formulation of biocomposites. They are often used in combination with other components in order to reduce their incompatibility with polymer matrices, facilitate processing, improve material performance according to specifications for different applications and consumer demands [1, 3, 4, 6].

The major component of wood is cellulose. This biopolymer plays an important role in wood-based composites as it confers stiffness to the material. Lignin, the second constituent of wood products, also contributes to the reinforcing action by acting as a binder for cellulose fibrils and allowing the stress transfer from the polymer to the fiber [111, 113, 126]. Wood particles may influence the mechanical behavior of composites by acting as nucleating agents toward the matrix. However, in some cases, the wood filler can also produce the embrittlement of the composite with a reduction of strength properties [1, 4, 23, 50, 85, 122, 127].

According to Teacà et al. [6], wood waste is mainly used for energy purposes and also in compost; however, its application in plastic formulations is of relevance as it can be further recycled or even submitted to biodegradation in controlled biological media or landfill sites. The final performance of wood composites can be explained on the basis of inherent material properties and processing effects. These topics can be resumed in three basic factors. The first is fiber damage during processing and, consequently, a poor wettability of fillers. The second is weak bonding between matrix and filler, entailing degradation processes at the interface. The third is high moisture absorption. Wood-based composites made of virgin materials may have low environmental impact if the wood content is significant. In this respect, wood based composite systems have been reported in literature, including wood flour or natural fibers [29, 34, 100, 128]. Avolio et al. [129] have used medium length wood pulp cellulose fiber as a reinforcing material in poly(butylene succinate-*co*-butylene adipate), resulting in increased mechanical properties in the biocomposites. *Acacia catechu* tree powder has been used as a reinforcing agent in polybenzoxazine resins. The addition of a low amount of filler (1–5 wt%) increased the impact strength of the matrix and reduced the curing temperature. In addition, it produced an enhancement of thermomechanical, tensile, and thermal properties [22]. Dash and Triphaty [1] have studied the use of teak wood dust as reinforcement in epoxy composites successfully developed using a hand-lay-up technique. They obtained an improvement in tensile strength.

9.6.2 Bamboo

Bamboo (*Bambusa vulgaris*) is an important plant fiber with a great potential for composite formulation [130]. The interest toward this biomass as a reinforcing agent stems from its outstanding mechanical properties; bamboo fibers show specific stiffness and specific strength even comparable to glass fibers [131]. Bamboo fibers have been applied in the manufacture of a new type of wood-like composite denoted parallel strand bamboo. Its preparation consists in flattening the bamboo strips into thin strands, impregnate them with phenolic resin, and gluing the impregnated strands under high pressure [89]. The final material shows toughness values higher than wood products commonly used in construction engineering.

Ismail et al. [8] have used bamboo mat to improve the properties of woven kenaf-reinforced epoxy composites. Woven kenaf/bamboo hybrid composites were fabricated by hand lay-up with different ratios of kenaf and bamboo fibers. With respect to pure kenaf composites, hybrid formulations showed a reduction in water absorption and thickness swelling along with an improvement in flexural properties. It was found that the composite with a 50/50 kenaf/bamboo ratio exhibited the best overall performance. This composition also showed the higher impact strength, even beyond the values of pure kenaf and bamboo composites. The studied materials are suitable for the manufacture of automotive components [8, 97, 132].

9.6.3 Flax

Flax (*Linum usitatissimum* L.) is one of the most frequently used natural fibers in composite materials due to its high specific mechanical properties and moderate cost. Flax fibers are characterized by a high cellulose content and high fiber length (10–65 mm) [133]. Figure 9.6 shows the SEM images of flax fibers reported by Tanguy et al. [133], where raw flax bundle cross section and longitudinal view of individual fibers are observed.

Bourmaud et al. [134] investigated the recycling of nonwoven PP–flax composites performed by injection and compression molding. The effects of repeated recycling cycles on the microstructure of the flax fiber and the tensile properties of the composites were studied. Their results showed that reprocessing produced a decrease in mechanical properties of the material, but the worsening effect was less pronounced for injection molding. In addition, this processing method also promotes good fiber dispersion, high volume fractions, and almost defect-free microstructures that are prerequisites for high strength materials.

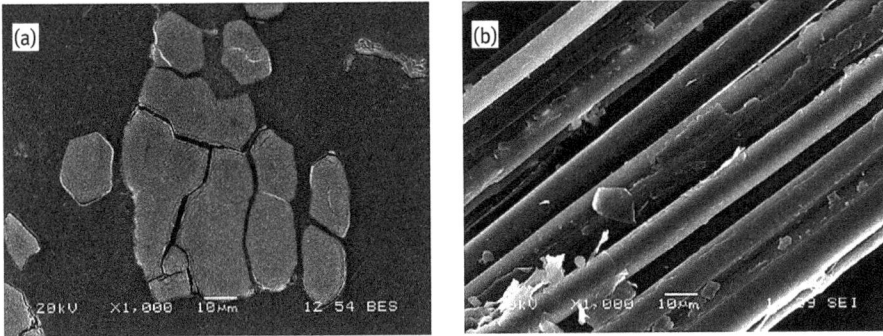

Figure 9.6: SEM images of (a) flax bundle cross section and (b) flax longitudinal bundle view reported by Tanguy et al. [133].

Tanguy et al. [133] studied the tensile properties of flax and jute fibers and their relationship with the mechanical performance of the corresponding PP-based composites prepared with 60 wt% filler content. The analysis was carried out on both unidirectional and short fiber composites; the latter were prepared by extrusion followed by injection molding. The experimental results evidenced that flax fiber exhibited higher tensile properties with respect to jute fiber (Figure 9.7(a)). The corresponding PP-based unidirectional composites showed a good correlation with the above results. However, the reverse relationship was found in the injection-molded samples (Figure 9.7(b)); in this instance, jute-based composites showed higher Young's modulus and strength with respect to the flax charged specimen. This difference in behavior was attributed to microstructural effects related to fiber orientation during the injection molding process.

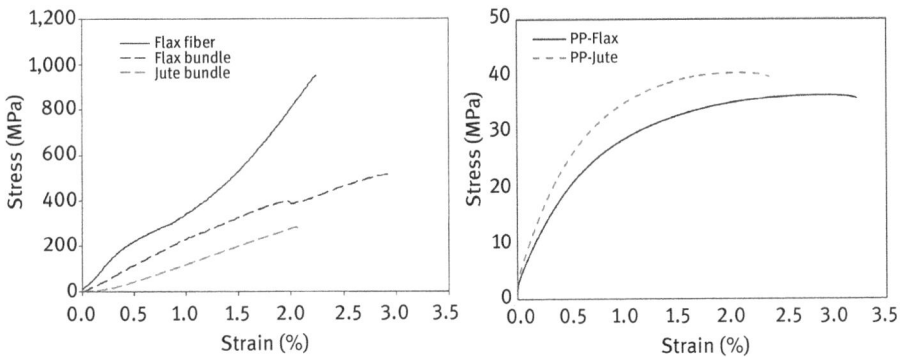

Figure 9.7: Tensile behavior of (a) flax fiber, flax fiber bundle, and jute bundle, and (b) injection-molded PP/flax and PP/jute composites, reported by Tanguy et al. [133].

9.6.4 Jute

Jute fiber is one of the most promising natural materials for composite applications. Similarly to flax, it is characterized by a high content of cellulose (61–72 wt%) but it also shows a larger lignin concentration (12–13 wt%) and a shorter fiber length (0.8–6 mm) [9, 133]. Raw jute fibers are organized in cohesive bundles where fiber connection is further improved by the relatively high lignin content of the material. Compared to other vegetal fibers, jute shows better tensile strength than bagasse but lower than flax. The same can be said for Young's modulus. Due to its good physical and mechanical properties and its availability in a number of forms, jute is used in many value-added applications from lamp shades to roof tiles [9, 26, 133, 135]. Figure 9.8 shows SEM images of jute fibers reported by Tanguy et al. [133], where the unidirectional tape layers are visible.

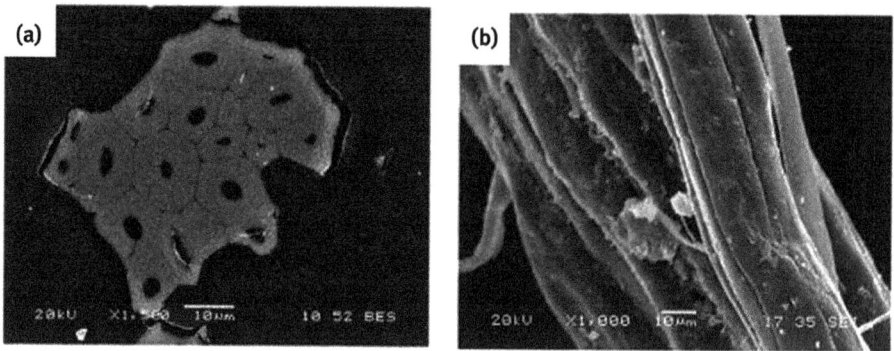

Figure 9.8: SEM images of (a) jute bundle cross section, and (b) jute longitudinal bundle view, as reported by Tanguy et al. [133].

Tripathi et al. [9] reported the use of jute in glass fiber–epoxy composites. They revealed that the addition of jute fiber improved mechanical properties of the hybrid composites. The prepared products showed better mechanical performance with respect to the corresponding composites based on bagasse fiber. They also found that by treating the organic fiber with a NaOH solution, a significant increase in strength of the prepared composites was reached. Such a finding was attributed to the removal of impurities, waxy substances, and natural oils from the fiber and to the increase in the number of free hydroxyl groups on its surface. These effects alter the exterior of the filler and improve the adhesion with the matrix resin. Sathishkumar et al. [135] reported the use of jute fiber (10 wt%) in hybrid epoxy/lignite fly ash hybrid composites. Among different formulations, hybrid compound prepared with 2 wt% of this cofiller showed the optimum performance in terms of thermal stability and mechanical strength.

9.6.5 Kenaf

Kenaf fiber is another natural filler with high tensile strength. This feature makes it suitable as a reinforcing agent for biocomposites, especially in the production of automotive components [8, 24, 60, 136]. Ramesh et al. [60] investigated the addition of montmorillonite clay to a kenaf-based PLA composite. Samples were prepared by using 30 wt% organic fiber and 0–3 wt% montmorillonite. An improvement of flexural and tensile strength was found for kenaf–PLA biocomposites. The addition of the inorganic charge produced a further enhancement in mechanical properties; this outcome was recorded at low content of cofiller and was attributed to the good matrix–fiber interaction, which improved load transfer capacity. The authors claimed that this adhesion effect is visible in the more uniform distribution of fibers recorded at low clay content (Figure 9.9(b)).

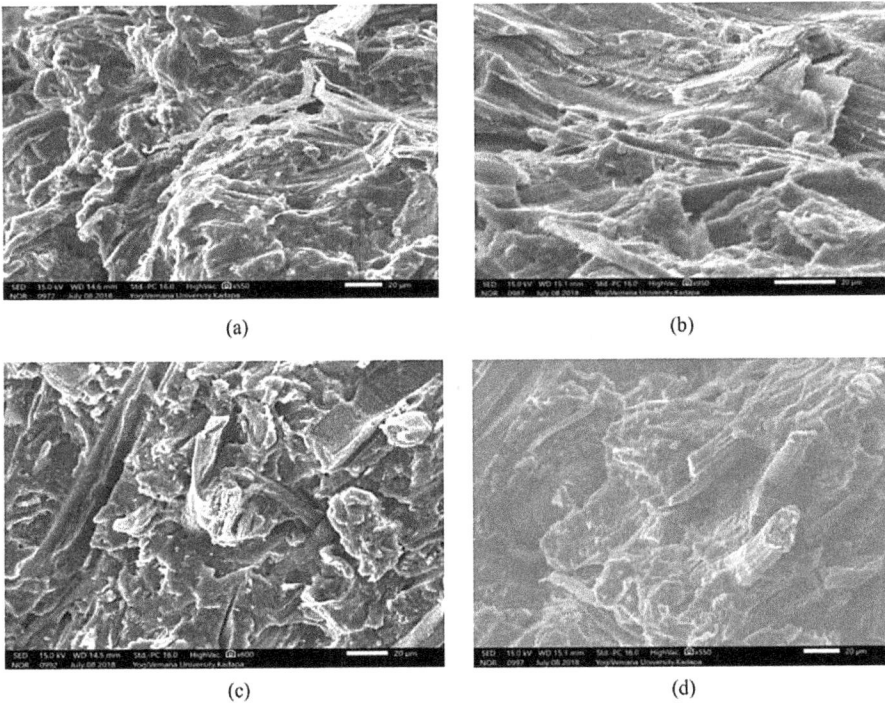

(a)

(b)

(c)

(d)

Figure 9.9: Micrographs for PLA hybrid composites charged with 30 wt% kenaf fiber and (a) 0 wt% clay, (b) 1 wt%, (c) 2 wt%, and (d) 3 wt% of montmorillonite clay. Reported by Ramesh et al. [60].

9.6.6 Hemp

Hemp plant (*Cannabis sativa* L.) belongs to the *Cannabis* species but, differently from *Cannabis indica* L., contains only low quantities of the psychoactive compound tetrahydrocannabinol (THC) [54, 137]. After jute, hemp is the largest grown bast fiber in the world [61]. The interest toward this natural fiber as a reinforcing agent in polymer composites is relatively new but, historically, hemp has been used in a great number of applications such as the production of paper, textiles, automotive components, and building materials [54, 137].

The effect of hemp fiber addition in polymeric matrices has been studied [54, 61]. Dayo et al. [7] found that polybenzoxazine composites charged with 30 vol% filler showed improved performance in flexural and impact strength. In addition, they demonstrated that by reducing fiber dimensions, a further increase in mechanical properties was achieved. A similar effect of fiber content and fiber dimensions was observed for water uptake; all the studied composites exhibited an enhancement in water absorption with the increase in vol% of fiber and the decrease in the fiber dimensions.

In a previous study, the same authors [61] investigated the effect of alkali-treated hemp fiber in curing behavior, and mechanical and thermal properties of polybenzoxazine composites. Samples were prepared by sandwiching a fiber film between two benzoxazine prepolymer layers. Next, the obtained compound was submitted to hot press curing. The dynamical mechanical characterization of the cured specimens showed that by increasing the vol% content of fiber, a regular enhancement in storage modulus and an increase in glass transition temperature were recorded. Hemp fiber also affected the flexural properties of the composites. In this instance, both flexural modulus and strength increased with fiber loading reaching a maximum of 20 vol%. After that, a slight decreasing trend was recorded (Figure 9.10). The reported effects were attributed to the ductile nature of natural fibers, the defibrillation of the fiber during the alkali treatment, and the strong hydrogen bonding effect between continuous phase and charge. This interaction improves the adhesion between the composite components and helps the transfer of the stress from matrix to the reinforcing agent.

9.6.7 Pecan nutshell

PNS (*Carya illinoinensis*) is a biowaste from pecan nut production, which is inexpensive and largely available. The potential of PNS as a source of fibers for the preparation of polymer composite materials was investigated in PLA-based systems [138, 139].

Álvarez-Chávez et al. [138] reported the characterization of PLA/PNS biocomposites, prepared by means of a single screw extruder, with a filler content up to

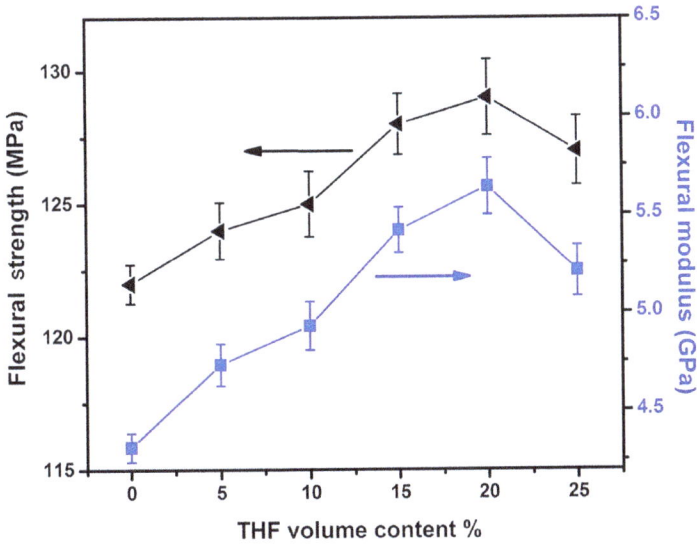

Figure 9.10: Variation in flexural strength and modulus of composites with different treated hemp fiber volume content [61].

7.5 wt%. The authors recorded that, compared to neat PLA, the obtained samples exhibited an increase in water sorption capability and improvement in thermal stability. Conversely, a decay in tensile strength and elongation at break was detected. The authors also investigated the effect of defatting of the biofiller, performed through organic solvent extraction, on the properties of the corresponding composites. They reported that the removal of solvent-soluble, nonstructural components had little effect on composite properties. Recently, the PLA/PNS system, in the same composition range, was studied by Sanchez-Acosta et al. [123]. In this case, samples were prepared by twin screw extrusion followed by injection molding. The tested composites, with respect to plain PLA, showed improvements in flexural and tensile modulus, along with enhancements in impact strength. In addition, the effect was further magnified if PNS was submitted to defatting. The same finding was also recorded through dynamical mechanical analysis, which showed a stiffening action of the defatted filler toward the matrix. These results, on the basis of the SEM analysis of the criofractured surfaces of the composites, were attributed to a better filler-matrix adhesion for samples charged with defatted PNS.

The potentiality of PNS as a biofiller for PLA was also investigated by Agustin-Salazar et al. [49]. They studied the reinforcing action of native PNS as well as the effect its structural components: HC and acid-insoluble lignin. According to the authors, DSC analysis of the formulated samples showed that native nutshell and HC acted as nucleating agents toward PLA by promoting its crystallization from the melt. Furthermore, they also affected the rheological behavior of the matrix producing an enhancement in its viscoelastic response. Finally, flexural tests showed that both biofillers led to an increase in flexural modulus and improved the stiffness of the composites.

Lignin, conversely, had no effect on the crystallization behavior and rheological properties of the studied composites. However, due to its amorphous nature, it influenced the flexural performance of prepared samples, leading to a decrease in modulus and an improvement in strength and elongation at break (Figure 9.11).

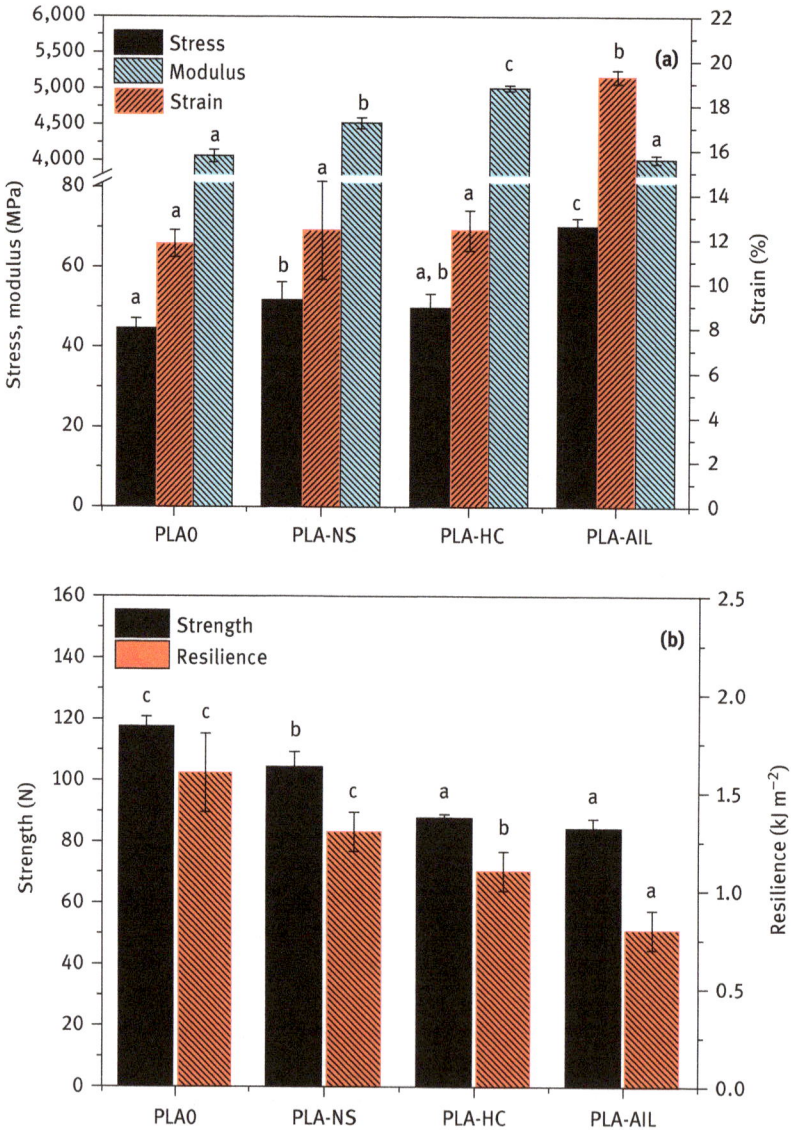

Figure 9.11: Mechanical properties, (a) flexural properties and (b) impact properties, of PLA biocomposites [49].

These results were attributed to the amorphous and heterogeneous nature of lignin, which render higher deformability to the material [5, 49, 113, 140].

9.7 Animal origin fibers

In the elaboration of composite materials, fibers of animal origin have also been used as reinforcement in polymeric matrices, and some applications are briefly described further. Animal hair, composed of proteins, lipids, water, and small amounts of trace elements, is a waste material in most parts of the world and its accumulation in waste streams causes environmental problems.

The use of hair as a reinforcement material is a new trend due to its capacity to resist stretching and compression [5, 46, 53, 55]. Human hair is mainly composed of 65–95% proteins, and remaining are water and lipid pigment [36]. In 2018, Srivastava and Sinha [36] reported the use of human hair fiber (HF) (treated and untreated with alkali) as a reinforcing filler in high-density polyethylene (HDPE) composites prepared by hot compression molding. Fiber morphology, before and after alkali treatment, was analyzed by SEM. Untreated HF showed a smooth surface (Figure 9.12(a)) that, according to the authors, was related to the presence of amino acids and cuticles. Upon treatment with 0.25 N alkali solution, an increase in surface roughness and porosity is detected (Figure 9.12(b)). By treating the fiber with alkali at higher concentration (0.5 N), cuticles damage was also recorded (Figure 9.12(c)).

Mechanical properties of the prepared composites were investigated through tensile and flexural tests. Modulus data, reported in Figure 9.13, show that filler addition produces a stiffening action toward the HDPE matrix. This outcome is recorded independently from fiber treatment and is more remarkable for HF treated with low concentration alkali (0.25 N). The stiffening effect is maximum at 15 wt% filler content while at higher HF loading a decay in modulus is recorded.

SEM analysis of the fractured surfaces was carried out in order to verify the effect of fiber treatment on the morphology of the composites (Figure 9.14). According to the authors, the fracture surface of the sample reinforced with untreated HF (15 wt% loading) was characterized by the presence of gaps and fiber-pullouts, which evidenced poor bonding between the matrix and the filler. These features were related to the smooth surface of the untreated fiber, which is evidenced in Figure 9.12(a), and the poor interlocking with the matrix (Figure 9.14(a)). The alkali treatment of HF improved the bonding between the matrix and the fiber, as evidenced by fiber damage associated with the fracture process (Figure 9.14(b)). However, some fiber-pullouts, related to scarce adhesion, are also present. These morphological evidences along with the results of mechanical properties are explained on the basis of the SEM analy-

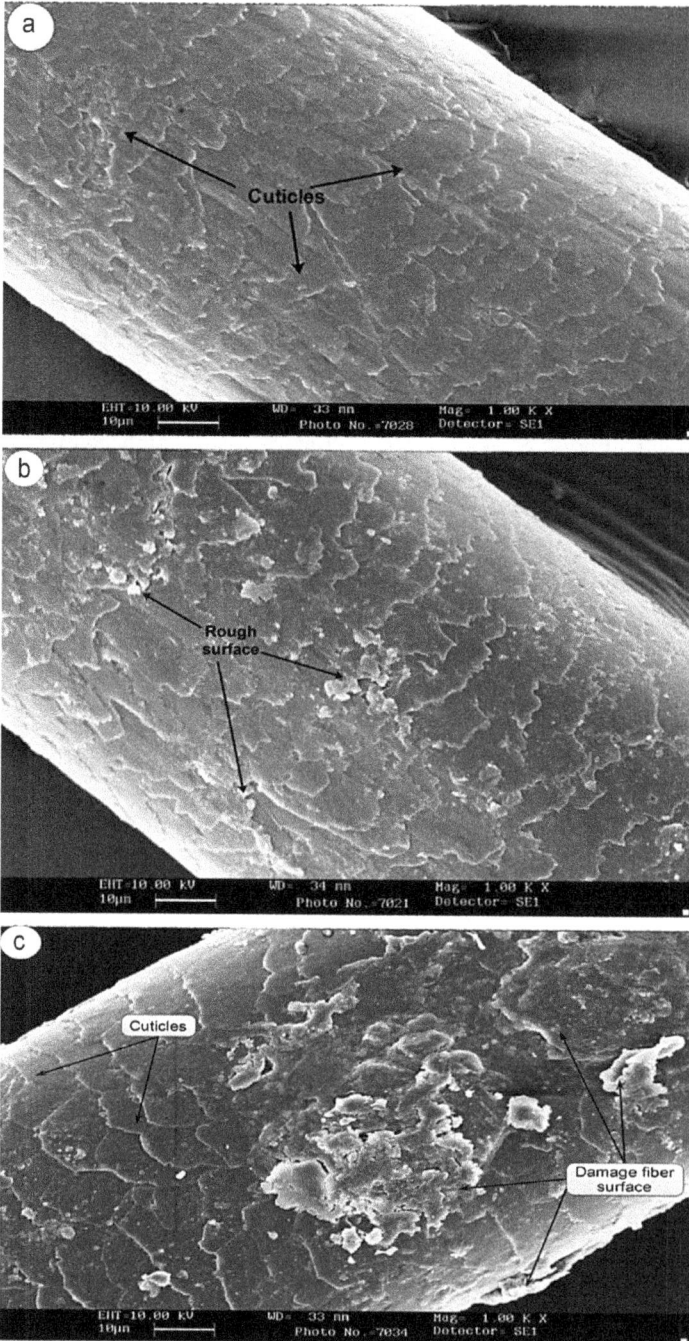

Figure 9.12: SEM micrographs of human hair: (a) untreated, (b) 0.25 N alkali-treated HF, and (c) 0.5 N alkali-treated HF [36].

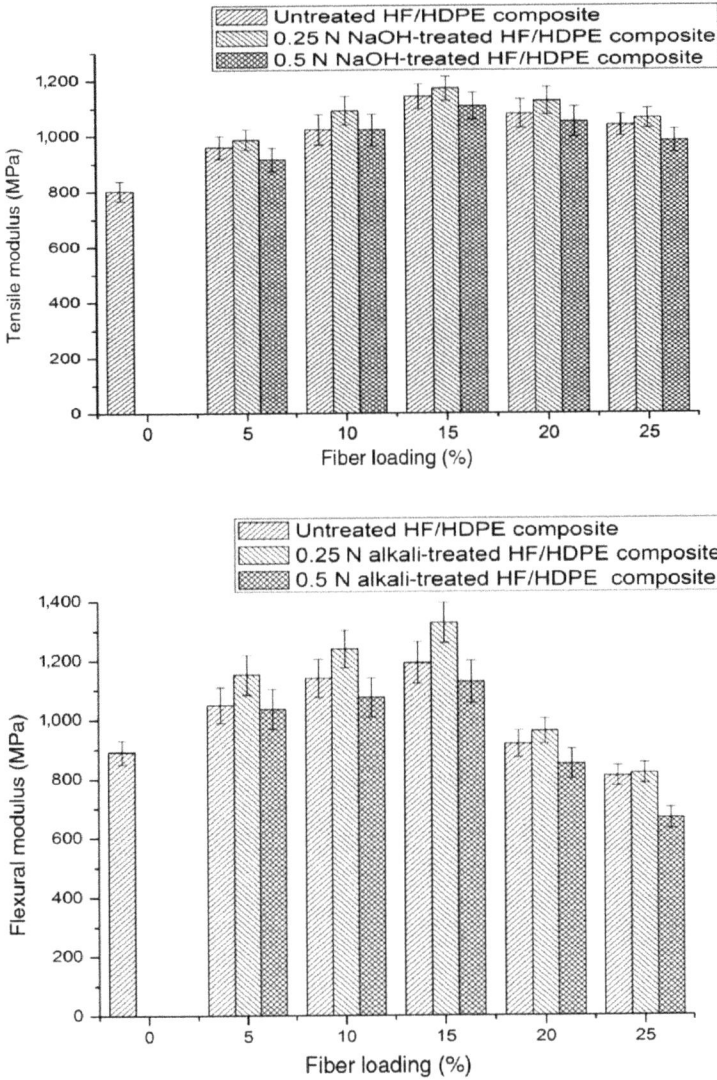

Figure 9.13: Effect of fiber loading on (a) the tensile modulus and (b) flexural modulus of human hair/high-density polyethylene reinforced composites [36].

sis of the 0.25 N alkali-treated HF (Figure 9.12(b)). The depicted micrograph evidences that alkali solution generates pits on the fiber surface, which also produces an increase in its roughness. These features bring about an interlocking effect between fiber and matrix as well as the enhancement of the mechanical properties of the composites [36].

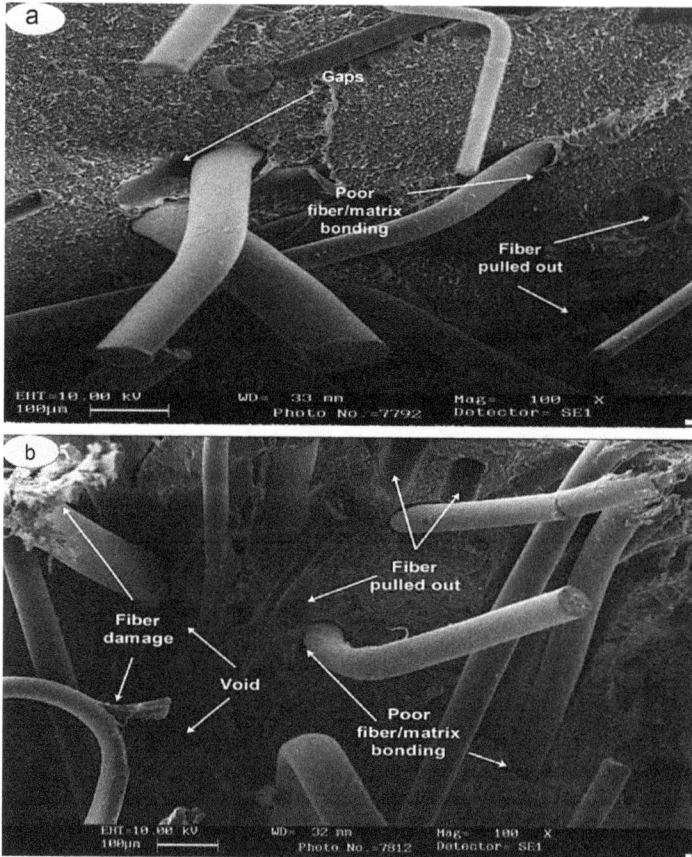

Figure 9.14: SEM micrographs (tensile fracture) of HDPE composites reinforced with (a) untreated HF and (b) 0.25 N alkali-treated HF [36].

Nanda and Satapathy [5] studied the reinforcement of epoxy composites with short human HF (6 wt%), revealing that human hair causes a substantial reduction in the effective thermal conductivity (9.91%), thereby improving the epoxy insulation capability. They also showed that the dielectric constant of the composites varies with the fiber content and the operating frequency.

Cow hair was used as a reinforcing agent in polyesters for the preparation of polymer composites [55]. The filler was applied as received as well as after modification by alkali treatment (0.5 M solution of NaOH at 50 °C). Compared to plain polyester, both untreated and treated cow hair produced an enhancement in flexural and tensile properties as well as an improvement in abrasion resistance. However, composites charged with unmodified fiber showed better mechanical performance with respect to composites based on the alkali-treated filler. This finding can be ascribed to the reduction in strength of the fiber after chemical treatment. The surface of the untreated fibers appears smooth. Conversely, in the NaOH-treated fiber, rough surfaces with cracks were observed, which are evidences of chemical damage.

Chicken feather has also been used in biocomposites due to its properties (hydrophobicity, high thermal insulation, and low density) and composition (91% protein and 1% lipids) [1, 26, 141]. Natural fiber chicken feather was used, along with teak wood dust, as reinforcement in epoxy composites [1]. The prepared samples were characterized through density measurements, tensile mechanical tests, and water absorption analysis. The biocomposites were charged with different amounts of wood dust (10, 15 and 20 wt%) while the chicken feather content was fixed at 5 wt%. The prepared composites showed a decrease in density and an improvement in tensile strength with the increase in wood flour content. The authors attributed the enhancement in mechanical behavior of composites to the good adhesion of the fiber to the matrix and to the effective transfer of the stress along the interface. Moisture absorption tests evidenced a decrease in water uptake of the composites with respect to the plain matrix. This effect envisages a higher hydrophobicity of the material and extends its application field in wet environments. Velmurugan et al. [26] studied the mechanical properties of epoxy hybrid composites based on eggshell, chicken feathers, and jute fiber as reinforcing additives. The authors analyzed only two samples prepared with the same chicken feather content (5 vol%) and different eggshell concentrations (10 and 15 vol%). The composite at 10 wt% of eggshell content showed better tensile properties with respect to the sample at 15 wt% concentration. The impact test, conversely, recorded an opposite trend, as the specimen charged with 15 wt% of cofiller exhibited higher impact strength than the 10 wt% sample.

Spider dragline silk is a fibrous material that exhibits excellent biodegradability and outstanding physical properties like high strength, high ductility, good extensibility, and the highest toughness among natural and synthetic fibers. This is due to its higher order structures based on N- and C-terminal domains, forming hard β-sheet crystallites [93]. Composite films have been fabricated using dragline silk protein from *Araneus diadematus* with polyalanine derivatives. The protein was used to synthesize the silk doped with polyalanine, improving its mechanical properties.

9.8 Biodegradability of biocomposite materials

The accumulation of plastic wastes in the environment fosters the research for new, cost-effective, energetically efficient, and environmentally friendly methods for their elimination. Recycling has been proposed as a sustainable and economically valuable technique to avoid the disposal of plastic materials at the end of their life cycle. However, polymer reprocessing usually brings about thermo-oxidative degradation that produces a decay in material performances and, consequently, only downgraded applications are possible [142]. These issues are much more relevant for bioplastics, usually based on aliphatic polyesters characterized by a relatively low thermodegradative stability. For this class of polymeric materials, biodegradation represents a valuable alternative to recycling for their management at the end of their life cycle. This disposal methodology can be applied to a number of biodegradable polymers as well as to their biocomposites. It allows the reincorporation of plastic materials into the carbon cycle [143] and is particularly valuable in fields where a short lifetime of the product is required such as packaging and mulching.

Biodegradation of biocomposites in biotic conditions has been reported in a number of papers [144–148]. In most cases, the presence of the biofiller promotes the biodegradation process. The effect has been recorded for PHB/wood fiber under composting conditions and PHA/*Posidonia oceanica* in marine environment [144, 145]. The same outcome has been reported for a PLA/*Phormium tenax* submitted to a composting test [149], biopol/jute and BAK/jute biocomposites [148]. The effect of biofillers on the biodegradation process is due to their inherent biodegradability and also to their hydrophilicity. The latter promotes water penetration and helps microorganism migration inside the biocomposite bulk. In this way, the contact surface between matrix polymer and degrading agent is widened. However, biocomposite biodegradation also depends on other factors; a few of them are discussed in the following.

The nature of the matrix [150] and the biofiller [146] also plays a role in determining the biodegradation properties. Teramoto et al. [150] studied, through soil burial test, the biodegradation in soil of a series of composites based on different aliphatic polyesters charged with 10 wt% abaca fiber. They showed, by means of weight loss data along with SEM and visual analysis, that the PCL-based system was scarcely affected by the presence of the abaca fiber. Conversely, the biofiller affected the behavior of Poly(hydroxybutyrate-co-hydroxyvalerate) (PHBV), PBS, and PLA composites. The effect was particularly remarkable for the PBS/abaca fiber system. In this case, as shown in Figure 9.15, after 120 days of soil burial neat PBS shows a mere 5 wt% weight loss. The corresponding abaca composite, conversely, exhibited nearly 45 wt% weight loss after 60 days, and no further measurement was possible for complete deterioration of the sample. Similar results were obtained for PHBV composites. The different behavior of the PCL system was attributed to the inherent biodegradability of the latter that cannot be further improved. Biofiller can also limit matrix biodegradability.

The addition of coupling agents, or fiber treatments aimed to improve matrix/filler adhesion, in most cases, produces a reduction in biodegradability. This effect was reported for aliphatic polyester/acetylated abaca fiber biocomposites tested by soil burial [150] and in a system based on PHA charged with tetraethoxysilane (TEOS)-modified plant fibers [151]. Similar results were obtained by Mohanty et al. for biopol/jute and polyester amide/jute systems tested in composting conditions [152, 153]. The authors modified the jute fiber by grafting acrylonitrile and methyl methacrylate. They found that the corresponding composites showed a reduced biodegradability with

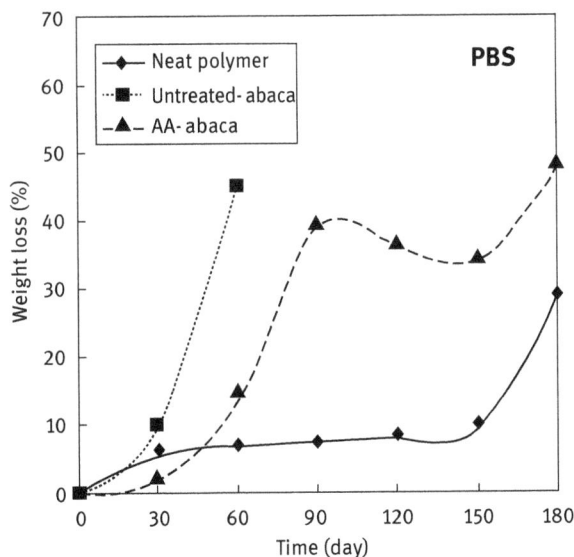

Figure 9.15: Weight loss of neat PBS and PBS/abaca composites during the burial in soil was published by Teramoto et al. [150]. Untreated abaca refers to the composite containing untreated filler. AA abaca refers to composite containing acetic anhydride–treated abaca filler.

respect to the nonmodified systems. They attributed the reported effects to the improvement in fiber matrix adhesion, which limits water diffusion and to the nonbiodegradability of the grafted acrylic oligomers.

Other fiber treatments, such as alkali extraction, also influence the biodegradation of the corresponding composites but the effect is not always predictable. Alkali extraction, removing the more hydrophilic and more biodegradable components from the biomass, is expected to limit water uptake and microorganism penetration. However, the alkali action is not selective and can involve other, more hydrophobic, components. Joyyi et al. [136] revealed that alkali treatment of kenaf fiber produced a reduction of biodegradability, tested by soil burial, of the relative PHA-based composites. The same finding was described by Mohanty et al. [154] for biopol/jute fabric composites and was attributed to the modification of fiber–matrix interactions. Conversely, other authors do not report remarkable effects of alkali treatments of the biofiller on composite biodegradability [148, 153].

9.9 Conclusions

This chapter presented a general view on the use of the most important biobased structural additives in the manufacturing of polymer biocomposites. As reported in the chapter, a number of literature instances that dealt with the development of composites reinforced with naturally occurring filler are ever increasing, in line with the continuously expanding application of biobased and biodegradable polymers. So far, the overall performances of the composites filled with biobased structural additives are still lower than those achieved by conventional heavy duty fillers. Therefore, further work is required to improve the properties of natural fibers, by employing suitable and cost-effective selection, purification, and chemical modification approaches, in order to make fully biobased and biodegradable composites a technological breakthrough.

References

[1] Dash, A., and Tripathy, S. (2018). Mechanical characteristics of chicken feather teak wood dust epoxy filled composite. IOP Conf. Ser. Mater. Sci. Eng. 377, 012111. https://doi.org/10.1088/1757-899X/377/1/012111.

[2] Mittal, V., Saini, R., and Sinha, S. (2016). Natural fiber-mediated epoxy composites – a review. Compos. Part B Eng. 99, 425–435. https://doi.org/10.1016/j.compositesb.2016.06.051.

[3] Krause, K. C., Müller, M., Militz, H., and Krause, A. (2017). Enhanced water resistance of extruded wood–polypropylene composites based on alternative wood sources. Eur. J. Wood Wood Prod. 75, 125–134. https://doi.org/10.1007/s00107-016-1091-5.

[4] Karimi, A., Nazari, S., Ghasemi, I., Tajvidi, M., and Ebrahimi, G. (2006). Effect of the delignification of wood fibers on the mechanical properties of wood fiber–polypropylene composites. J. Appl. Polym. Sci. 102, 4759–4763. https://doi.org/10.1002/app.23967.

[5] Nanda, B. P., and Satapathy, A. (2018). Effect of fiber content on the thermal conductivity and dielectric constant of hair fiber reinforced epoxy composite. IOP Conf. Ser. Mater. Sci. Eng. 338, 012040. https://doi.org/10.1088/1757-899X/338/1/012040.

[6] Teacă, C.-A., Tanasà, F., and Zànoagà, M. (2018). Multi-component polymer systems comprising wood as bio-based component and thermoplastic polymer matrices – an overview. BioResources 13, 4728–4769.

[7] Dayo, A. Q., Wang, A., Kiran, S., Wang, J., Qureshi, K., le Xu, Y., Zegaoui, A., Derradji, M., Babar, A.A., and bin Liu, W. (2018). Impacts of hemp fiber diameter on mechanical and water uptake properties of polybenzoxazine composites. Ind. Crops Prod. 111, 277–284. https://doi.org/10.1016/j.indcrop.2017.10.039.

[8] Ismail, A. S., Jawaid, M., Sultan, M. T. H., and Hassan, A. (2019). Physical and mechanical properties of woven kenaf/bamboo fiber mat reinforced epoxy hybrid composites. Bioresources 14, 1390–1404.

[9] Tripathi, P., Kumar Gupta, V., Dixit, A., Kumar Mishra, R., and Sharma, S. (2018). Development and characterization of low cost jute, bagasse and glass fiber reinforced advanced hybrid epoxy composites. AIMS Mater. Sci. 5, 320–337. https://doi.org/10.3934/matersci.2018.2.320.

[10] Väisäne, T., Das, O., and Tomppo, L. (2017). A review on new bio-based constituents for natural fiber-polymer composites. J. Clean. Prod. 149, 582–596. https://doi.org/10.1016/j.jclepro.2017.02.132.

[11] Shi, D., Hua, J., Zhang, L., and Chen, M. (2015). Synthesis of bio-based poly(lactic acid-co-10-hydroxy decanoate) copolymers with high thermal stability and ductility. Polymers (Basel) 7, 468–483. https://doi.org/10.3390/polym7030468.

[12] le Duigou, A., Merotte, J., Bourmaud, A., Davies, P., Belhouli, K., and Baley, C. (2017). Hygroscopic expansion: a key point to describe natural fibre/polymer matrix interface bond strength. Compos. Sci. Technol. 151, 228–233. https://doi.org/10.1016/j.compscitech.2017.08.028.

[13] Hadjadj, A., Jbara, O., Tara, A., Gilliot, M., Malek, F., Maafi, E. M., and Tighzert, L. (2016). Effects of cellulose fiber content on physical properties of polyurethane based composites. Compos. Struct. 135, 217–223. https://doi.org/10.1016/j.compstruct.2015.09.043.

[14] Sullins, T., Pillay, S., Komus, A., and Ning, H. (2017). Hemp fiber reinforced polypropylene composites: the effects of material treatments. Compos. Part B Eng. 114, 15–22. https://doi.org/10.1016/j.compositesb.2017.02.001.

[15] Li, Z., Ye, L., Shen, J., Xie, K., and Li, Y. (2018). Strain-gauge sensing composite films with self-restoring water-repellent properties for monitoring human movements. Compos. Commun. 7, 23–29. https://doi.org/10.1016/j.coco.2017.12.007.

[16] Agustin-Salazar, S., Gamez-Meza, N., Medina-Juárez, L. Á., Malinconico, M., and Cerruti, P. (2017). Stabilization of polylactic acid and polyethylene with nutshell extract: efficiency assessment and economic evaluation. ACS Sustain. Chem. Eng. 5, 4607–4618. https://doi.org/10.1021/acssuschemeng.6b03124.

[17] Pellis, A., Herrero Acero, E., Ferrario, V., Ribitsch, D., Guebitz, G. M., and Gardossi, L. (2016). The closure of the cycle: enzymatic synthesis and functionalization of bio-based polyesters. Trends Biotechnol. 34, 316–328. https://doi.org/10.1016/j.tibtech.2015.12.009.

[18] Tumbic, J., Romo-uribe, A., Boden, M., and Mather, P. T. (2016). Hot-compacted interwoven webs of biodegradable polymers. Polymer (Guildf) 101, 127–138. https://doi.org/10.1016/j.polymer.2016.08.057.

[19] Belkhir, K., Jegat, C., and Taha, M. (2016). Thiol-ene grafting from polylactic acid, polycaprolactone, and polyhydroxybutyrate. React. Funct. Polym. 101, 82–89. https://doi.org/10.1016/j.reactfunctpolym.2016.02.009.

[20] Hu, Y., Daoud, W. A., Cheuk, K. K. L., and Lin, C. S. K. (2016). Newly developed techniques on polycondensation, ring-opening polymerization and polymer modification: focus on poly (lactic acid). Materials (Basel) 9. https://doi.org/10.3390/ma9030133.

[21] Hamad, K., Kaseem, M., Yang, H. W., Deri, F., and Ko, Y. G. (2015). Properties and medical applications of polylactic acid: a review. Express Polym. Lett. 9, 435–455. https://doi.org/10.3144/expresspolymlett.2015.42.

[22] Dayo, A. Q., kun Ma, R., Kiran, S., Zegaoui, A., an Cai, W., Shah, A. H., Wang, J., Derradji, M., and bin Liu, W. (2018). Reinforcement of economical and environment friendly Acacia catechu particles for the reduction of brittleness and curing temperature of polybenzoxazine thermosets. Compos. Part A Appl. Sci. Manuf. 105, 258–264. https://doi.org/10.1016/j.compositesa.2017.12.004.

[23] Labidi, K., Cao, Z., Zrida, M., Murphy, A., Hamzaoui, A. H., and Devine, D. M. (2019). Alfa fiber/polypropylene composites: influence of fiber extraction method and chemical treatments, J. Appl. Polym. Sci. 136, 47392. https://doi.org/10.1002/app.47392.

[24] Peter, C., Rani, M. N. A., Yunus, M. A., Zin, M. S. M., Kalam, A., and Shah, M. A. S. A. (2019). Inverse method for material properties identification of a kenaf reinforced composite. 020058. https://doi.org/10.1063/1.5086001.

[25] Sood, M., and Dwivedi, G. (2018). Effect of fiber treatment on flexural properties of natural fiber reinforced composites: a review. Egypt. J. Pet. 27, 775–783. https://doi.org/10.1016/j.ejpe.2017.11.005.

[26] Velmurugan, K., Sanjaikumar, A., Saravanaraj, V., and Kumar, P. R. (2017). Studies on mechanical properties of bio material filler and jute fiber-feathers reinforced hybrid composite. Int. J. Sci. Res. Sci. Technol. 3, 310–316.

[27] Vaidya, A. A., Collet, C., Gaugler, M., and Lloyd-Jones, G. (2019). Integrating softwood biorefinery lignin into polyhydroxybutyrate composites and application in 3D printing. Mater. Today Commun. 19, 286–296. https://doi.org/10.1016/j.mtcomm.2019.02.008.

[28] Yang, J., Ching, Y. C., and Chuah, C. H. (2019). Applications of lignocellulosic fibers and lignin in bioplastics: a review. Polymers (Basel) 11, 1–26. https://doi.org/10.3390/polym11050751.

[29] Castro-Aguirre, E., Iñiguez-Franco, F., Samsudin, H., Fang, X., and Auras, R. (2016). Poly(lactic acid) – mass production, processing, industrial applications, and end of life. Adv. Drug Deliv. Rev. 107, 333–366. https://doi.org/10.1016/j.addr.2016.03.010.

[30] Benyahia, H., Tarfaoui, M., Datsyuk, V., El Moumen, A., Trotsenko, S., and Reich, S. (2017). Dynamic properties of hybrid composite structures based multiwalled carbon nanotubes. Compos. Sci. Technol. 148, 70–79. https://doi.org/10.1016/j.compscitech.2017.05.021.

[31] Otto, G. P., Moisés, M. P., Carvalho, G., Rinaldi, A. W., Garcia, J. C., Radovanovic, E., and Fávaro, S. L. (2017). Mechanical properties of a polyurethane hybrid composite with natural lignocellulosic fibers. Compos. Part B Eng. 110, 459–465. https://doi.org/10.1016/j.composi tesb.2016.11.035.

[32] Manteghi, S., Mahboob, Z., Fawaz, Z., and Bougherara, H. (2017). Investigation of the mechanical properties and failure modes of hybrid natural fiber composites for potential bone fracture fixation plates. J. Mech. Behav. Biomed. Mater. 65, 306–316. https://doi.org/10.1016/j.jmbbm.2016.08.035.

[33] Vytejčková, S., Vápenka, L., Hradecký, J., Dobiáš, J., Hajšlová, J., Loriot, C., Vannini, L., and Poustka, J. (2017). Testing of polybutylene succinate based films for poultry meat packaging. Polym. Test. 60, 357–364. https://doi.org/10.1016/j.polymertesting.2017.04.018.

[34] Ramesh, M., Palanikumar, K., and Reddy, K. H. (2017). Plant fibre based bio-composites: sustainable and renewable green materials. Renew. Sustain. Energy Rev. 79, 558–584. https://doi.org/10.1016/j.rser.2017.05.094.

[35] Väisänen, T., Das, O., and Tomppo, L. (2017). A review on new bio-based constituents for natural fiber-polymer composites. J. Clean. Prod. 149, 582–596. https://doi.org/10.1016/j.jclepro.2017.02.132.

[36] Srivastava, P., and Sinha, S. (2018). Effect of alkali treatment on hair fiber as reinforcement of HDPE composites: mechanical properties and water absorption behavior. IEEE J. Sel. Top. Quantum Electron. 25, 571–578. https://doi.org/10.1515/secm-2016-0198.

[37] Auras, R., Harte, B., and Selke, S. (2004). An overview of polylactides as packaging materials. Macromol. Biosci. 4, 835–864. https://doi.org/10.1002/mabi.200400043.

[38] Domínguez-Díaz, M., Meneses-Acosta, A., Romo-Uribe, A., Peña, C., Segura, D., and Espin, G. (2015). Thermo-mechanical properties, microstructure and biocompatibility in poly-β-hydroxybutyrates (PHB) produced by OP and OPN strains of Azotobacter vinelandii. Eur. Polym. J. 63, 101–112. https://doi.org/10.1016/j.eurpolymj.2014.12.002.

[39] Giro-Paloma, J., Rayón, E., Roa, J. J., Martínez, M., and Fernández, A. I. (2015). Effect of the filler on the nanomechanical properties of polypropylene in contact with paraffinic phase change material. Eur. Polym. J. 63, 29–36. https://doi.org/10.1016/j.eurpolymj.2014.11.029.

[40] Rychlý, J., Rychlá, L., Stloukal, P., Koutný, M., Pekařová, S., Verney, V., and Fiedlerová, A. (2013). UV initiated oxidation and chemiluminescence from aromatic-aliphatic co-polyesters and polylactic acid. Polym. Degrad. Stab. 98, 2556–2563. https://doi.org/10.1016/j.polymde gradstab.2013.09.016.

[41] Gurunathan, T., Mohanty, S., and Nayak, S. K. (2015). A review of the recent developments in biocomposites based on natural fibres and their application perspectives. Compos. Part A Appl. Sci. Manuf. 77, 1–25. https://doi.org/10.1016/j.compositesa.2015.06.007.

[42] Duval, A., and Lawoko, M. (2014). A review on lignin-based polymeric, micro- and nano-structured materials. React. Funct. Polym. 85, 78–96. https://doi.org/10.1016/j.reactfunctpolym.2014.09.017.

[43] Lasprilla, A. J. R., Martinez, G. A., Lunelli, B. H., Jardini, A. L., and Filho, R. M. (2011). Poly-lactic acid synthesis for application in biomedical devices – a review. Biotechnol. Adv. 30, 321–328. https://doi.org/10.1016/j.biotechadv.2011.06.019.

[44] McKeown, N. B. (2017). The synthesis of polymers of intrinsic microporosity (PIMs). Sci. China Chem. 60, 1023–1032. https://doi.org/10.1007/s11426-017-9058-x.

[45] Mandlekar, N., Cayla, A., Rault, F., Giraud, S., Salaün, F., Malucelli, G., and Guan, J.-P. (2018). An overview on the use of lignin and its derivatives in fire retardant polymer systems. In: Lignin – Trends Applications, InTech. https://doi.org/10.5772/intechopen.72963.

[46] Jayaseelan, J., Vijayakumar, K. R., Ethiraj, N., Sivabalan, T., and Andrew Nallayan, W. (2017). The effect of fibre loading and graphene on the mechanical properties of goat hair fibre epoxy composite. IOP Conf. Ser. Mater. Sci. Eng. 282. https://doi.org/10.1088/1757-899X/282/1/012018.

[47] Le Gall, M., Davies, P., Martin, N., and Baley, C. (2018). Recommended flax fibre density values for composite property predictions. Ind. Crops Prod. 114, 52–58. https://doi.org/10.1016/j.indcrop.2018.01.065.

[48] Scida, D., Bourmaud, A., and Baley, C. (2017). Influence of the scattering of flax fibres properties on flax/epoxy woven ply stiffness. Mater. Des. 122, 136–145. https://doi.org/10.1016/j.matdes.2017.02.094.

[49] Agustin-Salazar, S., Cerruti, P., Medina-Juárez, L. Á., Scarinzi, G., Malinconico, M., Soto-Valdez, H., and Gamez-Meza, N. (2018). Lignin and holocellulose from pecan nutshell as reinforcing fillers in poly (lactic acid) biocomposites. Int. J. Biol. Macromol. 115, 727–736. https://doi.org/10.1016/j.ijbiomac.2018.04.120.

[50] Zhang, X., Shi, J., Ye, H., Dong, Y., and Zhou, Q. (2018). Combined effect of cellulose nanocrystals and poly(butylene succinate) on poly(lactic acid) crystallization: the role of interfacial affinity. Carbohydr. Polym. 179, 79–85. https://doi.org/10.1016/j.carbpol.2017.09.077.

[51] Roselló-Soto, E., Poojary, M. M., Barba, F. J., Lorenzo, J. M., Mañes, J., and Moltó, J. C. (2018). Tiger nut and its by-products valorization: from extraction of oil and valuable compounds to development of new healthy products. Innov. Food Sci. Emerg. Technol. 45, 306–312. https://doi.org/10.1016/j.ifset.2017.11.016.

[52] Sanjay, M. R., Madhu, P., Jawaid, M., Senthamaraikannan, P., Senthil, S., and Pradeep, S. (2018). Characterization and properties of natural fiber polymer composites: a comprehensive review. J. Clean. Prod. 172, 566–581. https://doi.org/10.1016/j.jclepro.2017.10.101.

[53] Rao, P. D., Kiran, C. U., and Prasad, K. E. (2017). Effect of fiber loading and void content on tensile properties of keratin based randomly oriented human hair fiber composites. Int. J. Compos. Mater. 7, 136–143. https://doi.org/10.5923/j.cmaterials.20170705.02.

[54] Florentin, Y., Pearlmutter, D., Givoni, B., and Gal, E. (2017). A life-cycle energy and carbon analysis of hemp-lime bio-composite building materials. Energy Build. 156 293–305. https://doi.org/10.1016/j.enbuild.2017.09.097.

[55] Oladele, I. O., Agbeboh, N. I., Isola, B. A., and Daramola, O. O. (2018). Abrasion and mechanical properties of keratinous based polyester composites. J. Eng. Technol. 9, 1–15.

[56] Deshmukh, K., and Joshi, G. M. (2014). Thermo-mechanical properties of poly (vinyl chloride)/graphene oxide as high performance nanocomposites. Polym. Test. 34, 211–219. https://doi.org/10.1016/j.polymertesting.2014.01.015.

[57] Lu, X., and Mi, Y. (2015). Gradiently varied chain packing/orientation states of polymer thin films revealed by polarization-dependent infrared absorption. Eur. Polym. J. 63, 247–254. https://doi.org/10.1016/j.eurpolymj.2014.12.030.

[58] Dasdemir, M., Maze, B., Anantharamaiah, N., and Pourdeyhimi, B. (2015). Reactive compatibilization of polyamide 6/polyethylene nonwoven based thermoplastic composites. Eur. Polym. J. 63, 194–206. https://doi.org/10.1016/j.eurpolymj.2014.12.019.

[59] Bodîrlău, R., Teacă, C.-A., Roşu, D., Roşu, L., Varganici, C.-D., and Coroabă, A. (2013). Physico-chemical properties investigation of softwood surface after treatment with organic anhydride. Cent. Eur. J. Chem. 11, 2098–2106. https://doi.org/10.2478/s11532-013-0337-x.

[60] Ramesh, P., Prasad, B. D., and Narayana, K. L. (2019). Morphological and mechanical properties of treated kenaf fiber/MMT clay reinforced PLA hybrid biocomposites (p. 020035). https://doi.org/10.1063/1.5085606.

[61] Dayo, A. Q., Chang Gao, B., Wang, J., Bin Liu, W., Derradji, M., Shah, A. H., and Babar, A. A. (2017). Natural hemp fiber reinforced polybenzoxazine composites: curing behavior, mechanical and thermal properties. Compos. Sci. Technol. 144, 114–124. https://doi.org/10. 1016/j.compscitech.2017.03.024.

[62] Móczó, J., and Pukánszky, B. (2008). Polymer micro and nanocomposites: structure, interactions, properties. J. Ind. Eng. Chem. 14, 535–563. https://doi.org/10.1016/j.jiec.2008. 06.011.

[63] Yuan, X., Jayaraman, K., and Bhattacharyya, D. (2004). Effects of plasma treatment in enhancing the performance of woodfibre-polypropylene composites. Compos. Part A Appl. Sci. Manuf. 35, 1363–1374. https://doi.org/10.1016/j.compositesa.2004.06.023.

[64] Dányádi, L., Janecska, T., Szabó, Z., Nagy, G., Móczó, J., and Pukánszky, B. (2007). Wood flour filled PP composites: compatibilization and adhesion. Compos. Sci. Technol. 67, 2838–2846. https://doi.org/10.1016/j.compscitech.2007.01.024.

[65] Dányádi, L., Renner, K., Szabó, Z., Nagy, G., Móczó, J., and Pukánszky, B. (2006). Wood flour filled PP composites: adhesion, deformation, failure. Polym. Adv. Technol. 17, 967–974. https://doi.org/10.1002/pat.838.

[66] Beg, M. D. H., and Pickering, K. L. (2006). Fiber pretreatment and its effects on wood fiber reinforced polypropylene composites. Mater. Manuf. Process. 21, 303–307. https://doi.org/ 10.1080/10426910500464750.

[67] Anggono, J., Farkas, Á. E., Bartos, A., Móczó, J., Antoni,, Purwaningsih, H., and Pukánszky, B. (2019). Deformation and failure of sugarcane bagasse reinforced PP. Eur. Polym. J. 112, 153–160. https://doi.org/10.1016/j.eurpolymj.2018.12.033.

[68] Felix, J. M., and Gatenholm, P. (1991). The nature of adhesion in composites of modified cellulose fibers and polypropylene. J. Appl. Polym. Sci. 42, 609–620. https://doi.org/10. 1002/app.1991.070420307.

[69] Faludi, G., Dora, G., Renner, K., Móczó, J., and Pukánszky, B. (2013). Improving interfacial adhesion in pla/wood biocomposites. Compos. Sci. Technol. 89, 77–82. https://doi.org/10. 1016/j.compscitech.2013.09.009.

[70] Kokta, B. V., Maldas, D., Daneault, C., and Beland, P. (1990). Composites of polyvinyl chloride-wood fibers. I. Effect of isocyanate as a bonding agent. Polym. Plast. Technol. Eng. 29, 87–118. https://doi.org/10.1080/03602559008049836.

[71] Raj, R. G., Kokta, B. V., and Daneault, C. (1989). Effect of chemical treatment of fibers on the mechanical properties of polyethylene-wood fiber composites. J. Adhes. Sci. Technol. 3, 55–64. https://doi.org/10.1163/156856189X00056.

[72] Herrera-Franco, P., and Valadez-González, A. (2004). Mechanical properties of continuous natural fibre-reinforced polymer composites. Compos. Part A Appl. Sci. Manuf. 35, 339–345. https://doi.org/10.1016/j.compositesa.2003.09.012.

[73] Nam, T. H., Ogihara, S., Tung, N. H., and Kobayashi, S. (2011). Effect of alkali treatment on interfacial and mechanical properties of coir fiber reinforced poly(butylene succinate) biodegradable composites. Compos. Part B Eng. 42, 1648–1656. https://doi.org/10.1016/j.compositesb.2011.04.001.

[74] Rajesh, G., Ratna Prasad, A. V., and Gupta, A. (2015). Mechanical and degradation properties of successive alkali treated completely biodegradable sisal fiber reinforced poly lactic acid composites. J. Reinf. Plast. Compos. 34, 951–961. https://doi.org/10.1177/0731684415584784.

[75] Shukor, F., Hassan, A., Saiful Islam, M., Mokhtar, M., and Hasan, M. (2014). Effect of ammonium polyphosphate on flame retardancy, thermal stability and mechanical properties of alkali treated kenaf fiber filled PLA biocomposites. Mater. Des. 54, 425–429. https://doi.org/10.1016/j.matdes.2013.07.095.

[76] Fiore, V., Scalici, T., Nicoletti, F., Vitale, G., Prestipino, M., and Valenza, A. (2016). A new eco-friendly chemical treatment of natural fibres: effect of sodium bicarbonate on properties of sisal fibre and its epoxy composites. Compos. Part B Eng. 85, 150–160. https://doi.org/10.1016/j.compositesb.2015.09.028.

[77] Arfat, Y. A., Ahmed, J., Hiremath, N., Auras, R., and Joseph, A. (2017). Thermo-mechanical, rheological, structural and antimicrobial properties of bionanocomposite films based on fish skin gelatin and silver-copper nanoparticles. Food Hydrocoll. 62. https://doi.org/10.1016/j.foodhyd.2016.08.009.

[78] Cai, J., He, Y., Yu, X., Banks, S. W., Yang, Y., Zhang, X., Yu, Y., Liu, R., and Bridgwater, A. V. (2017). Review of physicochemical properties and analytical characterization of lignocellulosic biomass. Renew. Sustain. Energy Rev. 76, 309–322. https://doi.org/10.1016/j.rser.2017.03.072.

[79] Rizvi, S. M. A., Dwivedi, A., Raza, S. S., Awasthi, A., and Gupta, H. (2017). An investigation of thermal properties of reinforced coconut coir-bagasse fibres polymer hybrid composites. Int. J. Sci. Res. Sci. Eng. Technol. 3, 427–432.

[80] Vyazovkin, S., Chrissafis, K., Di Lorenzo, M. L., Koga, N., Pijolat, M., Roduit, B., Sbirrazzuoli, N., and Suñol, J. J. (2014). ICTAC Kinetics Committee recommendations for collecting experimental thermal analysis data for kinetic computations. Thermochim. Acta. 590, 1–23. https://doi.org/10.1016/j.tca.2014.05.036.

[81] Naseem, A., Tabasum, S., Zia, K. M., Zuber, M., Ali, M., and Noreen, A. (2016). Lignin-derivatives based polymers, blends and composites: a review. Int. J. Biol. Macromol. 93, 296–313. https://doi.org/10.1016/j.ijbiomac.2016.08.030.

[82] Poletto, M., Ornaghi, H., and Zattera, A. (2014). Native Cellulose: Structure, Characterization and Thermal Properties. Materials, 7(9), 6105–6119. https://doi.org/10.3390/ma7096105.

[83] Indran, S., Raj, R. E., and Sreenivasan, V. S. (2014). Characterization of new natural cellulosic fiber from Cissus quadrangularis root. Carbohydr. Polym. 110, 423–429. https://doi.org/10.1016/j.carbpol.2014.04.051.

[84] Spinacé, M. A. S., Lambert, C. S., Fermoselli, K. K. G., and De Paoli, M. (2009). Characterization of lignocellulosic curaua fibres. Carbohydr. Polym. 77, 47–53. https://doi.org/10.1016/j.carbpol.2008.12.005.

[85] Wang, J. P., Matthews, M. L., Williams, C. M., Shi, R., Yang, C., Tunlaya-Anukit, S., Chen, H. C., Li, Q., Liu, J., Lin, C. Y., Naik, P., Sun, Y. H., Loziuk, P. L., Yeh, T. F., Kim, H., Gjersing, E., Shollenberger, T., Shuford, C. M., Song, J., Miller, Z., Huang, Y. Y., Edmunds, C. W., Liu, B., Sun, Y., Lin, Y. C. J., Li, W., Chen, H., Peszlen, I., Ducoste, J. J., Ralph, J., Chang, H. M.,

Muddiman, D. C., Davis, M. F., Smith, C., Isik, F., Sederoff, R., and Chiang, V. L. (2018). Improving wood properties for wood utilization through multi-omics integration in lignin biosynthesis. Nat. Commun. 9. https://doi.org/10.1038/s41467-018-03863-z.

[86] Lang, Q., Guo, Y., Zheng, Q., Liu, Z., and Gai, C. (2018). Co-hydrothermal carbonization of lignocellulosic biomass and swine manure: hydrochar properties and heavy metal transformation behavior. Bioresour. Technol. 266, 242–248. https://doi.org/10.1016/j.biortech.2018.06.084.

[87] Williams, C. L., Westover, T. L., Petkovic, L. M., Matthews, A. C., Stevens, D. M., and Nelson, K. R. (2017). Determining thermal transport properties for softwoods under pyrolysis conditions. ACS Sustain. Chem. Eng. 5, 1019–1025. https://doi.org/10.1021/acssuschemeng.6b02326.

[88] Sanchis, M. J., Carsí, M., Gómez, C. M., Culebras, M., Gonzales, K. N., and Torres, F. G. (2017). Monitoring molecular dynamics of bacterial cellulose composites reinforced with graphene oxide by carboxymethyl cellulose addition. Carbohydr. Polym. 157, 353–360. https://doi.org/10.1016/j.carbpol.2016.10.001.

[89] Hu, B., Zhou, K., Liu, Y., Liu, A., Zhang, Q., Han, G., Liu, S., Yang, Y., Zhu, Y., and Zhu, D. (2018). Optimization of microwave-assisted extraction of oil from tiger nut (Cyperus esculentus L.) and its quality evaluation. Ind. Crops Prod. 115, 290–297. https://doi.org/10.1016/j.indcrop.2018.02.034.

[90] Ribeiro, V. P., Almeida, L. R., Martins, A. R., Pashkuleva, I., Marques, A. P., Ribeiro, A. S., Silva, C. J., Bonifácio, G., Sousa, R. A., Oliveira, A. L., and Reis, R. L. (2017). Modulating cell adhesion to polybutylene succinate biotextile constructs for tissue engineering applications. J. Tissue Eng. Regen. Med. 11, 2853–2863. https://doi.org/10.1002/term.2189.

[91] Sahoo, S., Seydibeyoğlu, M. Ö., Mohanty, A. K., and Misra, M. (2011). Characterization of industrial lignins for their utilization in future value added applications. Biomass Bioenergy 35, 4230–4237. https://doi.org/10.1016/j.biombioe.2011.07.009.

[92] Savy, D., and Piccolo, A. (2014). Physical-chemical characteristics of lignins separated from biomasses for second-generation ethanol. Biomass Bioenergy 62, 58–67. https://doi.org/10.1016/j.biombioe.2014.01.016.

[93] Tsuchiya, K., Ishii, T., Masunaga, H., and Numata, K. (2018). Spider dragline silk composite films doped with linear and telechelic polyalanine: effect of polyalanine on the structure and mechanical properties. Sci. Rep. 8, 1–9. https://doi.org/10.1038/s41598-018-21970-1.

[94] Farhat, W., Venditti, R. A., Hubbe, M., and Taha, M. (2017). A review of water-resistant hemicellulose-based materials: processing and applications. 305–323. https://doi.org/10.1002/cssc.201601047.

[95] John, M. J., and Thomas, S. (2008). Biofibres and biocomposites. Carbohydr. Polym. 71, 343–364. https://doi.org/10.1016/j.carbpol.2007.05.040.

[96] Scheller, H. V., and Ulvskov, P. (2010). Hemicelluloses. Annu. Rev. Plant Biol. 61, 263–289. https://doi.org/10.1146/annurev-arplant-042809-112315.

[97] Qian, S., and Sheng, K. (2017). PLA toughened by bamboo cellulose nanowhiskers: role of silane compatibilization on the PLA bionanocomposite properties. Compos. Sci. Technol. 148, 59–69. https://doi.org/10.1016/j.compscitech.2017.05.020.

[98] Mokhothu, T. H., and John, M. J. (2015). Review on hygroscopic aging of cellulose fibres and their biocomposites. Carbohydr. Polym. 131, 337–354. https://doi.org/10.1016/j.carbpol.2015.06.027.

[99] Foresti, M. L., Vázquez, A., and Boury, B. (2017). Applications of bacterial cellulose as precursor of carbon and composites with metal oxide, metal sulfide and metal nanoparticles: a review of recent advances. Carbohydr. Polym. 157, 447–467. https://doi.org/10.1016/j.carbpol.2016.09.008.

[100] Biagiotti, J., Puglia, D., and Kenny, J. M. (2004). A review on natural fibre-based composites – part II. J. Nat. Fibers. 1, 37–68. https://doi.org/10.1300/J395v01n03.

[101] Jackson, M. J., and Line, M. A. (1997). Organic composition of a pulp and paper mill sludge determined by FTIR, 13 C CP MAS NMR, and chemical extraction techniques. J. Agric. Food Chem. 45, 2354–2358. https://doi.org/10.1021/jf960946l.

[102] Haykiri-Acma, H., Yaman, S., and Kucukbayrak, S. (2010). Comparison of the thermal reactivities of isolated lignin and holocellulose during pyrolysis. Fuel Process. Technol. 91, 759–764. https://doi.org/10.1016/j.fuproc.2010.02.009.

[103] Dhyani, V., and Bhaskar, T. (2017). A comprehensive review on the pyrolysis of lignocellulosic biomass. Renew. Energy. 129, 695–716. https://doi.org/10.1016/j.renene. 2017.04.035.

[104] Fundador, N. G. V., Enomoto-Rogers, Y., Takemura, A., and Iwata, T. (2012). Acetylation and characterization of xylan from hardwood kraft pulp. Carbohydr. Polym. 87, 170–176. https:// doi.org/10.1016/j.carbpol.2011.07.034.

[105] Sun, R., Fang, J. M., Tomkinson, J., and Jones, G. L. (1999). Acetylation of wheat straw hemicelluloses in N, N-dimethylacetamide/LiCl solvent system. 10, 209–218.

[106] Klemm, D., Heublein, B., Fink, H.-P., and Bohn, A. (2005). Cellulose: fascinating biopolymer and sustainable raw material. ChemInform. 36, 36. https://doi.org/10.1002/chin. 200536238.

[107] Lessa, E. F., Gularte, M. S., Garcia, E. S., and Fajardo, A. R. (2017). Orange waste: a valuable carbohydrate source for the development of beads with enhanced adsorption properties for cationic dyes. Carbohydr. Polym. 157, 660–668. https://doi.org/10.1016/j.carb pol.2016.10.019.

[108] Jústiz-Smith, N. G., Virgo, G. J., and Buchanan, V. E. (2008). Potential of Jamaican banana, coconut coir and bagasse fibres as composite materials. Mater. Charact. 59, 1273–1278. https://doi.org/10.1016/j.matchar.2007.10.011.

[109] Kim, J. H., Park, S., Kim, H., Kim, H. J., Yang, Y., Kim, Y. H., Jung, S., Kan, E., and Lee, S. H. (2017). Alginate/bacterial cellulose nanocomposite beads prepared using Gluconacetobacter xylinus and their application in lipase immobilization. Carbohydr. Polym. 157, 137–145. https://doi.org/10.1016/j.carbpol.2016.09.074.

[110] Peng, S., Fan, L., Wei, C., Liu, X., Zhang, H., Xu, W., and Xu, J. (2017). Flexible polypyrrole/ copper sulfide/bacterial cellulose nanofibrous composite membranes as supercapacitor electrodes. Carbohydr. Polym. 157, 344–352. https://doi.org/10.1016/j.carbpol.2016.10.004.

[111] Nuruddin, M., Hosur, M., Uddin, M. J., Baah, D., and Jeelani, S. (2016). A novel approach for extracting cellulose nanofibers from lignocellulosic biomass by ball milling combined with chemical treatment. J. Appl. Polym. Sci. 133, 1–10. https://doi.org/10.1002/app.42990.

[112] Krishnamachari, P., Hashaikeh, R., and Tiner, M. (2011). Modified cellulose morphologies and its composites; SEM and TEM analysis. Micron. 42, 751–761. https://doi.org/10.1016/j. micron.2011.05.001.

[113] Laurichesse, S., and Avérous, L. (2014). Chemical modification of lignins: towards biobased polymers. Prog. Polym. Sci. 39, 1266–1290. https://doi.org/10.1016/j.progpo lymsci.2013.11.004.

[114] Brígida, A. I. S., Calado, V. M. A., Gonçalves, L. R. B., and Coelho, M. A. Z. (2010). Effect of chemical treatments on properties of green coconut fiber. Carbohydr. Polym. 79, 832–838. https://doi.org/10.1016/j.carbpol.2009.10.005.

[115] Wilson, R. H., Smith, A. C., Kacuráková, M., Saunders, P. K., Wellner, N., and Waldron, K. W. (2000). The mechanical properties and molecular dynamics of plant cell wall polysaccharides studied by Fourier-transform infrared spectroscopy. Plant Physiol. 124, 397–405. https://doi.org/10.1104/pp.124.1.397.

[116] Peredo, K., Escobar, D., Vega-Lara, J., Berg, A., and Pereira, M. (2016). Thermochemical properties of cellulose acetate blends with acetosolv and sawdust lignin: a comparative study. Int. J. Biol. Macromol. 83, 403–409. https://doi.org/10.1016/j.ijbiomac.2015.11.022.

[117] Harshvardhan, K., Suri, M., Goswami, A., and Goswami, T. (2017). Biological approach for the production of vanillin from lignocellulosic biomass (Bambusa tulda). J. Clean. Prod. 149, 485–490. https://doi.org/10.1016/j.jclepro.2017.02.125.

[118] Holladay, J. E., Bozell, J. J., White, J. F., and Johnson, D. (2007). Top value added chemicals from biomass: volume II results of screening for potential candidates from biorefinery lignin.

[119] Franke, R., Hemm, M. R., Denault, J. W., Ruegger, M. O., Humphreys, J. M., and Chapple, C. (2002). Changes in secondary metabolism and deposition of an unusual lignin in the ref8 mutant of Arabidopsis. Plant J. 30, 47–59. http://www.ncbi.nlm.nih.gov/pubmed/11967092.

[120] Mottiar, Y., Vanholme, R., Boerjan, W., Ralph, J., and Mansfield, S. D. (2016). Designer lignins: harnessing the plasticity of lignification. Curr. Opin. Biotechnol. 37, 190–200. https://doi.org/10.1016/j.copbio.2015.10.009.

[121] Elizalde-González, M. P., and Hernández-Montoya, V. (2007). Characterization of mango pit as raw material in the preparation of activated carbon for wastewater treatment. Biochem. Eng. J. 36, 230–238. https://doi.org/10.1016/j.bej.2007.02.025.

[122] Agustin-Salazar, S., Gamez-Meza, N., Medina-Juárez, L. Á., Malinconico, M., and Cerruti, P. (2017). Stabilization of polylactic acid and polyethylene with nutshell extract: efficiency assessment and economic evaluation. ACS Sustain. Chem. Eng. 5, 4607–4618. https://doi.org/10.1021/acssuschemeng.6b03124.

[123] Gupta, A., Simmons, W., Schueneman, G. T., Hylton, D., and Mintz, E. A. (2017). Rheological and thermo-mechanical properties of poly(lactic acid)/lignin-coated cellulose nanocrystal composites. ACS Sustain. Chem. Eng. 5, 1711–1720. https://doi.org/10.1021/acssuschemeng.6b02458.

[124] Gadioli, R., Morais, J. A., Waldman, W. R., and De Paoli, M.-A. (2014). The role of lignin in polypropylene composites with semi-bleached cellulose fibers: mechanical properties and its activity as antioxidant. Polym. Degrad. Stab. 108, 23–34. https://doi.org/10.1016/j.polymdegradstab.2014.06.005.

[125] Ferry, L., Dorez, G., Taguet, A., Otazaghine, B., and Lopez-Cuesta, J. M. (2015). Chemical modification of lignin by phosphorus molecules to improve the fire behavior of polybutylene succinate. Polym. Degrad. Stab. 113, 135–143. https://doi.org/10.1016/j.polymdegradstab.2014.12.015.

[126] Wikberg, H., and Maunu, S. L. (2004). Characterisation of thermally modified hard- and softwoods by ^{13}C CPMAS NMR. Carbohydr. Polym. 58, 461–466. https://doi.org/10.1016/j.carbpol.2004.08.008.

[127] BioStruct. (2007). Nuevos materiales compuestos mejorados basados en madera y plástico (eWPC). http://www.biostructproject.eu/fileadmin/MEDIEN/Training/Material/1-ESP_New_eWPC_materials.pdf.

[128] Alekhina, M., Ershova, O., Ebert, A., Heikkinen, S., and Sixta, H. (2015). Softwood kraft lignin for value-added applications: fractionation and structural characterization. Ind. Crops Prod. 66, 220–228. https://doi.org/10.1016/j.indcrop.2014.12.021.

[129] Avolio, R., Graziano, V., Pereira, Y. D. F., Cocca, M., Gentile, G., Errico, M. E., Ambrogi, V., and Avella, M. (2015). Effect of cellulose structure and morphology on the properties of poly (butylene succinate-co-butylene adipate) biocomposites. Carbohydr. Polym. 133, 408–420. https://doi.org/10.1016/j.carbpol.2015.06.101.

[130] Khalil, H. P. S. A., Bhat, I. U. H., Jawaid, M., Zaidon, A., Hermawan, D., and Hadi, Y. S. (2012). Bamboo fibre reinforced biocomposites : a review. 42, 353–368. https://doi.org/10.1016/j.matdes.2012.06.015.

[131] Fazita, N., Rawi, M., and Jayaraman, K. (2013). A performance study on composites made from bamboo fabric and poly (lactic acid). https://doi.org/10.1177/0731684413498296.

[132] Gamon, G., Evon, P., and Rigal, L. (2013). Twin-screw extrusion impact on natural fibre morphology and material properties in poly(lactic acid) based biocomposites. Ind. Crops Prod. 46, 173–185. https://doi.org/10.1016/j.indcrop.2013.01.026.

[133] Tanguy, M., Bourmaud, A., Beaugrand, J., Gaudry, T., and Baley, C. (2018). Polypropylene reinforcement with flax or jute fibre; influence of microstructure and constituents properties on the performance of composite. Compos. Part B Eng. 139, 64–74. https://doi.org/10.1016/j.compositesb.2017.11.061.

[134] Bourmaud, A., Fazzini, M., Renouard, N., Behlouli, K., and Ouagne, P. (2018). Innovating routes for the reused of PP-flax and PP-glass non woven composites: a comparative study. Polym. Degrad. Stab. 152, 259–271. https://doi.org/10.1016/j.polymdegradstab.2018.05.006.

[135] Sathishkumar, G., Rajkumar, G., Srinivasan, K., and Umapathy, M. (2017). Structural analysis and mechanical properties of lignite fly-ash-added jute–epoxy polymer matrix composite. J. Reinf. Plast. Compos. 073168441773518. https://doi.org/10.1177/0731684417735183.

[136] Joyyi, L., Ahmad Thirmizir, M. Z., Salim, M. S., Han, L., Murugan, P., Ichi Kasuya, K., Maurer, F. H. J., Zainal Arifin, M. I., and Sudesh, K. (2017). Composite properties and biodegradation of biologically recovered P(3HB-co-3HHx) reinforced with short kenaf fibers. Polym. Degrad. Stab. 137, 100–108. https://doi.org/10.1016/j.polymdegradstab.2017.01.004.

[137] Decorte, T. (2011). Fibre hemp and marihuana: assessing the differences between distinct varieties. Int. Police Exec. Symp., Geneva Cen.

[138] Alvarez-Chavez. (2017). Characterization of extruded poly (lactic acid)/pecan nutshell biocomposites. 2017, 13–17. https://doi.org/https://doi.org/10.1155/2017/3264098.

[139] Sánchez, D., Arturo, A., Uribe, R., Álvarez, C. R., Amar, C., Misra, M., Cervantes, J. L., and Madera, T. J. (2019). Physicochemical characterization and evaluation of pecan nutshell as biofiller in a matrix of poly (lactic acid). J. Polym. Environ. 27, 521–532. https://doi.org/10.1007/s10924-019-01374-6.

[140] Maharana, T., Pattanaik, S., Routaray, A., Nath, N., and Sutar, A. K. (2015). Synthesis and characterization of poly(lactic acid) based graft copolymers. React. Funct. Polym. 93, 47–67. https://doi.org/10.1016/j.reactfunctpolym.2015.05.006.

[141] Bansal, G., Singh, V. K., Gope, P. C., and Gupta, T., (2017). Application and properties of chicken feather fiber (CFF) a livestock waste in composite material development. J. Graph. Era Univ. 5, 16–24.

[142] Badia, J. D., Gil-Castell, O., and Ribes-Greus, A. (2017). Long-term properties and end-of-life of polymers from renewable resources. Polym. Degrad. Stab. 137, 35–57. https://doi.org/10.1016/j.polymdegradstab.2017.01.002.

[143] Badia, J. D., Strömberg, E., Kittikorn, T., Ek, M., Karlsson, S., and Ribes-Greus, A. (2017). Relevant factors for the eco-design of polylactide/sisal biocomposites to control biodegradation in soil in an end-of-life scenario. Polym. Degrad. Stab. 143, 9–19. https://doi.org/10.1016/j.polymdegradstab.2017.06.004.

[144] Seggiani, M., Cinelli, P., Verstichel, S., Puccini, M., Vitolo, S., Anguillesi, I., and Lazzeri, A. (2015). Development of fibres-reinforced biodegradable composites. Chem. Eng. Trans. 43, 1813–1818. https://doi.org/10.3303/CET1543303.

[145] Seggiani, M., Cinelli, P., Mallegni, N., Balestri, E., Puccini, M., Vitolo, S., Lardicci, C., and Lazzeri, A. (2017). New bio-composites based on polyhydroxyalkanoates and Posidonia oceanica fibres for applications in a marine environment. Materials (Basel) 10. https://doi.org/10.3390/ma10040326.

[146] Angelini, S., Cerruti, P., Immirzi, B., Santagata, G., Scarinzi, G., and Malinconico, M. (2014). From biowaste to bioresource: effect of a lignocellulosic filler on the properties of poly(3-hydroxybutyrate). Int. J. Biol. Macromol. 71, 163–173. https://doi.org/10.1016/j.ijbiomac. 2014.07.038.

[147] Fortunati, E., Puglia, D., Santulli, C., Sarasini, F., and Kenny, J. M. (2012). Biodegradation of Phormium tenax/poly(lactic acid) composites. J. Appl. Polym. Sci. 125, E562–E572. https:// doi.org/10.1002/app.36839.

[148] Chander, K., Mohanty, A., and Joergensen, R. (2002). Decomposition of biodegradable packing materials jute, Biopol, BAK and their composites in soil. Biol. Fertil. Soils. 36, 344–349. https://doi.org/10.1007/s00374-002-0548-3.

[149] Fortunati, E., Peltzer, M., Armentano, I., Torre, L., Jiménez, A., and Kenny, J. M. (2012). Effects of modified cellulose nanocrystals on the barrier and migration properties of PLA nano-biocomposites. Carbohydr. Polym. 90, 948–956. https://doi.org/10.1016//j.carbpol.2012. 06.025.

[150] Teramoto, N., Urata, K., Ozawa, K., and Shibata, M. (2004). Biodegradation of aliphatic polyester composites reinforced by abaca fiber. Polym. Degrad. Stab. 86, 401–409. https:// doi.org/10.1016/j.polymdegradstab.2004.04.026.

[151] Wu, C. S. (2013). Preparation, characterization and biodegradability of crosslinked tea plant-fibre-reinforced polyhydroxyalkanoate composites. Polym. Degrad. Stab. 98, 1473–1480. https://doi.org/10.1016/j.polymdegradstab.2013.04.013.

[152] Mohanty, A. K., Misra, M., and Hinrichsen, G. (2000). Biofibres, biodegradable polymers and biocomposites: an overview. Macromol. Mater. Eng. 276–277, 1–24. https://doi.org/10. 1002/(SICI)1439-2054(20000301)276:1<1::AID-MAME1>3.0.CO;2-W.

[153] Mohanty, A. K., Khan, M. A., and Hinrichsen, G. (2000). Influence of chemical surface modification on the properties of biodegradable jute fabrics – polyester amide composites. Compos. Part A Appl. Sci. Manuf. 31, 143–150. https://doi.org/10.1016/S1359-835X(99) 00057-3.

[154] Mohanty, A. K., Khan, M. A., and Hinrichsen, G. (2000). Surface modification of jute and its influence on performance of biodegradable jute-fabric/Biopol composites. Compos. Sci. Technol. 60, 1115–1124. https://doi.org/10.1016/S0266-3538(00)00012-9.

Federico Olivieri, Cristina De Capitani, Andrea Sorrentino

10 Additive manufacturing for biodegradable polymers

Abstract: Additive manufacturing consists of a wide variety of techniques capable of responding to the needs of the several application sectors. From prototyping to direct production of parts, additive manufacturing is now widely used in industrial sectors such as biomedical, aerospace, automotive, architectural, and fashion. Thanks to its peculiar characteristics, it is often possible to increase sustainability and efficiency even of known and consolidated processes. The high flexibility of additive manufacturing allows to work with practically all types of known materials, although polymers are the most commonly used ones. Concerns about the environmental impact of the materials used in these new processes are growing more and more recently. Fortunately, polymeric materials can be produced from different sources, including renewable and low environmental impact ones. This chapter focuses on the description of the most used additive techniques. Particular attention is given to AD technologies that make use of biodegradable materials. The combination of AD production with biodegradable polymers provides the opportunity to achieve a true bio-based economy.

Keywords: Additive manufacturing, sustainability, biodegradable polymers, easy use, easy process, modelling

10.1 Introduction

Additive manufacturing (AM) is a group of manufacturing techniques that are defined as the process of joining materials layer upon layer to make objects from a 3D model [1]. For many years, AM was largely used in prototyping due to its ability to produce complex shapes and design new products [2].

Nowadays, the strong reduction in equipment costs and the increase in software capability have supported the rapid development and diffusion of these technologies also in production stages. These manufacturing technologies make possible to produce, repair, or replace products everywhere from the industry to the local shop, from the hospital to the school [3]. Everywhere, it is possible to download a file with the

Federico Olivieri, Institute for Polymers, Composites and Biomaterials, National Research Council of Italy, Portici, Italy
Cristina De Capitani, Andrea Sorrentino, Institute of Polymers, Composites and Biomaterials (IPCB-CNR), Lecco, Italy

https://doi.org/10.1515/9783110590586-010

part design and print it without particular competences. With limited efforts, it is possible to modify or personalize existing products by adding customized features [4, 5].

With respect to the conventional manufacturing processes, AM requires low startup costs, local production, and on-demand manufacturing (Figure 10.1). In addition, layer by layer deposition reduces largely the material waste and allows design freedom [6]. These technologies already find applications in biomedical, aerospace, automotive, engineering, and arts [7, 8].

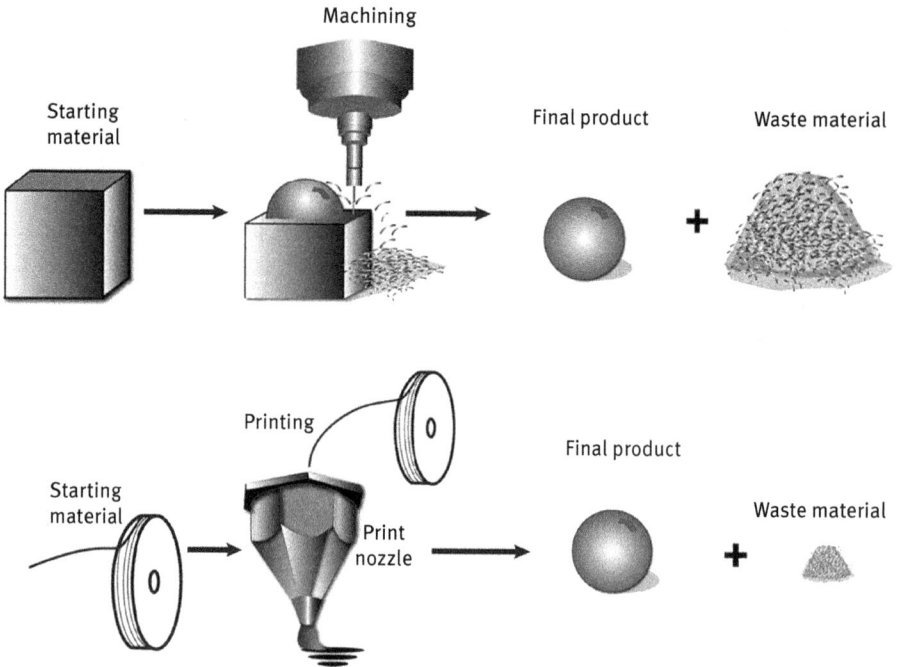

Figure 10.1: AM process versus conventional manufacturing process.

In the near future, AM promises to create a revolution in the modern manufacturing process [9]. Production and supply chain will be integrated in a more simple, efficient, and sustainable model. Mass production in factories and global logistics of both raw materials and products could become useless. Products will be transferred as files instead of materials or objects. This organization model will require only local logistics of raw materials and a capillary AM machines availability [10, 11]. Local productions will be favored for the lower environmental impact and greenhouse gas emissions [12]. In this context, objects produced from renewable resources provide the opportunity to realize a truly sustainable and circular economy [13, 14]. The limited availability of biobased materials will not represent a limiting factor as in the case of mass-produced parts and components. Biobased materials sourced from plant, animal, or marine origins will be readily used to replace fabrics, adhesives, reinforcement fibers, polymers, and other conventional materials [15–17]. They will be used alone or in combination with conventional materials to reduce mass and meet stricter fuel economy and emissions standards.

However, the replacement of conventional materials can be possible only if the biobased products meet or exceed current quality, performance, and price standards [18, 19]. It still requires the introduction of innovative manufacturing processes that are able to have complete control during the process [20]. From this point of view, the combination of biodegradable polymers with AM offers unique potentiality. The natural variability in biobased materials can be controlled quickly and efficiently by adjusting the manufacturing conditions [21]. Parts formed with AM can be designed to be stronger, lighter, or more functional than parts made with conventional manufacturing processes [22]. The geometries as well as the processing conditions can be changed for every single part produced, allowing an incredible flexibility. The perspective to process biodegradable materials in an easy way, keeping low costs and wastes, represents a very attractive and versatile approach for the production of sustainable products.

10.2 Rapid prototyping

Rapid prototyping starts with the development of a computer-aided design (CAD) that reproduces the real object [23]. The possibility to manipulate and optimize the CAD model represents probably the most powerful skill of rapid prototyping [24]. The rapid prototyping is then continued with transformation of the CAD model in the computer-aided manufacturing and computer numerical control file. At this point, specific techniques, such as subtractive or AM, can be utilized to produce the part [25, 26]. Different rapid prototyping techniques have been developed. They allow using various materials, such us metals, ceramics, sand, food, polymers, and even living cells. However, each technique has its processing limitations, in terms, for example, of energy need or geometric shape [27].

For the AM process, the CAD model is sliced into thousands of horizontal layers and then sent to the machine for the process development. Currently there are several technologies and variations that are considered as AM. These technologies are characterized by the fact that new material is continually added where it is necessary to form the required geometry. In Figure 10.2, a schematic view of all the possible classes of AM techniques is presented. Some of them are already suitable for biodegradable polymers, whereas for others many companies are working to offer new materials [28]. Among the biobased materials available for AM, there are a variety of materials made from natural fibers, sugars, algae, and even waste byproducts from coffee and beer production [29–31]. In the following sections, these AM approaches will be examined.

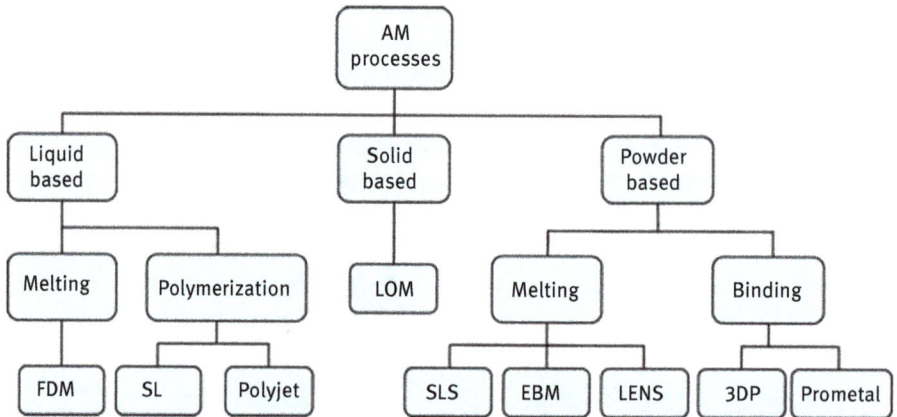

Figure 10.2: AM techniques: stereolithography (SL), fused deposition modeling (FDM), laminated object manufacturing (LOM), powder 3D printing (3DP), selective laser sintering (SLS), laminated engineered net shaping (LENS), and electron beam melting (EBM) [27].

10.2.1 Stereolithography

In 1986, Charles Hull deposited the first patent of a stereolithographic (SL) system. The process allows manufacturing 3D objects starting from a Ultraviolet (UV)-light responsive resin. A UV laser light is used to draw preprogrammed patterns onto the surface of the photopolymer [24]. In the right conditions, the resin is solidified and forms a single layer of the desired 3D object. This process is repeated for each layer of the design until the 3D object fabrication is complete [25, 27]. This process can fabricate high-resolution prototypes with adjustable micron layer density. In particular, layer thickness smaller than 10 µm can be obtained by micro-SL (µSL) [32].

Parameters such as exposure time, scan speed, and laser power can influence the printed object properties and the percentage of curing [33]. Two irradiation methods exist: laser writing and mask-based projection, which differ as for illumination method and build orientation [34]. While in the laser writing the CAD model is directly transferred to the machine, and in mask-based approach, the transfer of an image is obtained through the irradiation of a patterned mask (Figure 10.3).

In the mask-based approach, each layer is built at the same time. In this case, the print rate only depends on the exposure time required by the material. On the contrary, in laser writing method each layer is built point-by-point, thus the print rate is strongly dependent on the part geometry. However, the generation of a huge number of masks and its precise alignment in a very precise way make the mask-based approach more expensive and complex to manage.

In the SL machine, the plate used is transparent and nonadhering, in order to avoid interference with the light irradiation and facilitate the separation between

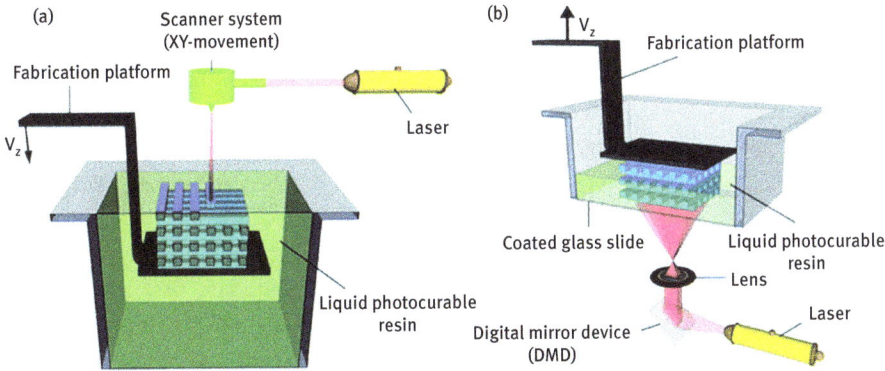

Figure 10.3: Laser writing (a) and mask-based (b) methods [35].

the plate and the object [36]. Once the process is ended, the object is not already cured and needs the removal of excess liquid resin from the external surface and the cleaning with solvents in order to remove uncured resin [2]. After that, a subsequent phase of postcuring occurs, whose duration depends on the material and object characteristics [37].

One of the main disadvantages of the SL process is the limited number of available resin materials [38, 39]. Nowadays, the most common used resins in SL are acrylic and epoxy resins. The starting material is generally composed of liquid compounds and a photocatalyst which is chemically activated by the energy source. One of the most difficult skills to control is the depth of the polymerization: in order to reach a good uniformity, UV absorbers can be added to the resin [40]. The presence of additives and the difficulty in controlling the extent of polymerization (linked to the possible cytotoxicity of the residual photoinitiator) make this technique not particularly suitable for biodegradable polymers. However, due to the high resolution and the versatility, SL hold great potential in biomedical applications, such as tissue engineering. Therefore, the study is focused on the formulation of nontoxic and biodegradable polymers and catalysts, with high attention on the degradation rate [41–45]. In the last decade, reactions to produce synthetic biodegradable polymers have been studied, and involve especially polyesters. Poly(propylene fumarate) (PPF), an unsaturated linear polyester introduced for the first time by Sanderson in 1988, is probably the most used biodegradable polymer in SL. Cooke et al. demonstrated that PPF mixed with other compounds can be used in SL for developing trabecular bones which have suitable mechanical properties [46]. Lan et al. obtained similar results using μSL [47]. Starting from these results, Lee et al. optimized the concentration of the compounds used [48]. Lactide polymers, such as polylactic acid (PLA), are also largely used [49, 50]. In particular, poly(D,L-lactide) (PDLLA) is one of the most common studied biodegradable polymers because its elastic modulus (about 3 GPa) is very close to that of some bones modulus' (ranging between 3 and 30 GPa) [51–53]. In order to reach a valid formulation to be used in SL, Jensen et al. studied a photopolymerization with a fumaric acid-based reagent and *N*-vinylpyrrolidone as a diluent. They obtained a polymer to be applied in tissue engineering. Melchels et al. developed a formulation without diluents [54] to obtain porous polylactide-based objects. Polycaprolactone (PCL) also finds applications due to its interesting characteristics; in particular the easy shaping combined to the tailorable degradation kinetics makes this polymer absolutely suitable in SL applied to biomedical engineering [55–59]. The first photocross-linkable PCL was developed by Kweon et al. [60]. Its PCL is also easy to functionalize with polar groups in order to improve the hydrophilicity, the adhesion, or biocompatibility [61, 62]. Furthermore, hydrogels are cross-linked polymers and they can be used in SL if the cross-linking is photoactivated [63]. Polyglycolic acid is one of the most largely used polymers in that sense, due to its easy functionalization with photoreactive groups [64, 65].

10.2.2 Selective laser sintering

The first selective laser sintering (SLS) system was patented by Carl Deckard in 1987 [66]. Its working principle includes the use of a laser (typically a CO_2 laser) which selectively fuses and sinters a powder material. The system includes a reservoir for the powder and a build platform, whereon the powder is placed and leveled by a blade or a roller (Figure 10.4).

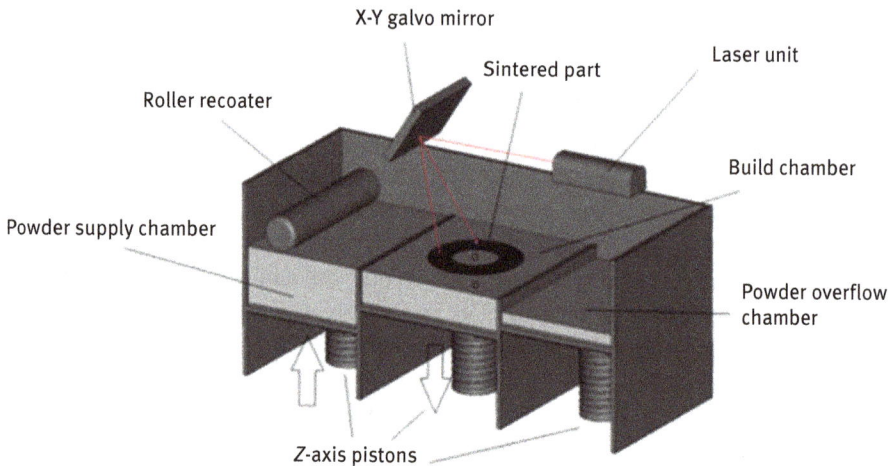

Figure 10.4: SLS setup [67].

This technique requires long starting time; the temperature in the camera and the bed temperature have to be controlled to guarantee a good result. Moreover, the bed temperature is heated gradually, even few degrees per minute, depending on the final temperature to reach (this phase can be 1–2 h long). The camera is isolated, but blade movements can influence the thermal stability in the camera; for this reason, the machine provides a heating system which works also during the printing (Figure 10.5). The layer deposition is guided by a z-direction translator linked to the platform; once a layer is completed, the translator lowers the platform allowing the formation of the next layer. The unsintered powder works as a support for the object, during its printing, and, at the end of the process, can be recovered. It means that the waste of material is very limited in SLS processing. Object properties can be affected by laser parameters, such as its power or exposure time, combined to the layer thickness and processing temperature [68].

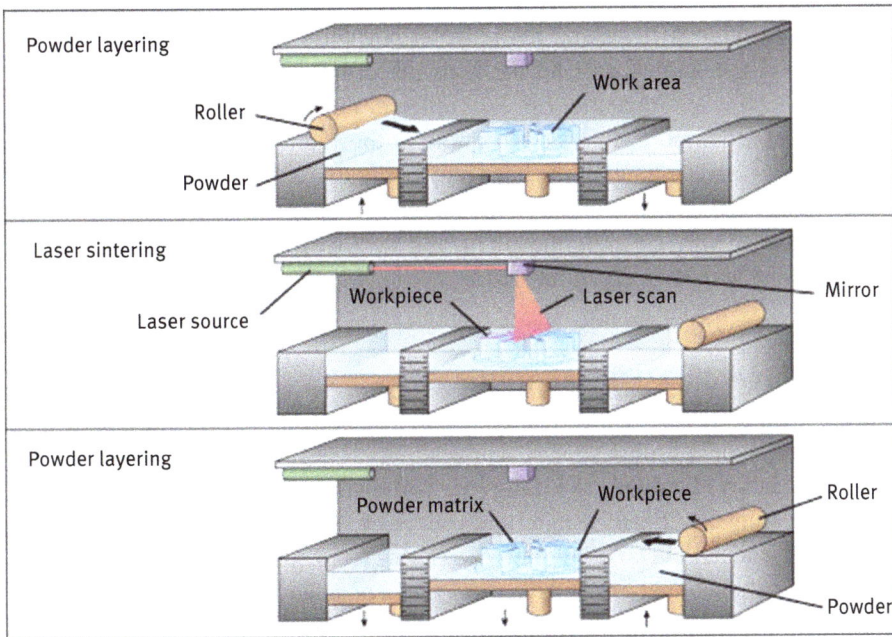

Figure 10.5: SLS working [69].

The great advantage of SLS technique is its flexibility: theoretically, all the materials that can be pulverized and sintered can be processed. This skill allows the process of polymers, ceramics, or metals. Biomedical industry is very interested in SLS, due to the possibility to work with several materials and, eventually, to combine different parts using the same technique [32, 70, 71]. Biodegradable polymers such as

PCL or PLLA are processed by SLS. These materials found application as polymeric scaffolds: Williams et al. used PCL [72], while Tan et al. exploited PLLA [73]. Other examples of SLS printed scaffolds provide the structuring of composite object, made by biodegradable polymers and biocompatible ceramics: Simpson et al. developed hydroxylapatite (HA) and β-tricalcium phosphate combined to poly(L-lactide-co-glycolide) (PLGA) [74] and Wiria et al. used PCL and HA [75]. Biopolymer functionalization is a powerful tool also for improving specific characteristics of the powder or of the final object. Nanofillers affect mechanical, optical, and thermal properties and allow the biodegradability and bioactivity of polymeric composites [76–78]. An example of nanofiller used to improve the mechanical properties is silica [79] or alumina [80]. Another interesting application point out the reduction of the required laser energy; nanofiller, if opportunely chosen, can absorb the energy more efficiently [81]. Several nanoparticles such as quartz, silica, talc, or graphite are used for this application. When mixed with polycarbonate, the obtained nanocomposites show lower energy requirement [82]. Obviously, it is crucial to guarantee a good adhesion between nanoparticles and the polymer powder [83, 84]. The easiest method consists of a mechanical mixing between the components but, especially when the particle sizes and densities are not uniform, the final properties could not be optimized. Furthermore, in order to improve the filler dispersion usually these nanoparticles are coated with the base polymer [85].

In biomedical application, the combination of biopolymers such as PLLA, chitosan, or collagen with calcium phosphate and hydroxyapatite as nanofillers or bioceramics were widely explored [86, 87].

10.2.3 Three-dimensional printing

Three-dimensional printing (3DP) is a powder-based technology that was developed in 1994 by Sachs et al. [88]. The difference with SLS is that here powder is not sintered by heating but "fixed" with a liquid adhesive [89, 90]. Theoretically, any powder reducible material could be printed with this technology. The process concerns the printing of the liquid in selective regions of a powder-coated platform [91]. Once a layer is completed, the platform shifts a relative displacement from the nozzle at a distance of the layer thickness imposed (Figure 10.6). At the end of the process, the unbounded powder is removed from the object and the removal from the support is relatively easy.

The quality of the printed object is strictly influenced by the interaction between powder and binder. Other relevant factors are the local amount of deposited binder, deposition rate, binder viscosity, and powder size [93, 94]. This technique is very flexible in terms of quantity of selectable materials; in fact, distilled water could also be used, as shown by Lam et al. [95]. However, 3DP does not allow to reach high resolution (about 300 µm), due to the difficulty in deposition binder control. Its flexibility

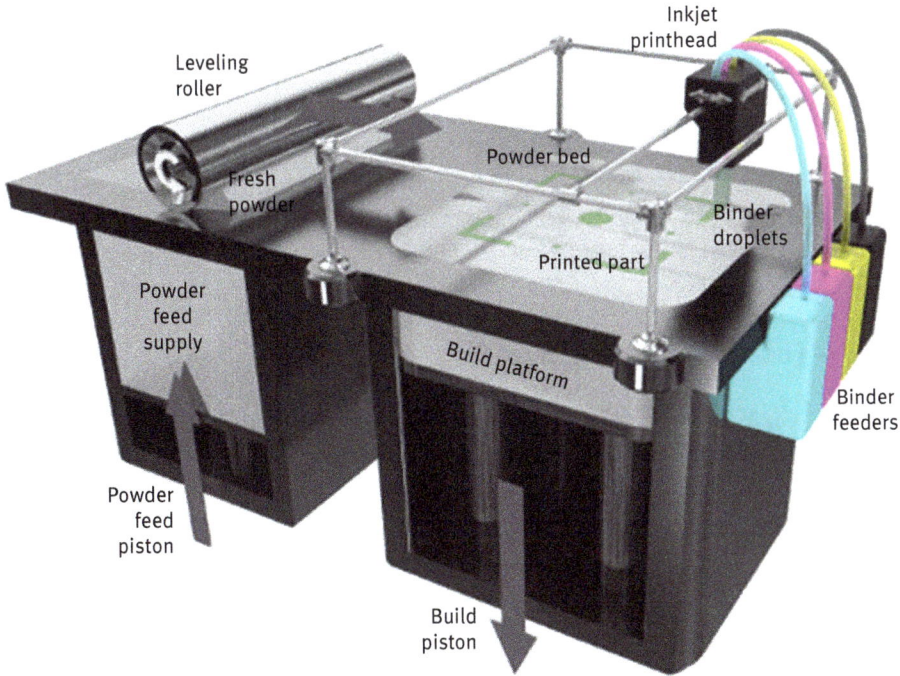

Figure 10.6: 3DP setup [92].

guarantees high microstructure control and ability to print overhangs and internal architectures [96]. The process occurs at room temperature: it is a great advantage when, for example, temperature sensitive filler are added to a matrix [91]. If the printer object shows very detailed regions or curved parts, it could be difficult in removing the unbounded powder [96].

Also in this case, the most used biodegradable polymers in 3DP are PDLLA, PLLA, PCL, PLGA, and so on. That is mainly due to the biomedical applications. Lam et al. combined water with natural polymers such as starch, dextran, and gelatin as binders [95]. Generic organic solvents have also been used as binders, but the choice of the solvent is very important. Commercial organic solvents already exist, but they are optimized only for few polymeric solutions causing usually an average reduction of the shape complexity [97]. Some examples concern the use of PCL, PLGA, and PLLA, as shown by Wu et al. [90], Kim et al. [96], and Zeltinger et al. [97]. The final application is the fabrication of scaffolds in which the printing of channels are supported by the unbounded powder [96]. Biodegradable polymer-based composites have also been developed for 3DP. Sherwood et al. designed a complex system in which a first region was composed of D,L-PLGA/L-PLA with 90% porosity, while a second region was composed of a L-PLGA/tricalcium phosphate (TCP) composite, in order to promote the bone and the cartilage regeneration [98]. Shanjani et al. studied and characterized polyvinylalcohol mixed with calcium polyphosphate tissue engineering structures [99]. An important approach, named indirected 3DP, consists of printing of a mold and its cavities are filled with biodegradable materials [100, 101]. Thanks to indirect 3DP, the limitations due to the use of biodegradable polymers are overcome: these polymers are cast in the printed mold. However, this approach involves a worsening in infill density capability and a restriction in design featuring and shaping, due to difficult demolding, causing a loss in architecture uniformity. PCL was used in indirect 3DP, combined to chitosan, exploiting a gelatin mold [102].

10.2.4 Fused deposition modeling

Due to the low machine costs in terms of machine and materials, nowadays fused deposition modeling (FDM) is probably the most common used AM technique [103]. It was developed in 1980s and commercialized by *Stratasys* in 1990. Its usage covers several fields, from industry to private hobby. Figure 10.7 shows the reference market of FDM (*Forbes Magazine*, 2018 [104]).

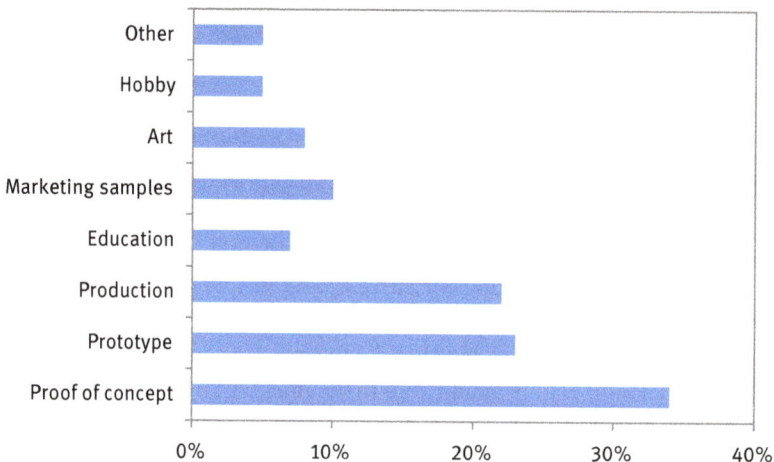

Figure 10.7: FDM applications [104].

FDM produces 3D models via the extrusion of a thermoplastic material through a small nozzle and its deposition onto a build platform. The thermoplastic material in filament shape is moved by two internal gears toward the heated nozzle tip of the extruder. The main mechanisms of a FDM system are illustrated in Figure 10.8.

The filament is conveyed through a wheel which clings and pulls the polymer into a heat room. Here melting occurs and the continuous wheel rotation presses the fused

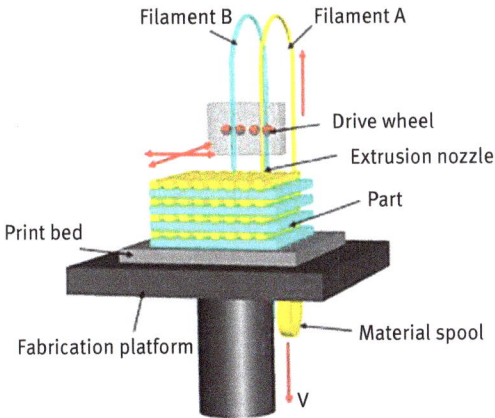

Figure 10.8: FDM setup [105].

filament into the extrusion nozzle [106]. Many skills are variable among the FDM print-ers, for example the building plate, the extrusion system, or the axis movements [107, 108]. The building plate is generally connected to a heat source, in order to im-prove the adhesion between the plate material and the extruded polymer [109]. Imposing the plate temperature at a value slightly higher than the T_g, the polymer keeps fluidity when it comes in contact with the plate [105]. It also promotes the adhesion between the polymer and the plate and facilitates the new layers deposi-tion. In fact, a typical problem in FDM is the bending of the object, due to the low adhesion with the plate during printing. Another solution consists of introducing a layer, typically having high roughness or high adhesive properties between the plate and the extruded polymer, in order to "trap" the first layer on the substrate. The ambient temperature is another important parameter to control in FDM. This is the reason why all professional systems generally have conditioned working chambers. Some FDM machines have more than one nozzle, which allows printing different materials simultaneously. The movement can be connected to the nozzle and/or the platform. Generally, the x-y movement is connected to the nozzle, while the z movement is connected to the platform.

Layer thickness depends on the nozzle diameter; reducing the layer thickness and the nozzle diameter lead to an increase in the part resolution but also to a higher proc-essing time. In addition, the polymer viscosity becomes really important. High viscosity can occlude the nozzle and avoid a good adhesion between polymer layers [110].

Infill percentage and print speed can be varied for reducing printing time. However, the mechanical properties and the final dimensions of the printed part are also affected by these parameters [2]. In particular, mechanical properties are affected by several parameters, such as infill percentage, layer thickness, printing orientation, processing temperature, and flow rate [111]. For particular shapes or detailed geometries, it is necessary to provide support generation during printing. They are printed, meanwhile the object layers are deposited. When smooth surfaces are required, eventually also due to the support presence, a postprocessing treatment is needed. Complex geometries or relatively big or high defined objects can require several working hours.

The most used polymers in FDM are PLA, acrylonitrile butadiene styrene, and thermoplastic polyurethane. Among the biodegradable materials, other suitable polymers for FDM are PCL [112, 113] and PLGA [114, 115]. It is also possible to use biodegradable composites having polymeric matrix, but the optimization of these composites is not easy. In fact, filler tends to increase the viscosity, which is a very crucial parameter to control. The filler distribution has to be very uniform in order to deposit uniform layers and keep the filler dispersion constant in the entire object. Commercially, filler reinforced filaments are largely available. Fillers as metallic or ceramic nanoparticles are used in order to improve specific properties, such as mechanical, electrical, or optical properties. Carbon nanotubes or graphene are able to deeply modify filament and 3D printed objects, as showed by Gnanasekaran et al. [116]. Biodegradable polymers processed by FDM are highly exploited in biomedical field. The polymer functionalization and polymer blends are widely studied in 3D scaffold printing. For instance, Kalita et al. developed TCP reinforced Polypropylene (PP) scaffold [117], while Woodfield et al. studied a scaffold obtained blending polyethyleneglycol and polybutylene terephthalate [118]. Rai et al. used PCL as a matrix and HA or TCP as fillers [119], while Korpela et al. combined PCL with a bioactive glass [120]. PLGA is as much used in scaffold: Kim et al. exploited a multinozzle system to combine PCL and PLGA [121], Shim et al. reinforced this blend with TCP [122], and Kim et al. mixed PLGA, TCP, and HA [115].

10.3 Conclusions

The combination of AM and biobased materials represents the next logical step in the current manufacturing process. It will help to produce and create a future without destroying the natural resources of our planet. In comparison to mass production, AM demonstrates to have less impact and more place for development.

AM generates less waste and uses less energy in comparison to the conventional manufacturing processes. The parts require less refining and assembly operations. The capillary distribution of small production plants allows removing the need of storing and transportation of the products before and during their sale. The potential in this technology allows saying that AM will continue developing and improving in every aspect, especially in terms of material sustainability. These techniques already have the potentiality to process biobased materials, powders, and resins eliminating the necessity of toxic and harmful materials. New greener products can be obtained by both using recycled materials and reducing or eliminating waste. However, these promises can be realized only if the right type of material will be developed or adjusted. Reuse and recycling of these materials on a local scale is equally important. The right combination of AM and biobased materials can reduce their manufacturing cost and finally encourage the complete replacement of traditional materials. AM is dramatically changing the way that products are developed, distributed, and acquired. Despite the great opportunities for this technology, uncertainties and speculations about its future developments remain. Further studies are therefore required to investigate these advantages and challenges. Several aspects related to the use of bio-based materials on AM have yet to be analyzed trying to remove the barriers that prevent their application.

Acknowledgments: The authors acknowledge the financial support from the "VINMAC" project, ID [139455], D.R. Lombardia n. 9559 – 02/08/2017, CUP: E67H16000980009.

References

[1]   Touri, M., Kabirian, F., Saadati, M., Ramakrishna, S., and Mozafari, M. (2019). Adv. Eng. Mater. 21, 1800511.
[2]   Wong, K. V., and Hernandez, A. (2012). ISRN Mech. Eng. 2012, 1–10.
[3]   González-Henríquez, C. M., Sarabia-Vallejos, M. A., and Rodriguez-Hernandez, J. Prog. Polym. Sci. doi:10.1016/j.progpolymsci.2019.03.001.
[4]   Molotnikov, A., Simon, G. P., and Estrin, Y. (2019). Architectured Materials in Nature and Engineering (pp. 257–285).
[5]   Goh, G. D., Yap, Y. L., Tan, H. K. J., Sing, S. L., Goh, G. L., and Yeong, W. Y. (2019). Crit. Rev. Solid State Mater. Sci. 1–21.
[6]   Wohlers, T. (2012). Manuf. Eng. 148, 55–63.
[7]   Bhushan, J., and Grover, V. (2019). Biomanufacturing (pp. 103–122). Springer International Publishing, Cham.
[8]   Gebhardt, A. (2011). Understanding Additive Manufacturing (Gebhardt, A., Ed., pp. 65–101). Hanser.
[9]   Jiang, R., Kleer, R., and Piller, F. T. (2017). Technol. Forecast. Soc. Change 117, 84–97.
[10] Lee, J.-Y., An, J., and Chua, C. K. (2017). Appl. Mater. Today 7, 120–133.
[11] Kietzmann, J., Pitt, L., and Berthon, P. (2015). Bus. Horiz. 58, 209–215.
[12] Weller, C., Kleer, R., and Piller, F. T. (2015). Int. J. Prod. Econ. 164, 43–56.
[13] Tofail, S. A. M., Koumoulos, E. P., Bandyopadhyay, A., Bose, S., O'Donoghue, L., and Charitidis, C. (2018). Mater. Today 21, 22–37.
[14] Ford, S., and Despeisse, M. (2016). J. Clean. Prod. 137, 1573–1587.
[15] Bioplastics, B.
[16] Top 8 alternative 3D printer filament – 3Dnatives. https://www.3dnatives.com/en/alterna tive-3d-printing-filament230820184/ (accessed 22 April 2019).
[17] 3ders.org – straw based 3D printer filament will cost half the price of PLA | 3D Printer News & 3D Printing News. https://www.3ders.org/articles/20140428-straw-based-3d-printer-filament-will-cost-half-the-price-of-pla.html (accessed 22 April 2019).

[18] Babu, R. P., O'Connor, K., and Seeram, R. (2013). Prog. Biomater. 2, 8.
[19] Gorrasi, G., and Sorrentino, A. (2015). Green Chem. 17, 2610–2625.
[20] Vietri, U., Sorrentino, A., Speranza, V., and Pantani, R. (2011). Polym. Eng. Sci. 51, 2542–2551.
[21] Gebhardt, A. (2011). Understanding Additive Manufacturing (Gebhardt, A., Ed., pp. 129–149). Carl Hanser Verlag GmbH & Co. KG, München.
[22] Vaneker, T. H. J. (2017). Procedia CIRP 60, 181–186.
[23] Ashley, S. (1991). Mech. Eng. 113, 34.
[24] Cooper, K. (2001). Rapid Prototyping Technology: Selection and Application. CRC Press.
[25] Chua, C. K., Leong, K. F., and Lim, C. S. (2010). Rapid Prototyping: Principles and Applications (with Companion CD-ROM). World Scientific Publishing Company.
[26] Kochan, A. (1997). Rapid Prototyp. J. 3, 150–152.
[27] Kruth, J.-P. (1991). CIRP Ann. 40, 603–614.
[28] Bentzen, N., and Laussen, E. (2018). Chalmers University of Technology.
[29] Gkartzou, E., Koumoulos, E. P., and Charitidis, C. A. (2017). Manuf. Rev. 4, 1.
[30] Nguyen, N. A., Barnes, S. H., Bowland, C. C., Meek, K. M., Littrell, K. C., Keum, J. K., and Naskar, A. K. (2018). Sci. Adv. 4, eaat4967.
[31] Li, T., Aspler, J., Kingsland, A., Cormier, L. M., and Zou, X. (2016). J. Sci. Technol. For. Prod. Process 5, 30.
[32] Halloran, J. W., Tomeckova, V., Gentry, S., Das, S., Cilino, P., Yuan, D., Guo, R., Rudraraju, A., Shao, P., Wu, T., Alabi, T. R., Baker, W., Legdzina, D., Wolski, D., Zimbeck, W. R., and Long, D. (2011). J. Eur. Ceram. Soc. 31, 2613–2619.
[33] Cho, Y. H., Lee, I. H., and Cho, D.-W. (2005). Microsyst. Technol. 11, 158–167.
[34] Wendel, B., Rietzel, D., Kühnlein, F., Feulner, R., Hülder, G., and Schmachtenberg, E. (2008). Macromol. Mater. Eng. 293, 799–809.
[35] Waheed, S., Cabot, J. M., Macdonald, N. P., Lewis, T., Guijt, R. M., Paull, B., and Breadmore, M. C. (2016). Lab Chip 16, 1993–2013.
[36] Vaezi, M., Seitz, H., and Yang, S. (2013). Int. J. Adv. Manuf. Technol. 67, 1721–1754.
[37] Gebhardt, A. (2011). Understanding Additive Manufacturing (Gebhardt, A., Ed., pp. 103–128). Hanser.
[38] Voet, V. S. D., Strating, T., Schnelting, G. H. M., Dijkstra, P., Tietema, M., Xu, J., Woortman, A. J. J., Loos, K., Jager, J., and Folkersma, R. (2018). ACS Omega 3, 1403–1408.
[39] Bertana, V., De Pasquale, G., Ferrero, S., Scaltrito, L., Catania, F., Nicosia, C., Marasso, S. L., Cocuzza, M., and Perrucci, F. (2019). Polymers (Basel) 11, 292.
[40] Heller, C., Schwentenwein, M., Russmueller, G., Varga, F., Stampfl, J., and Liska, R. (2009). J. Polym. Sci. Part A Polym. Chem. 47, 6941–6954.
[41] Ronca, A., Ambrosio, L., and Grijpma, D. W. (2013). Acta Biomater. 2013, 9, 5989–5996.
[42] Gill, A. A., and Claeyssens, F. (2011). pp. 309–321.
[43] Arcaute, K., Mann, B. K., and Wicker, R. B. (2011). Stereolithography (pp. 299–331). Springer, Boston.
[44] Melchels, F. P. W., Domingos, M. A. N., Klein, T. J., Malda, J., Bartolo, P. J., and Hutmacher, D. W. (2012). Prog. Polym. Sci. 37, 1079–1104.
[45] Yuan, D., Lasagni, A., Shao, P., and Das, S. (2008). Virtual Phys. Prototyp. 3, 221–229.
[46] Cooke, M. N., Fisher, J. P., Dean, D., Rimnac, C., and Mikos, A. G. (2003). J. Biomed. Mater. Res. B. Appl. Biomater. 64, 65–69.
[47] Lan, P. X., Lee, J. W., Seol, Y.-J., and Cho, D.-W. (2009). J. Mater. Sci. Mater. Med. 20, 271–279.
[48] Lee, K.-W., Wang, S., Fox, B. C., Ritman, E. L., Yaszemski, M. J., and Lu, L. (2007). Biomacromolecules 8, 1077–1084.

[49] Gunatillake, P. A., and Adhikari, R. (2003). Eur. Cell. Mater. 5, 1–16; discussion 16.
[50] Methachan, B., and Tanodekaew, S. (2013). The 6th 2013 Biomedical Engineering International Conference (pp. 1–3). IEEE.
[51] Pham, D., and Gault, R. (1998). Int. J. Mach. Tools Manuf. 38, 1257–1287.
[52] Osswald, T. A., and García-Rodríguez, S. (2011). Handbook of Applied Biopolymer Technology. Royal Society of Chemistry, Cambridge.
[53] Gao, M., and Zhu, P. (2012). Starch – Stärke 64, 97–104.
[54] Melchels, F. P. W., Feijen, J., and Grijpma, D. W. (2009). Biomaterials 30, 3801–3809.
[55] Ronca, A., Dessì, M., Guarino, V., Gloria, A., Raucci, M. G., Ambrosio, L., D'Amore, A., Acierno, D., and Grassia, L. (2008). AIP Conf. Proc.1042, 199–200.
[56] Ronca, A., Guarino, V., Raucci, M. G., Salamanna, F., Martini, L., Zeppetelli, S., Fini, M., Kon, E., Filardo, G., Marcacci, M., and Ambrosio, L. (2014). J. Biomater. Appl. 29, 715–727.
[57] Domingos, M., Intranuovo, F., Russo, T., De Santis, R., Gloria, A., and Ambrosio, L. (2013). J. Ciurana and P. Bartolo, Biofabrication, 5, 045004.
[58] Allen, C., Yu, Y., Maysinger, D., and Eisenberg, A. (1998). Bioconjug. Chem. 9, 564–572.
[59] Wei, X., Gong, C., Gou, M., Fu, S., Guo, Q., Shi, S., Luo, F., Guo, G., Qiu, L., and Qian, Z. (2009). Int. J. Pharm. 381, 1–18.
[60] Kweon, H., Yoo, M. K., Park, I. K., Kim, T. H., Lee, H. C., Lee, H.-S., Oh, J.-S., Akaike, T., and Cho, C. S, (2003). Biomaterials 24, 801–808.
[61] Zhu, Y., Gao, C., Liu, X., and Shen, J. (2002). Biomacromolecules 3, 1312–1319.
[62] Jiao, Y.-P., and Cui, F.-Z. (2007). Biomed. Mater. 2, R24–R37.
[63] Mao, Y., Miyazaki, T., Sakai, K., Gong, J., Zhu, M., and Ito, H. (2018). Polymers (Basel) 10, 1117.
[64] Arcaute, K., Mann, B. K., and Wicker, R. B. (2006). Ann. Biomed. Eng. 34, 1429–1441.
[65] Working, P. K., Newman, M. S., Johnson, J., and Cornacoff, J. B. (1997). Poly(ethylene glycol) (pp. 45–57). American Chemical Society.
[66] Selective Laser Sintering, Birth of an Industry – Department of Mechanical Engineering. https://www.me.utexas.edu/news/news/selective-laser-sintering-birth-of-an-industry (accessed 22 April 2019).
[67] Goodridge, R., and Ziegelmeier, S. (2017). Laser Additive Manufacturing (Brandt, M., Ed., pp. 181–204). Elsevier.
[68] Pilipović, A., Brajlih, T., and Drstvenšek, I. (2018). Polymers (Basel) 10, 1208.
[69] Koo, J. H., Ortiz, R., Ong, B., and Wu, H. (2017). Laser Additive Manufacturing (pp. 205–235). Elsevier.
[70] Tang, H.-H., Chiu, M.-L., and Yen, H.-C. (2011). J. Eur. Ceram. Soc. 31, 1383–1388.
[71] Salmoria, G. V., Paggi, R. A., Lago, A., and Beal, V. E. (2011). Polym. Test. 30, 611–615.
[72] Williams, J. M., Adewunmi, A., Schek, R. M., Flanagan, C. L., Krebsbach, P. H., Feinberg, S. E., Hollister, S. J., and Das, S. (2005). Biomaterials 26, 4817–4827.
[73] Tan, K. H., Chua, C. K., Leong, K. F., Cheah, C. M., Gui, W. S., Tan, W. S., and Wiria, F. E. (2005). Biomed. Mater. Eng. 15, 113–124.
[74] Simpson, R. L., Wiria, F. E., Amis, A. A., Chua, C. K., Leong, K. F., Hansen, U. N., Chandrasekaran, M., and Lee, M. W. (2008). J. Biomed. Mater. Res. Part B Appl. Biomater. 84B, 17–25.
[75] Wiria, F. E., Leong, K. F., Chua, C. K., and Liu, Y. (2007). Acta Biomater. 3, 1–12.
[76] Delogu, F., Gorrasi, G., and Sorrentino, A. (2017). Prog. Mater. Sci. 86, 75–126.
[77] Gorrasi, G., and Sorrentino, A. (2013). Polym. Degrad. Stab. 98, 963–971.
[78] Sorrentino, A., Altavilla, C., Merola, M., Senatore, A., Ciambelli, P., and Iannace, S. (2015). Polym. Compos. 36, 1124–1134.
[79] Chung, H., and Das, S. (2008). Mater. Sci. Eng. A 487, 251–257.

[80] Shi, W., Yan, Y., Wei, C., Wen, Q., and Zhu, S. (2015). Sci. Sin. Informationis 45, 204.

[81] Levy, G. N. (2010). Phys. Procedia 5, 65–80.

[82] Ho, H. C. H., Cheung, W. L., and Gibson, I. (2002). Rapid Prototyp. J. 8, 233–242.

[83] Sorrentino, A., Gorrasi, G., Tortora, M., and Vittoria, V. (2006). Polymer Nanocomposites (Yiu-Wing Mai, Z.-Z. Y., Ed., pp. 273–296). Elsevier, Cambridge.

[84] Sorrentino, A. (2011). Nanocoatings and Ultra-Thin Films (Makhlouf, A. S. H. and Tiginyanu, U.-T. F., Eds., pp. 203–234). Elsevier, Oxford.

[85] Jain, P. K., Pandey, P. M., and Rao, P. V. M. (2009). Polym. Compos. NA–NA.

[86] Duan, B., and Wang, M. (2010). Polym. Degrad. Stab. 95, 1655–1664.

[87] Zhou, W. Y., Lee, S. H., Wang, M., Cheung, W. L., and Ip, W. Y. (2008). J. Mater. Sci. Mater. Med. 19, 2535–2540.

[88] US5340656A, 1994.

[89] Cima, M. J., Sachs, E., Cima, L. G., Yoo, J., Khanuja, S., Borland, S. W., Wu, B., and Giordano, R. A. (1994). 1994 International Solid Freeform Fabrication Symposium.

[90] Wu, B. M., Borland, S. W., Giordano, R. A., Cima, L. G., Sachs, E. M., and Cima, M. J. (1996). J. Control. Release 40, 77–87.

[91] Billiet, T., Vandenhaute, M., Schelfhout, J., Van Vlierberghe, S., and Dubruel, P. (2012). Biomaterials 33, 6020–6041.

[92] Alexandrea, 3D Printing Through Powder Binding – 3Dnatives. https://www.3dnatives.com/en/powder-binding100420174/ (accessed 23 April 2019).

[93] Wu, B. M., and Cima, M. J. (1999). Polym. Eng. Sci. 39, 249–260.

[94] Utela, B., Storti, D., Anderson, R., and Ganter, M. (2008). J. Manuf. Process. 10, 96–104.

[95] Lam, C. X. F., Mo, X. M., Teoh, S.-H., and Hutmacher, D. W. (2002). Mater. Sci. Eng. C 20, 49–56.

[96] Kim, S. S., Utsunomiya, H., Koski, J. A., Wu, B. M., Cima, M. J., Sohn, J., Mukai, K., Griffith, L. G., and Vacanti, J. P. (1998). Ann. Surg. 228, 8.

[97] Zeltinger, J., Sherwood, J. K., Graham, D. A., Müeller, R., and Griffith, L. G. (2001). Tissue Eng. 7, 557–572.

[98] Sherwood, J. K., Riley, S. L., Palazzolo, R., Brown, S. C., Monkhouse, D. C., Coates, M., Griffith, L. G., Landeen, L. K., and Ratcliffe, A. (2002). Biomaterials 23, 4739–4751.

[99] Shanjani, Y., De Croos, J. N. A., Pilliar, R. M., Kandel, R. A., and Toyserkani, E. (2010). J. Biomed. Mater. Res. Part B Appl. Biomater. 93, 510–519.

[100] Lee, M., Wu, B. M., and Dunn, J. C. Y. (2008). J. Biomed. Mater. Res. Part A 87A, 1010–1016.

[101] Lee, M., Dunn, J. C. Y., and Wu, B. M. (2005). Biomaterials 26, 4281–4289.

[102] Lee, J.-Y., Choi, B., Wu, B., and Lee, M. (2013). Biofabrication 5, 045003.

[103] Jain, P., and Kuthe, A. M. (2013). Procedia Eng. 63, 4–11.

[104] Carausu, C., Mazurchevici, A., Ciofu, C., and Mazurchevici, S. (2018). IOP Conf. Ser. Mater. Sci. Eng. 400, 042008.

[105] Ngo, T. D., Kashani, A., Imbalzano, G., Nguyen, K. T. Q., and Hui, D. (2018). Compos. Part B Eng. 143, 172–196.

[106] Turner, B. N., Strong, R., and Gold, S. A. (2014). Rapid Prototyp. J. 20, 192–204.

[107] Dickson, A. N., Barry, J. N., McDonnell, K. A., and Dowling, D. P. (2017). Addit. Manuf., 16, 146–152.

[108] Parandoush, P., and Lin, D. (2017). Compos. Struct. 182, 36–53.

[109] Turner, B. N., and Gold, S. A. (2015). Rapid Prototyp. J. 21, 250–261.

[110] Tofangchi, A., Han, P., Izquierdo, J., Iyengar, A., and Hsu, K. (2019). Polymers (Basel) 11, 315.

[111] Sood, A. K., Ohdar, R. K., and Mahapatra, S. S. (2010). Mater. Des. 31, 287–295.

[112] Zein, I., Hutmacher, D. W., Tan, K. C., and Teoh, S. H. (2002). Biomaterials 23, 1169–1185.

[113] Hutmacher, D. W., Schantz, T., Zein, I., Ng, K. W., Teoh, S. H., and Tan, K. C. (2001). J. Biomed. Mater. Res. 55, 203–216.

[114] Park, S. H., Park, D. S., Shin, J. W., Kang, Y. G., Kim, H. K., Yoon, T. R., and Shin, J.-W. (2012). J. Mater. Sci. Mater. Med. 23, 2671–2678.

[115] Kim, J., McBride, S., Tellis, B., Alvarez-Urena, P., Song, Y.-H., Dean, D. D., Sylvia, V. L., Elgendy, H., Ong, J., and Hollinger, J. O. (2012). Biofabrication 4, 025003.

[116] Gnanasekaran, K., Heijmans, T., van Bennekom, S., Woldhuis, H., Wijnia, S., de With, G., and Friedrich, H. (2017). Appl. Mater. Today 9, 21–28.

[117] Kalita, S. J., Bose, S., Hosick, H. L., and Bandyopadhyay, A. (2003). Mater. Sci. Eng. C 23, 611–620.

[118] Woodfield, T. B. F., Malda, J., de Wijn, J., Péters, F., Riesle, J., and van Blitterswijk, C. A. (2004). Biomaterials 25, 4149–4161.

[119] Rai, B. (2004). Biomaterials 25, 5499–5506.

[120] Korpela, J., Kokkari, A., Korhonen, H., Malin, M., Närhi, T., and Seppälä, J. (2013). J. Biomed. Mater. Res. Part B Appl. Biomater. 101B, 610–619.

[121] Kim, J. Y., and Cho, D.-W. (2009). Microelectron. Eng. 86, 1447–1450.

[122] Shim, J.-H., Moon, T.-S., Yun, M.-J., Jeon, Y.-C., Jeong, C.-M., Cho, D.-W., and Huh, J.-B. (2012). J. Mater. Sci. Mater. Med. 23, 2993–3002.

Index

https://doi.org/10.1515/9783110590586-011

www.ingramcontent.com/pod-product-compliance
Lightning Source LLC
Chambersburg PA
CBHW061926190326
41458CB00009B/2667